Civil Engineering Problems and Solutions

Civil Engineering Problems and Solutions

Editor:

Dr. Tanjina Nur, Ph.D.

ArclerPress

www.arclerpress.com

Civil Engineering Problems and Solutions

Editor: Dr. Tanjina Nur, Ph.D.

708 3rd Avenue, 6th Floor
New York
NY 10017
United States of America
www.arclerpress.com

ISBN: 978-1-68094-373-3

Library of Congress Control Number: 2016950949

About the Editor

Dr. Tanjina Nur, Ph.D.

Tanjina finished her PhD in Civil and Environmental Engineering in 2014 from University of Technology Sydney (UTS). Now she is working as Post-Doctoral Researcher in the Centre for Technology in Water and Wastewater (CTWW) and published about eight International journal papers with 80 citations. Her research interest is wastewater treatment technology using adsorption process.

Contents

List of Contributors

Jurgita Antucheviciene
Department of Construction Technology and Management, Faculty of Civil Engineering, Vilnius Gediminas Technical University, Saulėtekio Alėja 11, LT-10223 Vilnius, Lithuania

Zdeněk Kala
Department of Structural Mechanics, Faculty of Civil Engineering, Brno University of Technology, Veveří Street 95, 602 00 Brno, Czech Republic

Mohamed Marzouk
Department of Structural Engineering, Faculty of Engineering, Cairo University, Giza 12613, Egypt

Egidijus Rytas Vaidogas
Department of Labour Safety and Fire Protection, Faculty of Civil Engineering, Vilnius Gediminas Technical University, Saulėtekio Alėja 11, LT-10223 Vilnius, Lithuania

Edmundas Kazimieras Zavadskas
Department of Construction Technology and Management, Vilnius Gediminas Technical University, Saulėtekio Alėja 11, LT-10223 Vilnius, Lithuania

Jurgita Antucheviciene
Department of Construction Technology and Management, Vilnius Gediminas Technical University, Saulėtekio Alėja 11, LT-10223 Vilnius, Lithuania

Seyed Hossein Razavi Hajiagha
Department of Management, Khatam Institute of Higher Education, No. 30, Hakim A'zam Alley, Mollasadra Street, North Shiraz Avenue, Tehran 19941633356, Iran

Shide Sadat Hashemi
Department of Management, Islamic Azad University, Kashan Branch, Kashan 8715998151, Iran

Yan Han
Institute of Geographic Sciences and Natural Resources Research, Chinese Academy of Sciences, Beijing 100101, China
State Key Laboratory of Simulation and Regulation of Water Cycle in River Basin, China Institute of Water Resources and Hydropower Research, Beijing 100038, China

Yuefei Huang
State Key Laboratory of Hydroscience and Engineering, Tsinghua University, Beijing 100084, China

Shaofeng Jia
Institute of Geographic Sciences and Natural Resources Research, Chinese Academy of Sciences, Beijing 100101, China

Jiahong Liu
State Key Laboratory of Simulation and Regulation of Water Cycle in River Basin, China Institute of Water Resources and Hydropower Research, Beijing 100038, China

Danxia Wang
Department of Mathematics, Taiyuan University of Technology, Taiyuan 030024, China

Jianwen Zhang
Department of Mathematics, Taiyuan University of Technology, Taiyuan 030024, China

Yinzhu Wang
Department of Mathematics, Taiyuan University of Science and Technology, Taiyuan 030024, China

Sufang Zhang
Department of Mathematics, Taiyuan University of Technology, Taiyuan 030024, China

Engin C. Seyhan
Department of Civil Engineering, Civil Engineering Faculty, Istanbul Technical University, Ayazaga, Sariyer, Istanbul 34469, Turkey

Caglar Goksu
Department of Civil Engineering, Civil Engineering Faculty, Istanbul Technical University, Ayazaga, Sariyer, Istanbul 34469, Turkey

Ahmet Uzunhasanoglu
Department of Civil Engineering, Civil Engineering Faculty, Istanbul Technical University, Ayazaga, Sariyer, Istanbul 34469, Turkey

Alper Ilki
Department of Civil Engineering, Civil Engineering Faculty, Istanbul Technical University, Ayazaga, Sariyer, Istanbul 34469, Turkey

Engin Gücüyen
Department of Civil Engineering, Celal Bayar University, 45140, Muradiye, Manisa, Turkey

R. Tuğrul Erdem
Department of Civil Engineering, Celal Bayar University, 45140, Muradiye, Manisa, Turkey

Erkan Kantar
Department of Civil Engineering, Celal Bayar University, 45140, Muradiye, Manisa, Turkey

Muhiddin Bağcı
Department of Civil Engineering, Celal Bayar University, 45140, Muradiye, Manisa, Turkey

Muhiddin Bağcı
Civil Engineering Department, Celal Bayar University, Manisa, Turkey

A. Barari
Departments of Civil Engineering and Mechanical Engineering, Mazandaran University of Technology, Babol, Iran

M. Omidvar
Technical and Engineering Faculty, Gorgan University of Agricultural Sciences and Natural Resources, Gorgan, Iran

D. D. Ganji
Departments of Civil Engineering and Mechanical Engineering, Mazandaran University of Technology, Babol, Iran

Abbas Tahmasebi Poor
Departments of Civil Engineering and Mechanical Engineering, Mazandaran University of Technology, Babol, Iran

Egidijus Rytas Vaidogas
Vilnius Gediminas Technical University, 11 Sauletekio Avenue, LT-10223 Vilnius, Lithuania

Jurgita Šakėnaitė
Vilnius Gediminas Technical University, 11 Sauletekio Avenue, LT-10223 Vilnius, Lithuania

L. X. Peng
College of Civil Engineering and Architecture, Guangxi University, Nanning 530004, China
Guangxi Key Laboratory of Disaster Prevention and Engineering Safety, Guangxi University, Nanning 530004, China

Murat Sari
Department of Mathematics, Faculty of Art and Science, Pamukkale University, 20070 Denizli, Turkey

Artur Andrzejczak
Department of Microelectronics and Computer Science, Lodz University of Technology, Wólczańska 221/223, 90-924 Lodz, Poland

PaweB Awczycki
Department of Microelectronics and Computer Science, Lodz University of Technology, Wólczańska 221/223, 90-924 Lodz, Poland

MichaB Wojtera
Department of Microelectronics and Computer Science, Lodz University of Technology, Wólczańska 221/223, 90-924 Lodz, Poland

Piotr Pietrzak
Department of Microelectronics and Computer Science, Lodz University of Technology, Wólczańska 221/223, 90-924 Lodz, Poland

Bartosz PwkosBawski
Department of Microelectronics and Computer Science, Lodz University of Technology, Wólczańska 221/223, 90-924 Lodz, Poland

Andrzej Napieralski
Department of Microelectronics and Computer Science, Lodz University of Technology, Wólczańska 221/223, 90-924 Lodz, Poland

Linsheng Huo
Faculty of Infrastructure Engineering, Dalian University of Technology, Dalian 116024, China

Chunxu Qu,
Faculty of Infrastructure Engineering, Dalian University of Technology, Dalian 116024, China

Hongnan Li
Faculty of Infrastructure Engineering, Dalian University of Technology, Dalian 116024, China

Francesco Soldovieri
Institute for Electromagnetic Sensing, National Research Council, Via Diocleziano 328, 80124 Napoli, Italy

Jean Dumoulin
Monitoring, Assessment and Computational Science Department (MACS), French Institute of Science and Technology for Transport, Development and Networks (IFSTTAR), Route Bouaye, CS4, 44344 Bouguenais, France

Nicola Masini
Institute for Archaeological and Monumental, CNR, C.da S. Loja, 85050 Tito Scalo (PZ), Italy

Erica Utsi
Utsi Electronics Ltd., Cambridge CB22 7NZ, UK

Baki Ozturk
Department of Civil Engineering, Faculty of Engineering, Hacettepe University, 06800 Ankara, Turkey

Safa Bozkurt Coskun
Department of Civil Engineering, Faculty of Engineering, Kocaeli University, 41380 Kocaeli, Turkey

Lina Wu
College of Traffic, Northeast Forestry University, Harbin 150080, China
College of Automobile and Traffic Engineering, Heilongjiang Institute of Technology, Harbin 150050, China

Jiangwei Chu
College of Traffic, Northeast Forestry University, Harbin 150080, China

Yusheng Ci
School of Transportation Science and Engineering, Harbin Institute of Technology, Harbin 150090, China

Shumin Feng
School of Transportation Science and Engineering, Harbin Institute of Technology, Harbin 150090, China

Xingwang Liu

Research Institute of Highway, Ministry of Transport, Beijing 100088, China

Yashon O. Ouma
Department of Civil and Structural Engineering, Moi University, Eldoret 30100, Kenya

J. Opudo
Department of Civil and Structural Engineering, Moi University, Eldoret 30100, Kenya

S. Nyambenya
Department of Civil and Structural Engineering, Moi University, Eldoret 30100, Kenya

Preface

This book presents review examines of methods developed for dealing with and applied to solve problems of civil engineering. The book is focused on formal methods of solving different parts of civil engineering problems using numerical methods and also fuzzy and uncertainty methods. Civil Engineering is concerned with such projects as roads, bridges, railways, harbours and docks, water supply and wastewater treatment, dams, tunnels and underground construction, power projects, offshore structures, and commercial and industrial buildings. The effective practice of civil engineering necessitates a strong background in the mathematical and physical sciences, an understanding of the properties of construction materials such as steel, concrete, timber and soils. Firstly, we discuss current state and future research methods for solving civil engineering problems by means of fuzzy and stochastic methods and the interval-valued intuitionistic fuzzy method. Then we present an interval-parameter fuzzy linear programming with stochastic vertices model for water resources management under uncertainty and attractor of beam equation with structural damping under nonlinear boundary conditions. Seismic behavior of substandard RC columns retrofitted with embedded aramid fiber reinforced polymer (AFRP) reinforcement and determination of the impact behavior of concrete and reinforced concrete beams are also discussed.

In the middle part of the book, we focus on neural network model for moment-curvature relationship of reinforced concrete sections and an approximate solution for boundary value problems in structural engineering and fluid mechanics. Then solving the problem of multiple-criteria building design and free vibration analysis of symmetrically laminated folded plate structures are also discussed. We also try to present solution of the porous media equation by a compact finite difference method and architecture of wireless vehicle weight measurement system for structural health monitoring in Civil Engineering application.

The last chapters of the book describe parametric optimization for the vibration control of building structures and non-invasive sensing techniques and geophysical methods for cultural heritage and civil infrastructures monitoring. We also try to find out analytical solution for free vibration analysis of beam on elastic foundation with different support conditions and engineering solutions to enhance traffic safety performance on two-lane highways. At the end, we describe comparison of fuzzy AHP and fuzzy TOPSIS for road pavement maintenance.

Editor
Dr. Tanjina Nur, Ph.D.

INTRODUCTION

Civil engineering is a technology that includes numerous other disciplines that produce useful facilities for the human beings, including roads, dams, waste disposal and other facilities that are used in our daily life. Civil engineering is progressing at a fast pace as are other technologies.

Civil engineering is considered as the first discipline of the various branches of engineering after military engineering, and includes the designing, planning, construction, and maintenance of the infrastructure. The works include roads, bridges, buildings, dams, canals, water supply and numerous other facilities that affect the life of human beings. Civil engineering is intimately associated with the private and public sectors, including the individual homeowners and international enterprises.

Civil engineering has a significant role in the life of every human being, though may not truly sense its importance in our daily routine. The function of civil engineering commences with the start of the day when we take a shower, since the water is delivered through a water supply system including a well-designed network of pipes, water treatment plant and other numerous associated services. The network of roads on which we drive while proceeding to school or work, the huge structural bridges we come across and the tall buildings where we work, all have been designed and constructed by civil engineers. Even the benefits of electricity we use are available to us through the contribution of civil engineers who constructed the towers for the transmission lines. In fact, no sphere of life may be identified that does not include the contribution of civil engineering. Thus, the importance of civil engineering may be determined according to its usefulness in our daily life.

Scope of Different Fields of Civil Engineering

Civil Engineering may be divided into the following fields:

- Surveying
- Building Materials
- Construction Technology
- Structural Engineering
- Geotechnical Engineering

- Hydraulics
- Water Resources and Irrigation Engineering
- Transportation Engineering
- Environmental Engineering and
- Architecture and Town planning

Scope of each one of these is discussed below.

Surveying

Surveying is the science of map making. To start any development activity in an area the relative positions of various objects in the horizontal and vertical directions are required. This is approved by surveying the area. Earlier, the conventional instruments like chain, tape and levelling instruments were used. In this electronic era, modern equipments like distance meters and total stations are used to get more accurate results easily. The modern technologies like photogrammetry and remote sensing have made surveying easier.

Building Materials

Shelter is the basic need of civilization. To get good shelter continuous efforts are going on right from the beginning of civilization. Stones, bricks, timber, lime, cement, sand, jellies and tiles are the traditional building materials. Use of steel, aluminium, glass, glazed tiles, plaster of paris, paints and varnishes have improved the quality of buildings. The appropriate mixture of binding materials like lime and cement with sand is known as mortar. The mixture of cement, sand and jelly (crushed stones) with water is known as concrete. The use of concrete with steel bars placed in appropriate position has helped in building strong and durable tall structures. The composite material of concrete and steel is called reinforced cement concrete which is popularly known as R C C. A civil engineer must know the properties of all the building materials so that they can be used appropriately. Improved versions of many building materials appear in the market. A good civil engineer will make use of them at the earliest.

Construction Technology

Construction is the major activity of civil engineering which is continuously improving. As land cost is going up there is demand for tall structures in urban areas while in rural areas need is for low cost constructions. One has to develop technology using locally available materials.

Structural Engineering

Load acting on a structure is ultimately transferred to ground. In doing so, various components of the structure are subjected to internal stresses. For example, in a building, load acting on a slab is transferred by slab to ground through beams, columns and footings. Assessing the internal stresses in the components of a structure is known as Structural Analysis and finding the suitable size of the structural component is known as design of structure. The structure to be analyzed and designed may be of masonry, R. C. C. or steel. Upto midsixties considerable improvements were seen in classical analysis. With the advent of computers numerical methods emerged and analysis and design packages are becoming popular. Matrix Method of analysis and Finite Elements Analysis have helped in the analysis of complex structures. A civil engineer has not only to give a safe structure but he has to give economical sections. To get economical section mathematical optimization techniques are used. Frequent earthquakes in the recent years have brought, importance of analysis of the structure for earthquake forces. Designing earthquake resistant structures is attracting lot of researches. All these aspects fall under structural engineering field.

Geotechnical Engineering

Soil property changes from place to place. Even in the same place it may not be uniform at various depths. The soil property may vary from season to season due to variation in moisture content. The load from the structure is to be safely transferred to soil. For this, safe bearing capacity of the soil is to be properly assessed. This branch of study in Civil Engineering is called as Geotechnical Engineering. Apart from finding safe bearing capacity for foundation of buildings, geotechnical engineering involves various studies required for the design of pavements, tunnels, earthen dams, canals and earth retaining structures. It involves study of ground improvement techniques also.

Hydraulics

Water is an important need for all living beings. Study of mechanics of water and its flow characteristics is another important field in Civil Engineering and it is known as hydraulics.

Water Resources and Irrigation Engineering

Water is to be supplied to agriculture field and for drinking purposes. Hence suitable water resources are to be identified and water is to be stored. Identifying, planning and building water retaining structures like tanks and dams and carrying stored water to fields is known as water resources and irrigation engineering.

Transportation Engineering

Transportation facility is another important need. Providing good and economical roads is an important duty of civil engineers. It involves design of base courses, suitable surface finishes, cross drainage works, road intersections, culverts, bridges, tunnels etc. Railway is another important long-way transport facility. Design, construction and maintenance of railway lines, signal system are part of transportation engineering. There is need for airports and harbors. For proper planning of these transportation facility, traffic survey is to be carried out Carrying out traffic survey, design, construction and maintenance of roads, bridges, railway, harbor and airports is known as transportation engineering.

Environmental Engineering

Proper distribution of water to rural areas, towns and cities and disposal of waste water and solid waste are another field of civil engineering. Industrialization and increase in vehicular traffic are creating air pollution problems. Environmental engineering while tackling all these problems provides healthy environment to public.

Architecture and Town Planning

Aesthetically good structures are required. Towns and cities are to be planned properly. This field of engineering has grown considerably and has become a course separate from Civil Engineering.

Types of Infrastructure

Infrastructure facilities involve various civil engineering amenities, electricity, telephone, internet facility, educational and healthcare facilities.

Civil engineering amenities in the infrastructure developments are:

- A good town planning and developing sites
- Providing suitable roads and network of roads
- Railway connection to important places
- Airports of national and international standards
- Assured water supply to towns, cities and rural areas
- A good drainage and waste disposal system
- Pollution free environment

Effect of Infrastructure Facilities

Connecting producing center to marketing places minimizes exploitation from middlemen. Both producer and consumers are benefitted. Imports and exports become easy as a result of which whole world becomes a village. The infrastructure development generates scope for lots of industries. Manpower is utilized for the benefit of mankind. Antisocial activities come under control. Improved education and healthcare give rise to skilled and healthy work force. Quality of life of the people is improved. In case of natural calamities assistance can be extended easily and misery of affected people is reduced. Infrastructure facility improves defense system and peace exists in the country. Improved economical power of the country brings a respectable status in the world. The world has realized that a government should not involve itself in production and distribution but should develop infrastructure to create an atmosphere for economical development.

Role of Civil Engineers in the Infrastructure Development

A civil engineer has to conceive, plan, estimate, get approval, create and maintain all civil engineering infrastructure activities. He has to carry out research and training programmes to improve the technology.

Civil engineer has a very important role in the development of the following infrastructures:

- Town and city planning
- Build suitable structures for the rural and urban areas for various utilities.

- Build tanks, dams to exploit water resources.
- Purify the water and supply water to needy areas like houses, schools, offices, and agriculture field.
- Provide good drainage system and purification plants.
- Provide and maintain communication systems like roads, railways, harbors and airports.
- Monitor land, water and air pollution and take measures to control them.

CHAPTER 1

SOLVING CIVIL ENGINEERING PROBLEMS BY MEANS OF FUZZY AND STOCHASTIC MCDM METHODS

Jurgita Antucheviciene, Zdeněk Kala, Mohamed Marzouk, and Egidijus Rytas Vaidogas

INTRODUCTION

Decision-making is applied in different areas of human activities. In the case of existence of at least two possible options, a person (i.e., a decision-maker) has to make a decision and to select the one which is best suited for his demands. Complex problems in science, engineering, technology, or management are characterised by multiple criteria. Usually they are hardly measurable, conflicting or interacting with each other. Decision-making (DM) problems based on multiple criteria are objects of MCDM.

MCDM is a discipline concerned with the theory and methodology for handling problems common in everyday life. They arise in such areas as business, engineering, social organisation, and so forth [1]. MCDM has grown as a part of operation research pertaining to the design of computational and mathematical tools for supporting the subjective evaluation of performance criteria by decision-makers [2].

As it is mentioned in a review paper [3], the origins of MCDM methods can be dated over 270 years ago. As an individual scientific discipline, MCDM has been widely spreading since the middle of the previous century. Numerous works on MCDM are summarized in a number of review papers [4]. Three main types of review papers related to MCDM can be distinguished: reviews of developments and extensions of a particular method (e.g., [5]) as well as its applications (e.g., [6, 7]); reviews of different approaches to modern MCDM methods (e.g., DM under uncertain and incomplete information and DM for groups of persons [2, 8]); reviews of applications of different MCDM methods for a particular problem [9–11].

Most, if not all, decisions in engineering are made under uncertainty. Accordingly, classical crisp methods, involving deterministic conditions under certain decision environments, are losing their relevance and new extensions for handling various types of uncertainties in MCDM problems appear. In most cases, uncertain elements of MCDM problems express subjective expert opinions and less often they represent stochastic nature of input data. Understanding the significance and impact of different types of uncertainties in input information and its relation to reality can increase the quality of DM process.

The present review strives to examine the current state of uncertainty modelling in MCDM and to identify several problems which decision-makers may face in the immediate future. The review is organized as follows. Section 2 presents the scope of the review and identifies several problems related to MCDM under uncertainty. Section 3 presents the latest references on applications of MCDM approaches in uncertain decision environments of civil engineering. Section 4 summarizes the applications of sensitivity analysis (SA) in fuzzy and stochastic MCDM techniques. Section 5 analyses published work on introducing measures of risk and reliability in the MCDM related to civil engineering. Section 6 attempts to identify the need for future research. This section presents a brief discussion on potential applications of MCDM under uncertainty in the areas which currently receive much attention in research and practice. Finally, conclusions and prospective research trends are presented in Section 7.

Citation: Jurgita Antucheviciene, Zdeněk Kala, Mohamed Marzouk, and Egidijus Rytas Vaidogas, "Solving Civil Engineering Problems by Means of Fuzzy and Stochastic MCDM Methods," Mathematical Problems in Engineering, vol. 2015, Article ID 362579, 16 pages, 2015. doi:10.1155/2015/362579.

The Scope of the Review

Generally, a MCDM problem is defined as follows. Let a = (a_1, a_2,...,,...,$_m$) ^{T} be a vector of decision alternatives and c = (c_1, c_2,...,c_j,...,c_n) a set of criteria, according to which suitability of alternatives a_i is to be judged. The problem is stated as $am \times n$ decision-making matrix with the elements c_{ij}. The value c_{ij} expresses an impact of the criterion cj on the alternative ai. Values of c related to the alternatives a_i are the row vectors c_i = (c_{i1}, c_{i2},...,,...,c_{in}) and, with these vectors, the matrix C is formulated as [c_1, c_2,..., c_j,..., c_m] ^{T}. In many MCDM methods, the importance of the criteria c_j is expressed by the weights wj which sum up to unity and are usually grouped into the vector w = (w_1, w_2,...,w_j,...,w_n). The purpose of DM is to determine the most preferable alternative among ai with respect to all criteria or to rank the alternatives. The result of ranking is expressed by a preference sequence, for example, π: $a_x \succ a_y \succ a_z \succ \cdots$.

A large variety of MCDM methods have been developed to date. A universal and comprehensive classification of these methods does not exist. For the purposes of the present review, MCDM methods will be classified according to the nature of the criteria grouped into the decision-making matrix C (Figure 1). In the previous five decades, a development of MCDM was focused mainly on solving MCDM problems with crisp elements of C. In parallel with this process, fuzzy MCDM methods have been developed since 1970s [12]. These methods allow solving MCDM problems with a decision-making matrix $\overleftrightarrow{C}$, the elements of which are modelled by fuzzy sets. Development and application of fuzzy MCDM methods significantly increased during the last two-three decades. One of the possible representations of cij by fuzzy sets is grey numbers based on interval arithmetic. Several recent publications devoted to a use of grey numbers for decision-making in engineering can be cited here [13– 15]. Finally, in the recent time, several MCDM methods were created and applied to a solution of decision-making problems, in which elements of C are modelled by means of probability distributions. Such a matrix will be denoted by the symbol $\tilde{C}$.

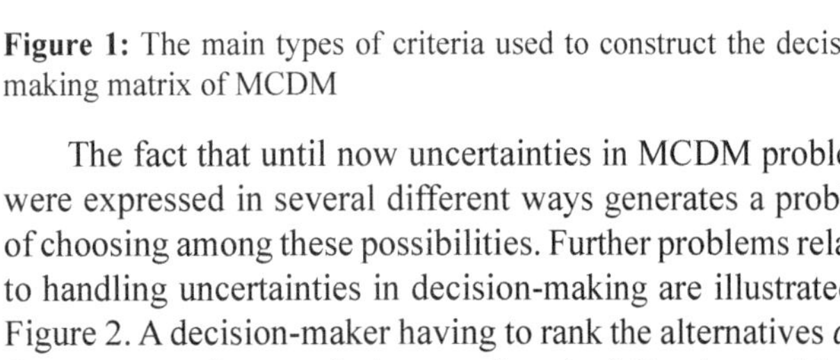

Figure 1: The main types of criteria used to construct the decision-making matrix of MCDM

The fact that until now uncertainties in MCDM problems were expressed in several different ways generates a problem of choosing among these possibilities. Further problems related to handling uncertainties in decision-making are illustrated in Figure 2. A decision-maker having to rank the alternatives a_i in the presence of uncertainties may face the following problems:

(1) The problem of choice among different representations of uncertainty related to criteria values c_{ij} and weights w_j.

(2) Specification of the weights w_j in the case where they are uncertain quantities.

(3) The problem of choice among available MCDM methods for solving a problem with the decision-making matrices $\overleftrightarrow{C}$ or $\tilde{C}$.

(4) Analysis of sensitivity of MCDM results to the changes in all elements of input data given by the matrices $\overleftrightarrow{C}$ or $\tilde{C}$ and the vectors w, $\overleftrightarrow{w}$ or $\tilde{w}$. Input uncertainties can cause possible permutations in the sequences of preferences, π_k, obtained by means of a specific MCDM method k.

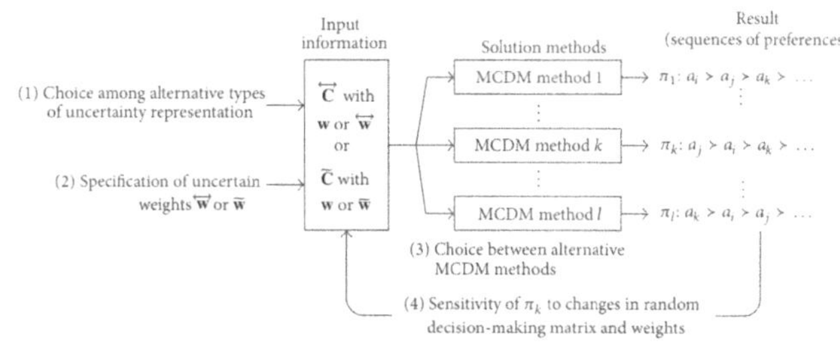

Figure 2: Four problems related to handling uncertainties in MCDM.

The first problem has its origin "outside" the domain of MCDM. Discussions of the type "fuzzy versus probabilistic" or "no probabilistic versus probabilistic" last many years and do not seem to be finished until now (e.g., [16]). In most applications of MCDM under uncertainty, authors do not trouble themselves to explain why a particular method of uncertainty quantification was preferred over others. The fuzzy logic prevails over probabilistic modelling by the number of MCDM methods developed to date and the number of MCDM applications to practical problems. Currently, one can state the obvious that probability distributions can be specified in Fisherian format for c_{ij} in the presence of sufficient amount of statistical data on c_{ij}. A representation of c_{ij} by fuzzy sets is better suited for c_{ij} expressing subjective expert opinions. However, it is important to remember that the Bayesian approach to probabilistic modelling has excellent means of quantifying and updating subjective judgments (e.g., [17]). This approach does not break down when data on c_{ij} is sparse or absent. MCDM problems can be solved when uncertainties in c_{ij} are represented in any of the aforementioned formats. A true problem will arise when the decision-making matrix will be a "mixture" of the elements c_{ij} expressed by different means of uncertainty quantification, probabilistic and nonprobabilistic.

The second problem of specifying elements of the vector of weights, w, is obviously an "internal" problem of MCDM. There is a body of literature devoted to specifying values of w_j. Methods used to determine criteria weights are classified into subjective, objective, and hybrid or integrated ones [18–20]. However, the task of assigning specific values to the components w_j will always be subjective or at least partly subjective, no matter what is the degree of mathematical sophistication behind this assignment. A really intriguing question is how w_j can be interpreted as uncertain quantities. The experts who usually specify the values of w_j can be uncertain (vague) regarding their opinions on w_j. Should weights be modelled by a fuzzy vector $\overleftrightarrow{w}$ or a vector $\tilde{w}$ with components expressed by probability distributions?

The third problem reflects the well-known decisionmaking paradox which was first identified by Triantaphyllou and Mann in 1989 [21]. The paradox was exhibited by many MCDM methods developed to deal with the crisp decisionmaking matrix C. Switching to uncertain matrices $\overleftrightarrow{C}$ or $\tilde{C}$ will not resolve this paradox. It will be likely to persist also in the field of decision-making under uncertainty. The present review will not address this tricky issue

The fourth problem arises naturally, because a MCDM method is in essence a mathematical model relating the input information expressed by C and w to the output information given by the preference sequence π. A sensitivity of π to changes in elements of C and w can be estimated (analysed) by standard mathematical means of general use. However, a sensitivity analysis (SA) of a MCDM problem with uncertain matrices $\overleftrightarrow{C}$ or $\tilde{C}$ and/or uncertain vectors $\overleftrightarrow{w}$ or $\tilde{w}$ is a nontrivial task.

The four problems of MCDM under uncertainty listed and briefly discussed above are of methodological nature. This kind of DM can face also problems of different nature, namely, an application in such areas of civil engineering as development of sustainable and energy efficient buildings, building information modelling, assurance of security, and safety of built property. Solving MCDM problems in these areas will face the necessity to model uncertainties related to long-term predictions, vague information available in the process of building design, and possibility of rare but extremely damaging events. A particular need for modelling uncertainties in MCDM problems arises at a design of safety critical objects of civil engineering. Failures of such objects or damage to them can cause severe consequences to society and environment. Many of safety critical objects are assessed (designed) by applying methods of reliability theory or probabilistic risk assessment. Measures of reliability and components of risk estimates can be uncertain quantities. An application of MCDM to the design of safety critical objects requires including these quantities into the decision-making matrix C. In most cases, uncertainties related to reliably and risk are modelled by means of probability distributions. Therefore, the decision-making matrix will be formulated as the matrix $\tilde{C}$. A combination of SA and MCDM applied to the design of the safety critical objects will allow revealing the most important MCDM criteria related to safety.

The following sections of this review will consider the problems raised above in greater detail. The attention will be focused on both results achieved until now and problematic issues which may require attention in the future.

MCDM TECHNIQUES UNDER UNCERTAINTY IN CIVIL ENGINEERING

Decision-making in the field of civil engineering is increasingly complex and is associated with situations where robust decisions are required to be taken. These decisions are made in different stages of civil engineering projects. For example, decision-making takes place during feasibility study stage prior to design, procurement, and construction stages in order to determine the viability of project undertaken by an investor. Also it often faces the need to deal with hazardous phenomena, including industrial accidents able to damage built property, structural failures, extreme natural phenomena, and human acts affecting security. Proper decisions made by architects and civil engineers may reduce the risk posed by the aforementioned phenomena. Decision-making in this field can be facilitated by an application of formal methods, such as methods of multiple criteria decision-making. Factors relevant to decisions can be identified using the methods of uncertainty and SA of mathematical model outputs.

The result of the evaluation of an alternative according to a given criteria is a single value (e.g., number or verbal expressions) for unambiguous deterministic information. For ambiguous information, the result of the evaluation of variants according to a given criteria is a random variable if the information is stochastic in nature or a fuzzy variable if the information is nonstochastic in nature. The stochastic nature of a phenomenon is associated with the unlimited repeatability of the phenomenon. Unrepeatable phenomena or phenomena with uncertain knowledge significance are nonstochastic in nature.

Fuzzy MCDM

The fuzzy framework is the most common approach to describe and handle uncertainty in MCDM. Considering the imprecisions and uncertainties, that is, fuzziness of the available data and the decision-making procedures, fuzzy set theory can be applied [22]. In a fuzzy MCDM, the elements *cij* of C are characterised by fuzzy sets. In what follows, such elements will be denoted by the symbol $\overleftrightarrow{c_{ij}}$ and a decision-making matrix containing fuzzy elements will be $\overleftrightarrow{C}$. The fuzzy set theory can be used for a MCDM, whenever probability distributions of c_{ij} are unknown due to lack of statistical data or there is no wish to express subjective judgments about values of c_{ij} in a probabilistic way. The weights *wj* can also be modelled as fuzzy variables $\overleftrightarrow{w_j}$ and so the vector w can be a fuzzy one, $\overleftrightarrow{w}$

In MCDM problems, three types of fuzziness can be observed: the ratings of each alternative with respect to each criterion are uncertain or imprecise and weights are crisp numbers$(\overleftrightarrow{c_{ij}}, w_j)$; the ratings of alternatives are crisp numbers while fuzzy numbers are used to assess the weights of all criteria$(c_{ij}, \overleftrightarrow{w_j})$; both the ratings and the weights are fuzzy$(\overleftrightarrow{c_{ij}}, \overleftrightarrow{w_j})$. Each type of uncertainty has its own characteristics and is appropriate for special cases.

A lot of fuzzy extensions of MCDM methods have been proposed and applied in engineering, technology, or management. Fuzzy extensions of MCDM involve application of basic fuzzy logic, using triangular fuzzy numbers and arithmetic operations, also trapezoidal fuzzy numbers, as well as intuitionistic fuzzy relations, interval-valued intuitionistic fuzzy relations, type-2 fuzzy sets, and hesitant fuzzy sets concepts. Nonfuzzy uncertain decision methods include reliability theory and probabilistic and grey-valued formulations for handling incomplete or imprecise information. The most widely applied fuzzy extensions of MCDM methods are fuzzy AHP (analytic hierarchy process), with the origins dated to 1983 [23] and various numerous extensions of fuzzy TOPSIS (technique for order preference by similarity to ideal solution) [2, 24]. Several latest extensions of more recently developed methods are worth to be mentioned: extension of weighted aggregated sum product assessment in an intervalvalued intuitionistic fuzzy environment (WASPAS-IVIF) [25], a complex proportional assessment method extended with interval-valued intuitionistic fuzzy numbers (COPRASIVIF), and suitable for group decision-making [26]. The most recent extended versions of MULTIMOORA (multiobjective optimization by ratio analysis plus full multiplicative form) utilize the basic fuzzy logic [27, 28] or are based on the interval 2-tuple linguistic variables [29], intuitionistic fuzzy numbers [30], or hesitant fuzzy numbers [31]. ARAS under fuzzy environment was also presented [32] and further applied [33, 34].

Numerous researches utilize not a single fuzzy MCDM (FMCDM) method but combine several of the methods. Two groups of researches can be distinguished: either integrating two or more available techniques and proposing so-called hybrid methods or employing several methods for a solution of a problem and comparing ranking results. The study [2] reviews papers on development and applications of MCDM during the last two decades by various aspects. The analysis covers 1081 papers related to MCDM and fuzzy MCDM. While a detailed survey is made on 403 papers published in peer review journals and entirely devoted to fuzzy decisions, involving 217 papers in a field of engineering, it is obvious that engineering applications cover over 50 percent of overall applications. Accordingly, it can be stated that fuzzy MCDM is useful and applicable technique for complex decisionmaking in civil engineering under uncertainty.

The increasing importance of the methods can be based on the findings that the number of developments and applications is increasing every year. On the basis of collected data [2], annual distribution of papers in engineering field is explored. The period is divided into several stages. The first stage covers 1994–2005, when a number of papers are small or even absent in certain years. The later period starting from 2006, when a number of researches began increasing, is grouped in three years. A chart (Figure 3) shows increase of FMCDM applications in engineering, measured by a number of publications included in the review paper [2] on FMCDM techniques and applications during two decades

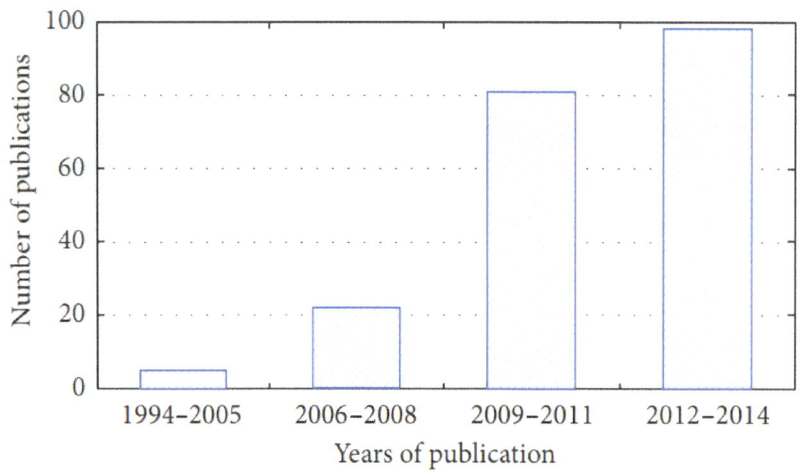

Figure 3: Increase of FMCDM applications in engineering

The last period from the above four covers 46 percent of publications issued during 20 years. The largest number of researches on the whole is observed in 2014.

The latest FMCDM approaches are applied in various subfields of civil engineering field. Subfields which are the most numerously supported by considered approaches are related to sustainability and building lifecycle assessment, supply chain management in construction, technology and inventory selection in construction industries, location selection and infrastructure modelling, and knowledge management.

Environmental, social and economic aspects of sustainable building were incorporated when evaluating mining projects and their impact on environment [35], assessing building energy performance [36] or performance of pavements with emphasis on sustainability [37]. Supply chain management is handled through proper supplier selection considering multiple criteria simultaneously. Various approaches are applied for supplier evaluation and selection, starting from the most common fuzzy AHP and hybrid methods integrating AHP and TOPSIS [38], AHP and PROMETHEE [39], and so forth. Also novel extensions based on interval type-2 fuzzy sets [40] or intuitionistic fuzzy information [41] are utilized. Selection of technologies, inventory, or materials in construction is also widely supported by FMCDM. Examples of the most recent applications are for selecting material handling equipment [42, 43], inventory classification [44], and selection of robots for automated technology operations [45].

Fuzzy decision-making methods are extremely important for handling vagueness in special-purpose building projects and their location. Fuzzy multiple criteria approaches successfully applied for nuclear power plant site selection [46], deep-water see port selection [47], and health monitoring of tunnels [48]. Regarding DM methodology in civil engineering applications, it can be concluded that the most common approach is observed to be fuzzy AHP-TOPSIS hybrid technique [38, 49], also combining fuzzy AHP with other classical methods as PROMETHEE [50] and DEMATEL [51, 52]. Another group of researches applies several FMCDM methods simultaneously and compares ranking results. An example can be provided of material selection applying a number of hybrid approaches, namely, FAHP-VIKOR, FAHPPROMETHEE, FAHP-TOPSIS, and FAHP-ELECTRE [53], technology selection applying FAHP, TOPSIS-F, VIKOR-F, and COPRAS-G [45], and so forth.

Stochastic MCDM

The probabilistic framework is the second approach to describe and handle uncertainty in MCDM. In this framework, the elements c_{ij} of the decisionmaking matrix C are modelled by random variables $\tilde{c}_{ij}$ expressing uncertainty in possible values of c_{ij}. The uncertainty is quantified by means of probability distributions which can be specified in the format Fisherian or Bayesian statistics. Presence of $\tilde{c}_{ij}$ in C generates a matrix $\tilde{C}$, all or some elements of which are random. Stochastic MCDM methods are used to solve the decision problems with the matrix $\tilde{C}$.

MCDM solved with an emphasis on stochastic uncertainty (SMCDM) is focused on decision problems of the selection of alternatives from several criteria that are mathematically described neither as crisp numbers nor as fuzzy numbers or linguistic variables but as random variables [54]. Stochastic methods for the determination of weights from various types of information on the character of the significance of criteria are effectively implemented especially in the AHP method. The classical AHP method lacks probability values for the distinction of adjacent alternatives in the final ranking [55]. Vargas [56] considered the case where members of the pairwise comparison matrix were random variables.

It should be noted that although random variables are considered, purely stochastic uncertainty of input data and decision-making procedures is rare. The occurrence of a random phenomenon is almost always accompanied by a certain degree of personal belief; therefore, the term subjective probability is sometimes used. The reason for the implementation of random variables in MCDM is the attempt to use methods of the theory of probability and mathematical statistics for the analysis of uncertainty of the results of the decision-making process. Advanced methods of stochastic SA, the equivalent of which is unknown in fuzzy MCDM, are available for stochastic MCDM [57].

There are numerous SMCDM methods and applications in real world situations [58, 59]. The classical Monte Carlo (MC) method or its improved variants can be used to tackle most SMCDM problems [60, 61]. The theoretical application of the Monte Carlo method is very extensive; however, the results of stochastic analysis are very sensitive to the laws of probability density functions and type of dependencies between input variables resulting in fuzzy (epistemic) uncertainty in decision-making tasks. Analysis of forecast uncertainty based on fuzzy stochastic approaches and its application to a number of problems in civil engineering and related fields is described in detail, for example, in the book [62]. However, the complexity of mathematical approaches and the interpretation of results limit the scientific ingenuity of the application of fuzzy stochastic SA methods.

SENSITIVITY ANALYSIS APPLIED TO MCDM

Two types of SA are most often mentioned in the literature: local SA and global SA (e.g., [63, 64]). Local SA determines a contribution of a given input parameter of a mathematical model to its output. Local methods do not attempt to fully explore the input parameter space. They examine small perturbations, usually one parameter at a time. Global SA examines a mathematical model in the presence of uncertain input parameters. SA of this type apportions output uncertainty to different sources of uncertainty in the input. Input and output uncertainties are expressed by probability distributions. Global SA applies perturbations of the entire space of input parameters. It is also argued that global SA should be used "in tandem" with uncertainty analysis and the latter should precede the former in practical applications [65, 66]. The aim of uncertainty analysis is to estimate variation in model output and SA apportions this variation to input parameters.

Local SA was applied in the recent past for better understanding of MCDM methods. Global SA was used for improving models which are related MCDM but do not fit strictly into its scheme. MCDM constitutes a special class of mathematical models. An application of SA to MCDM rises specific problems. However, any methodological specificity of performing SA for applications of MCDM in civil engineering is not known to us. In addition, combined applications of MCDM and SA to problems of civil engineering are few in number and deal mainly with selecting locations of buildings [67, 68]. A certain relation to civil engineering has applications of combined MCDM and SA to geographic information systems (see Malczewski and Rinner [69] and references cited therein). Therefore, the remainder of this section will be a general discussion on SA applications to a better understanding of MCDM methods.

Applications of SA in Deterministic MCDM and MCDM under Uncertainty

In the light of the aforementioned SA definitions, a deterministic (crisp) MCDM method can be interpreted as follows: elements of the matrix C and components of the vector w represent input parameters, the preference sequence π is model output, and the procedure relating π toCandw is a mathematical model. A sensitivity of π versus C and w is called ranking sensitivity or sensitivity to ranking stability (e.g., [70, 71]). With the deterministic input C and w, a sensitivity of π versus C and w can be determined by means of local SA methods.

Triantaphyllou and Sanchez applied local SA to determine the criterion c_j and the decision matrix element c_{ij} which is most critical to the ranking expressed by π [72, 73]. They applied sensitivity measures based on minimum changes in the weights w_j and matrix elements c_{ij} which cause changes between ranks of the alternatives a_i. Bevilacqua and Braglia used a simple wide-range variation of the weights w_j to explore shifts in the ranking π obtained by means of a deterministic AHP [74]. Any numerical sensitivity measure or graphical representation of SA results was not suggested in this study. Chang et al. and Wu et al. carried out SA of AHP results based on increasing values of w_j up to 35% [67, 75]. They expressed SA results graphically. This approach was applied to SA of results obtained with fuzzy decisionmaking matrix $\tilde{C}$ [71]. The results were produced by means of fuzzy TOPSIS and fuzzy AHP methods.

Another kind of perturbations used to reveal sensitivity of π versus w_j is exchanging positions of the weights within the vector w. Choudhary and Shankar used such perturbations for results obtained with a combined fuzzy AHP and TOPSIS method [76]. A total of np perturbations produces a set of preference sequences π_p (p = 1, 2,...,) which are treated as SA result and expressed graphically. In a series of methodologically similar articles, Awasthi et al. suggested using the sequences

π_p for a final ranking of the alternatives a_i [68, 77–79]. The sequences π_p were obtained for a problem with fuzzy weights $\overleftrightarrow{w}$ and fuzzy matrix $\overleftrightarrow{c}$. The alternatives ai were ranked by counting scores for each a_i according to its position in π_p. These authors also tried to use this scoring for a qualitative assessment of ranking sensitivity. However, any quantitative sensitivity measure was not suggested.

A quantitative measure of ranking sensitivity used in several studies is known as an average shift in ranks (ASR). Some of these studies fit into the scheme of MCDM and some have common elements with this field. In line with the notations used herein, ASR is expressed as the mean $m^{-1}\sum_{i=1}^{m}|r_i^{\varsigma} - r_i^{\text{ref}}|$ where r_i^{ς} is the rank of the alternative ai related to a perturbation ς and r_i^{ref} is the rank of ai in a reference (base) ranking. Saisana et al. used ASR as a model output for a global SA [80]. ASR was computed on the basis of a composite indicator. The reference ranks r_i^{ref} were obtained from one specific application of the indicator. Ben-Arieh used ASR in the format of MCDM for sort of a local SA [81]. ASR was applied to pairwise comparisons of alternative rankings obtained with different linguistic quantifiers (probabilities). They were used for calculating the weights w_j. The ranks r_i^{ς} and r_i^{ref} were obtained using pairs of different linguistic quantifiers. Later on, the same approach was used by other authors for assessing ranking sensitivity to fuzzy linguistic quantifiers [82–84]. ASR was applied also to a global SA related to MCDM. Ligman-Zielinska suggested using ASR as a scalar representation of output ranking and computing a sensitivity index based on variance of ASR [65]. Unfortunately, values of ASR depend on the choice of the reference ranks r_i^{ref} and this introduces certain arbitrariness in the process of SA. Apart from ASR, an alternative and well-elaborated scalar measure expressing a sensitivity of permutations within π to input of MCDM models is not known to us.

Sensitivity to Model Selection

One of the key problems of MCDM is a selection of method (model) relating input information expressed by C and w to the output ranking sequence π. If the MCDM method requires normalisation of the initial matrix C, the analyst will also face the problem of choosing among several normalisation rules (formulas) (e.g., [85]). Using different normalisation rules may also contribute to the variability of ranks within the sequence π.

An application of fuzzy numbers as elements of the matrix $\overleftrightarrow{c}$ and components of the vector $\overleftrightarrow{w}$ will introduce an additional problem of model selection. Membership functions of fuzzy numbers, $\mu_{\overleftrightarrow{w}_j}(w_j)$ and $\mu_{\overleftrightarrow{c}_{ij}}(c_{ij})$ may not necessarily be triangular as in the fuzziest MCDM applications (Links 1 and 2, Figure 4). The choice of the type of fuzzy numbers in $\overleftrightarrow{c}$ and components of the vector $\overleftrightarrow{w}$ is a subjective exercise and introduces arbitrariness into MCDM process. Similar statements can be made about the use of random variables in $\tilde{C}$ and $\tilde{w}$. A specification of MCDM model input will require selecting specific types of probability densities $f_{\tilde{c}_{ij}}(c_{ij} \mid \mu_{ij}, \sigma_{ij})$ and $f_{\tilde{w}_j}(w_j \mid \alpha_j, \beta_j)$ for decision matrix elements $\tilde{c}_{ij}$ and criteria weights $\tilde{w}j$ (Links 3 and 4, Figure 4). In case of $\tilde{w}_j$, the selection of distribution type and then specification of its parameters α_j and β_j will be purely subjective task. However, in some cases, statistical data on the random components $\tilde{c}_{ij}$ of $\tilde{C}$ can be available. In such cases, distribution type of $\tilde{c}_{ij}$ will be dictated by this data. Consequently, we can speak about three SA problems:

(1) Sensitivity of the ranking sequence π to the selection of MCDM method.

(2) Sensitivity of π to the choice of normalisation rule.

(3) Sensitivity of π to the selection of membership function type in case of fuzzy MCDM and probability density type in case of stochastic MCDM.

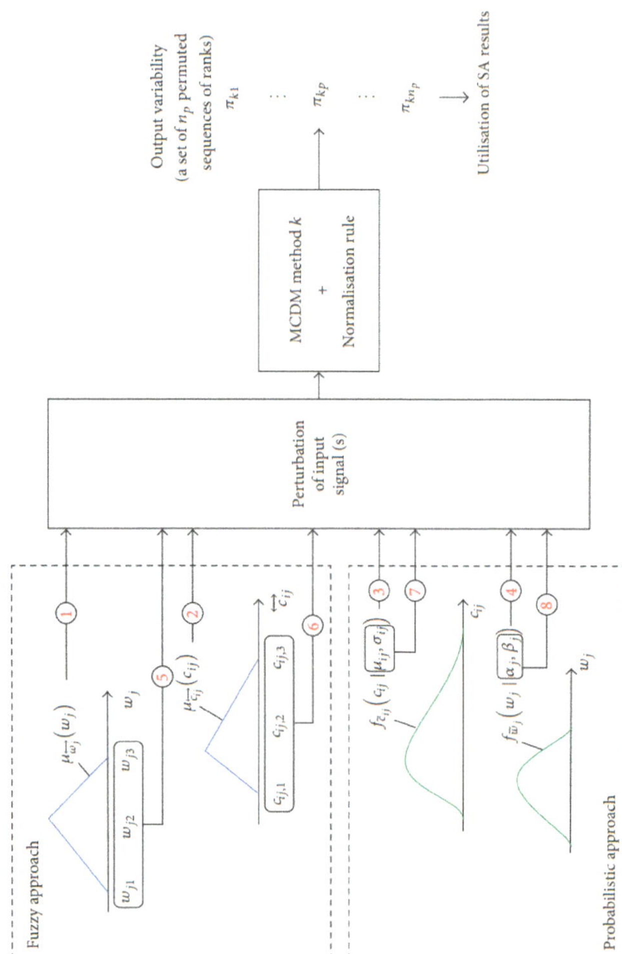

Figure 4: Possible SA links related to the input space of a MCDM problem and a specific MCDM method.

The first two problems will be present in both deterministic MCDM and MCDM under uncertainty. The third problem will arise with the need to introduce uncertainties in MCDM process. To the best of our knowledge, a systematic, in-depth solution of these three problems is not available in the MCDM literature to date.

Apart from model selection problem, the use of the membership functions $\mu_{\overleftrightarrow{w}_j}(w_j)$ and $\mu_{\overleftrightarrow{c}_{ij}}(c_{ij})$ and the density functions $f_{\tilde{c}_{ij}}(c_{ij} \mid \mu_{ij}, \sigma_{ij})$ and $f_{\tilde{w}_j}(w_j \mid \alpha_j, \beta_j)$ will give rise to a problem of assessing sensitivity of the ranking sequence π to parameters of these functions (Links 5 to 8, Figure 4). An association of all or some w_j and c_{ij} with two or more parameters of membership function or density function may substantially increase the dimensionality of input space. This can encumber SA process, especially in case of global SA. It will be necessary to assign additional distributions to this parameter in order to carry out Monte Carlo analysis of a MCDM model under study

MEASURES OF RISK AND RELIABILITY IN THE MCDM RELATED TO CIVIL ENGINEERING

Civil engineering systems can be damaged by deliberate assaults, actions induced during industrial accidents and extremes of nature. Damage to components of civil engineering systems induces mechanical and thermal actions called often abnormal or accidental ones. Incidents with abnormal actions are relatively rare, short-lasting, and usually unexpected events. They can cause serious harm and sometimes catastrophic consequences [86]. In terms of civil engineering, such incidents are called "abnormal situations" or "accidental situations." The latter term is used in the widely known standards ISO 2394 and ENV 1991-1. An adequate design of civil engineering systems for abnormal situations will result in resilient buildings and infrastructure able to avoid or absorb damage without undergoing a complete failure [87– 89].

MCDM and Safety in Civil Engineering

The design of components of safety critical civil engineering systems for abnormal situations requires, among other things, comparing alternative design solutions of these components. In terms of MCDM, they can be called alternative designs or simply alternatives a_i (i = 1, 2,...,). Interalternative comparisons of ai must include criteria cij expressing safety of ai or, alternatively, risk posed by a_i. The row vector of criteria, $c_i = (c_{i1}, c_{i2}, ..., ..., c_{in})$, will include also elements which are not necessarily related to safety, for instance, economic, functional, and aesthetic criteria.

The adequacy of the design of critical objects of civil engineering for abnormal situations is assured by applying methods of reliability theory and probabilistic risk assessment (PRA) [90]. These two fields of engineering are closely related and the criteria c_{ij} can be specified by applying methods developed in each of them.

Values of c_{ij} can be probabilities of failure or quantitative estimates of risk related to the designs a_i [91]. Practical applications of MCDM to the design for abnormal situations will face the problem of uncertainty related to failure probabilities and risk estimates. Quantitative measures of this uncertainty can be introduced into MCDM problems. A wellestablished platform of uncertainty modelling in PRA is the Bayesian statistical theory or, in brief, the Bayesian approach [92]. In line with this approach, the uncertainty in potential safety-related criteria c_{ij} is divided into two kinds: stochastic (aleatory) and state-of-knowledge (epistemic) uncertainty [93–96]. Uncertainties of either kind can be accommodated in MCDM problems.

An Integration of Reliability Measures and Related Quantities into MCDM

A failure probability pfi characterising a particular alternative design a_i can be used as a MCDM criterion c_{ij} [85, 91]. A value of the failure probability p_{fi} assigned to the alternative a_i accounts for the possibility of its potential failures. The probability pfi can be "mechanically" included into a MCDM problem as one of components of the vector c_i. However, in many cases the probabilities p_{fi} will be uncertain in the epistemic sense. The uncertainty in p_{fi} may stem primarily from sparse information on accidental actions. Quantitatively this uncertainty can be expressed by modelling the probabilities p_{fi} as epistemic random variables $\tilde{p}_{fi}$ [97]. If the uncertain failure probability $\tilde{p}_{fi}$ of a_i is taken as a MCDM criterion, the vectors c_i will contain at least one random component and can be replaced by the stochastic vectors $\tilde{C}_i$ given by $(\tilde{p}_{fi1}, \tilde{p}_{fi2}, \ldots, \tilde{p}_{fik_i}, c_{k_i+1}, c_{k_i+2}, \ldots, c_{in})$, where ki is the number of failure modes of ai. Some or all criteria in the vector just mentioned can be uncertain in the stochastic (aleatory) sense. For instance, such criteria can be time of construction or cost of ai. Uncertainty related to the stochastic components of $\tilde{C}_i$ can be expressed by the random variables $\tilde{c}ij$. With the random vectors $\tilde{C}_i$, a MCDM problem will have to be solved by applying a decision-making matrix $\tilde{C}$ some or all elements of which are random variables. Stochastic MCDM methods must be applied to deal with the matrix $\tilde{C}$.

Vaidogas and Zavadskas suggested introducing the failure probabilities p_{fil} ($1 \leq l \leq k_i$) into a MCDM problem indirectly, through comparison of the total (life-cycle) utilities $u_{tot,}$ related to the alternatives a_i [91]. The utility $u_{tot,}$ is expressed as a difference between expected benefit from ai and a total cost of a_i. The failure probabilities p_{fil} are incorporated into the total cost of ai through the cost of failures expressed by the sum $\sum_{l=1}^{k_i} p_{fil} c_{fl}$, where c_{fl} is the anticipated cost of failure according to the failure mode l.

Vaidogas et al. applied reliability-oriented MCDM for a selection among alternative construction projects of a building [85]. Reliability of the alternative projects a_i was expressed as a probability that a specified construction time will not be exceeded. A further application of MCDM was a ranking of designs of a reinforced concrete slab with different floorings. Probabilities of two failure modes of the slab were used as the criteria c_{ij}: probability of collapse and probability of excessive deflection.

An Integration of Risk Estimates into MCDM

The alternatives a_i can represent hazardous industrial objects which pose risk to people and environment in the form of industrial accidents. The magnitude of consequences of such accidents can range between minor and catastrophic consequences [98]. In the European Union, most of hazardous objects are regulated by Seveso Directives (currently Seveso III Directive) [99]. Such objects are assessed by means of formal methods developed in the field of PRA [93, 100].

A risk related to the design ai is a very informative characteristic suitable for inclusion into a MCDM problem [98]. In line with PRA, the risk related to a_i is expressed by the set $\{(\lambda_{is}, c_{is}, \mathrm{m}_{is}), s = 1, 2, \ldots, n_{ai}\}$, in which λ_{is} and c_{is} are likelihood-consequence pairs, m_{is} are the vectors of magnitudes (severities) of c_{is}, and n_{ai} is the number of accident scenarios related to a_i. The vector m_{is} is given by a certain number n_s of magnitudes m_{isj} ($j = 1, 2..._s$). Zavadskas and Vaidogas suggested expressing the criteria c_{ij} in the form of expected magnitudes $\overline{m}_{ij}$ [97]. The component c_{ij} of the decision-making matrix C represented by $\overline{m}_{ij}$ is computed as the sum $\sum_{s=1}^{n_{ai}} \lambda_{is} m_{isj}$. As in the case of uncertain failure probabilities $\tilde{p}_{fi}$, the likelihoods λis can be uncertain quantities modelled by epistemic random variables $\tilde{\lambda}_{is}$. In most instances, the variables $\tilde{\lambda}_{is}$ will represent uncertain annual frequencies of the consequences c_{is}. The presence of the random likelihood $\tilde{\lambda}_{is}$ will stochastise the expected severities $\overline{m}_{ij}$ and they will turn into random variables, for example, $\overline{\tilde{m}}_{ij}$. Consequently, a MCDM problem will turn into a stochastic one with a random decision-making matrix $\tilde{C}$. The row $\tilde{c}_i$ of $\tilde{C}$ will be expressed as $(\overline{\tilde{m}}_{i1}, \overline{\tilde{m}}_{i2}, \ldots, \overline{\tilde{m}}_{in_s}, c_{i,n_s+1}, c_{i,n_s+2}, \ldots, c_{in})$ with $n - n_s \geq 1$, where the components denoted by the letter "c" can be either deterministic or stochastically uncertain quantities. With the vectors $\tilde{c}i$ including the risk-related components $\overline{\tilde{m}}_{ij}$, the matrix $\tilde{C}$ can be expressed as a two-block matrix $[\tilde{C}_1 \mid C_2]$. The $m \times ns$ matrix $\tilde{C}_1$ reflects risk estimates of the alternatives ai, whereas the $m \times (n - n_s)$ matrix C_2 includes criteria which are not directly related to the risk. Stochastic MCDM methods will be necessary to solve the MCDM problem with the matrix $[\tilde{C}_1 \mid C_2]$.

Vaidogas and Sak ˇ enait ˙ e applied the risk-based MCDM to ˙ a choice among alternative sprinkler systems [101]. Zhou et al. used a safety-oriented MCDM for solving decision-making problems of hydropower construction project management [102]. Catrinu and Nordgard applied PRA and MCDM to a ° management of electricity distribution system asset [103].

In recent years, a fairly large number of publications considered an application of MCDM methods for handling managerial risk related to construction projects and running built facilities. Although risk of this type differs by nature from the "pure" risk posed by (to) physical objects, assessments of managerial and "pure" risk are related through a need to deal with uncertainties in risky objects or processes. Nieto-Morote and Ruz-Vila used fuzzy AHP method for assessing building projects and selection of contractors [104, 105]. Xiang et al. applied fuzzy AHP for assessing risk arising at a construction of submerged floating tunnels [106]. Wang et al. used AHP in combination with other decision-making methods to assess risk posed by exploitation of bridges [107]. El-Abbasy et al. applied AHP method together with Monte Carlo simulation for selecting contractors of a highway project [108]. The studies just listed involve elements of a nonprobabilistic uncertainty quantification based on fuzzy sets. As the "pure" risk is always a part of managerial risk, uncertain criteria $\tilde{c}_{ij}$ specified by means of probabilistic methods of PRA can be included into the decision matrix C alongside with "fuzzy" criteria $\overleftrightarrow{c}_{ij}$. However, MCDM methods which allow a simultaneous juggling of "probabilistic" and "fuzzy" criteria $\tilde{c}_{ij}$ and $\overleftrightarrow{c}_{ij}$ do not exist at present, to the best of our knowledge.

MCDM and Fire Protection of Civil Engineering Objects

Fire is a prevailing hazard in most objects of civil engineering. Fire accidents often occur on construction sites [109, 110]. As regards fire protection, MCDM methods were used until now mainly for ranking attributes expressing fire safety of completed buildings. AHP method was applied for developing weights of fire safety attributes in the socalled Edinburgh study [111]. A stochastic AHP was used by Zhao et al. to rank attributes of building fire safety [112]. Wong et al. used attributes of fire detection and alarm systems among a fairly large number of characteristics of an intelligent building. They applied two MCDM methods, AHP and ANP, to rank these characteristics [113, 114]. Vaidogas and Sak ˇ enait ˙ e formulated a number of MCDM problems, ˙ in which building fire safety is considered with respect to economics of fire protection: selection among existing buildings, building projects, and construction materials [115]. Vaidogas and Linkute considered also problems of decision- ˙ making in the design of structures used for protection of built property against accidental explosions [116].

MCDM IN INNOVATIVE AREAS OF CIVIL ENGINEERING: A LOOK AT DECISION-MAKING UNDER UNCERTAINTY

Developing Sustainable and Energy Efficient Building

Sustainability is a natural subject of MCDM, because it automatically includes three subsets of criteria, involving economics, environmental, and social aspects. When solving problems of sustainable building, the fourth subset of criteria, involving engineering-technological dimensions, is also necessary. One of the innovative themes in sustainable construction is related to using materials of low embodied energy and energy efficient applications. However, such things as future and real building cost, environmental impact, and future social status of a constructed facility are very uncertain if considered in a long sight. For instance, large built areas in Hamburg (Germany) lost a lot of image due to a social downgrade of inhabitants. Also a lot of industrial and farming buildings having perfect infrastructure were left abandoned due to political and respective economic changes in post-Soviet states in Eastern Europe [117]. These buildings and territories make a great potential for further redevelopment as the recent trends in construction emphasize rehabilitation instead of occupying new territories, wasting building materials, and so forth. Building rehabilitation should be performed in accordance with principles of sustainable development, thus combining a number of usually conflicting and hardly measurable aspects.

The usefulness and even necessity of application of decision-making methods under uncertainty for aforementioned problems are summarized below. It is worth mentioning that DM under uncertainty is more characteristic to rehabilitation than to new construction. New construction is more regulated by technical norms, standards, and comprehensive planning. However, aspiration to redevelop a building in the most proper way is certainly a multiple criteria DM problem. Problem related to upgrading abandoned or depreciated buildings as well as physically and morally deteriorated built environment can generate several potential alternatives as demolishing depreciated building and building new structures (technically sound approach but contradicting to principles of sustainability), dismantling, reusing, or recycling of building materials (partly meeting principles of sustainability), renewal according to up-to-date requirements and using a building for previous purposes, or upgrading a building with intention to use for other purpose (e.g., conversion of industrial buildings to commercial or apartments). When searching for the best alternative solution from several potential decision alternatives a_i, a set of criteria c_j should be defined and the matrix C can be formulated. The main challenge is in setting criteria values c_{ij} related to the alternatives *ai* and criteria relative significances expressed by the weights w_j.

The four problems related to uncertainty measures could be identified. The first group of criteria has to assess features of building or surrounding environment, regional peculiarities using hardly measurable, vague, or incomplete data. Examples of criteria associated with vague, incomplete, or hardly measurable information, having no fixed values or expressed by linguistic scale, are architectural or historical value of a building (in terms of importance of preserving a building or its elements in a time of redevelopment), aesthetic appearance of a building and harmony with the environment after implementing a specific upgrading alternative, reflection of period norms, technologies, and so forth (in a case of upgrading vernacular buildings to contemporary building norms [14]). Examples of criteria associated with varying information, either related to a specific terms of a particular project or based on periodical or territorial measures, and expressed as interval data are building redevelopment costs (depending on particular material costs, labour costs in a locality, tax rates, etc.), energy savings (embodied energy, different technologies, and varying costs of energy), material and foreign investments in the area, changes in level of unemployment of local population or population activity index, and characterising social-economic value of a civil engineering project. The aforementioned uncertainty could be determined as fuzziness and managed with the help of fuzzy sets. Then, the elements c_{ij} of C are characterised by fuzzy sets and denoted as $\overleftrightarrow{c_{ij}}$. In that case, a decision-making matrix containing fuzzy elements $\overleftrightarrow{C}$ could be composed.

Another group of criteria should assess development possibilities and should be able to reflect probable changes over the time, involving long-term prediction. Examples of criteria could be risk related to political and respective economic changes, technology innovations in civil engineering and related areas, state/region business foresights, cash flow and net present value of a potential civil engineering project, life quality parameters and a real value of a building, state income from business, and property taxes, and so forth. In the current case, risk factors should be taken into account and probabilistic modelling should be applied when the elements c_{ij} of the decision-making matrix C are modelled by random variables $\tilde{c}_{ij}$ expressing uncertainty in criteria values, usually in Fisherian or Bayesian format. A matrix $\tilde{C}$ is subject to stochastic DM.

The third problem is based on combination of rather different criteria into a single solution. Combination of criteria is analysed in two senses. At first, a problem is in establishing relative significances of criteria in a particular task, based on decision-makers' preferences and peculiarities of the analysed problem. It is usually rather subjective and uncertain process. How could you compare significance of a local (e.g., cost of thermal insulation material) and a global criterion (e.g., energy savings, CO_2 emissions, and their impact on global warming) for different stakeholders? How could you express uncertain or linguistic decisionmakers' preferences in a numerical format? What could be better specification of the weights *wj* in the case they are of uncertain quantities: $\overleftrightarrow{W}$ or $\tilde{w}$?

The next issue as the fourth problem is combining fuzzy and stochastic uncertainties in a single task. Based on the above assumptions on criteria $\overleftrightarrow{c_{ij}}$ and $\tilde{c}_{ij}$ as well as relative significances of criteria ($\overleftrightarrow{W}$ or $\tilde{w}$), there is the need to combine fuzzy and probabilistic representation of uncertainties in one initial decision-making matrix. Next comes the problem of choice among available MCDM methods or their extensions to be able to deal with fuzzy and stochastic data simultaneously. Thus, development of sustainable construction can be solved adequately by combining fuzzy and probabilistic representation of uncertainties, applying MCDM methods able to deal with fuzzy and stochastic input data, and including SA, that is, analysing effect related to the input data on the optimal solution of a MCDM problem.

Possibilities to Apply MCDM Methods within BIM Process

Building information modelling (BIM) goes through a series of levels of development (LODs) which represent an increasing level of detailing [118, 119]. The "fuzzy" detailing in such levels as LOD100 or LOD200 naturally introduces uncertainties in a DM process. They must be taken into account as long as MCDM is applied in this early design stage.

In the context of building design organised as a successive passing of LODs, a comparison of the alternatives a_i will make sense if *ai* will represent the same LOD. In principle, branching of the design process into the alternatives a_i is possible on the conceptual level LOD100, as long as it is possible to characterise the vague information expressed by conceptual designs by some qualitative or quantitative criteria c_j. Most of them will express subjective opinions of architect and client, because very little can be measured quantitatively and objectively at this stage. This may require to express the opinions by fuzzy numbers or probability distributions and to solve a MCDM problem with respective matrices $\overleftrightarrow{C}$ and $\tilde{C}$.

Higher levels of detail represented by LODs from 200 to 400 give better opportunity to formulate and solve a MCDM problem. Branching of the design process into the alternatives a_i can be done on each of these levels. However, uncertainties may still be present because the designs will not be fully finished. Some characteristics of a_i will be uncertain independently of available LOD. For instance, the final and precise construction cost and duration related to a_i will be known only after completion of construction process.

A development of a large number of the alternatives *ai* and characteriation of each of them by the criteria c_j can encumber the design process. In essence, a solution of a MCDM problem will require to prepare *m* different designs of building related to the LOD reached in the design process. Thus, one can say that there is a "mega-uncertainty" related to the number of LODs at which an application of MCDM will be most efficient.

Deciding on Security and Safety of Buildings

Security and safety are things and activities which are done in order to keep people and organisation safe from attack, harm, and damage. Three primary areas of security are physical security, information security, and personnel security [120]. Safety is achieved by protecting objects of civil engineering against fire, lightning, earthquakes, extreme winds, and contamination events which can negatively affect indoor air quality.

Measures and procedures of physical security are used to improve protection of an organisation against threats and vulnerabilities. Physical security of built property includes elements related to civil engineering: access control, barriers, doors, fencing, perimeter security, vaults, walls, and windows. Physical security is also interrelated with safety of built property. Malicious acts of people committed outside or inside perimeters of built objects may not only induce structural damage but also cause severe harm to their occupants. Examples of such acts are bombings, shootings, deliberate vehicular impacts, and arsons. The saddest events of this kind are September 11 attacks on the world trade centre in New York.

A higher investment in security systems may reduce damaging potential of malicious acts. If such acts will happen in spite of all security measures, damage caused by them can be limited by a better design for safety. Such design involves architectural and structural solutions enhancing safety and provision of better safety systems (alarms, fire and evacuation control systems, monitors of indoor air, etc.). Effectiveness of security and safety can be characterised by at least two MCDM criteria (one for security and one for safety). Further criteria related to them may be costs of security and safety systems. We can add criteria which measure impact of architectural and structural solutions on security and safety. Presence of a number of criteria makes out a case for an application of MCDM to design of buildings with respect to security and safety.

In a MCDM oriented towards security and safety, the alternatives a_i may represent different configurations of physical security systems and different solutions of safety systems. The alternatives a_i can also represent different architectural and structural solutions influencing security and safety. New structural solutions of tall buildings suggested since September 11 can serve as an example [121]. The architectural and structural solutions may seriously influence security and safety. For instance, a low-rise building provides better opportunities for escape from fire and firefighting than a tall one. However, a low-rise building will have much longer external perimeter to be guarded than a tall building with the same floor area. On the other hand, a tall building is a better target for shooting or an attack by a plane.

A simultaneous consideration of security and safety as well as architectural and structural aspects in the same MCDM problem may pose also several problems at generating the set of a_i. The number of a_i can be large due to a wide variety of security and safety systems available on the market. This number can be also increased by a large number of architectural and design solutions of the same building. A generation of the alternatives a_i can be influenced by legal and regulatory restrictions which must be regarded simultaneously at a formulation of the designs represented by a_i.

The criteria c_j used for evaluating the alternative triplets security-safety-architecture can be simple: amount and cost security equipment, number of security personnel, number of escape roots, minimum distance from a guarded object at which unauthorised persons must be stopped, and so forth. Such criteria are obvious and their values can be specified with relative ease. However, they do not reflect in-depth effectiveness of security and safety systems.

The effectiveness of safety systems can be expressed by applying methods of PRA as discussed in Section 5.2. Safety related criteria c_j can be either likelihood or expected severities of accident scenarios. These accidents can be initiated by security breaches. For instance, an arsonist, after an unauthorised access to secured areas of a building, can initiate fires in different places. They will lead to different fire scenarios. The higher is an estimated level of safety and robustness of the building represented by a_i, and the smaller should be likelihood and consequences of potential accidents. The possibility of different accident scenarios does not allow to measure the level of safety by a single MCDM criterion.

The effectiveness of a physical security system can also be expressed by several criteria c_j reflecting different identifiable scenarios of security breach. Examples of such scenarios are unauthorised access to secured areas, armed attack from outside of object perimeter, an attempt to hit a protected object with a vehicle, external disruption of power supply, and gas attack from outside the building. As in the case of PRA, likelihood measures can be estimated and consequences of security breaches assessed. The likelihood and consequence severities can be used as MCDM criteria c_j.The higher is level of physical security, and the smaller must be values of such criteria.

Attacks and adverse physical events impairing security and safety of built property are in most cases rare events. Data on such events is generally sparse. Therefore, estimating the likelihood of security and safety events will almost inevitably face the problem of quantifying epistemic uncertainties in the manner of PRA. This will lead to an inclusion of uncertain criteria into MCDM which are uncertain in the epistemic sense. In other words, problems of MCDM, in which security and safety of built property are regarded, can be solved adequately by composing the random decisionmaking matrix $\tilde{C}$.

CONFLICT OF INTERESTS

The authors declare that there is no conflict of interests regarding the publication of this paper.

ACKNOWLEDGMENT

The part of the paper done by the author from Czech Republic has been worked out under Project no. LO1408 "AdMAs UP."

REFERENCES

1. M. M. Wiecek, M. Ehrgott, G. Fadel, and J. R. Figueira, "Multiple criteria decision making for engineering," Omega, vol. 36, no. 3, pp. 337–339, 2008.
2. A. Mardani, A. Jusoh, and E. K. Zavadskas, "Fuzzy multiple criteria decision-making techniques and applications—two decades review from 1994 to 2014," Expert Systems with Applications, vol. 42, no. 8, pp. 4126–4148, 2015.
3. J. J. H. Liou and G.-H. Tzeng, "Comments on 'Multiple criteria decision making (MCDM) methods in economics: an overview'," Technological and Economic Development of Economy, vol. 18, no. 4, pp. 672–695, 2012.
4. E. K. Zavadskas, Z. Turskis, and S. Kildiene, "State

of art surveys of overviews on MCDM/MADM methods," Technological and Economic Development of Economy, vol. 20, no. 1, pp. 165–179, 2014.

5. T. Baležentis and A. Baležentis, "A survey on development and applications of the multi-criteria decision making method MULTIMOORA," Journal of Multi-Criteria Decision Analysis, vol. 21, no. 3-4, pp. 209–222, 2014.
6. M. Behzadian, R. B. Kazemzadeh, A. Albadvi, and M. Aghdasi, "PROMETHEE: a comprehensive literature review on methodologies and applications," European Journal of Operational Research, vol. 200, no. 1, pp. 198–215, 2010.
7. M. Behzadian, S. Khanmohammadi Otaghsara, M. Yazdani, and J. Ignatius, "A state-of the-art survey of TOPSIS applications," Expert Systems with Applications, vol. 39, no. 17, pp. 13051–13069, 2012.
8. R. Ureña, F. Chiclana, J. A. Morente-Molinera, and E. Herrera-Viedma, "Managing incomplete preference relations in decision making: a review and future trends," Information Sciences, vol. 302, no. 1, pp. 14–32, 2015.
9. J. Chai, J. N. K. Liu, and E. W. T. Ngai, "Application of decision-making techniques in supplier selection: a systematic review of literature," Expert Systems with Applications, vol. 40, no. 10, pp. 3872–3885, 2013.
10. G. Kabir, R. Sadiq, and S. Tesfamariam, "A review of multi-criteria decision-making methods for infrastructure management," Structure and Infrastructure Engineering: Maintenance, Management, Life-Cycle Design and Performance, vol. 10, no. 9, pp. 1176–1210, 2014.
11. D. Jato-Espino, E. Castillo-Lopez, J. Rodriguez-Hernandez, and J. C. Canteras-Jordana, "A review of application of multi-criteria decision making methods in construction," Automation in Construction, vol. 45, pp. 151–162, 2014.
12. C. Kahraman, "Multi-criteria decision making methods and fuzzy sets," in Fuzzy Multi-Criteria Decision Making, C. Kahraman, Ed., vol. 16 of Springer Optimization and Its Applications, pp. 1–18, Springer, New York, NY, USA, 2008.
13. J. Tamošaitiene and E. Gaudutis, "Complex assessment of structural systems used for high-rise buildings," Journal of Civil Engineering and Management, vol. 19, no. 2, pp. 305–317, 2013.
14. E. Šiožinyte, J. Antuchevičiene, and V. Kutut, "Upgrading the old vernacular building to contemporary norms: multiple criteria approach," Journal of Civil Engineering and Management, vol. 20, no. 2, pp. 291–298, 2014.
15. J. L. Hougaard and T. Baležentis, "Fuzzy efficiency without convexity," Fuzzy Sets and Systems, vol. 255, pp. 17–29, 2014.
16. L. A. Zadeh, "Fuzzy set theory and probability theory: what is the relationship?" in International Encyclopedia of Statistical Science, pp. 563–566, Springer, 2014.
17. D. Kelly and C. Smith, Bayesian Inference for Probabilistic Risk Assessment: A Practitioner's Guidebook, Springer, New York, NY, USA, 2011.
18. J.-J. Wang, Y.-Y. Jing, C.-F. Zhang, and J.-H. Zhao, "Review on multi-criteria decision analysis aid in sustainable energy decision-making," Renewable and Sustainable Energy Reviews, vol. 13, no. 9, pp. 2263–2278, 2009.
19. Y.-M. Wang and Y. Luo, "Integration of correlations with standard deviations for determining attribute weights in multiple attribute decision making," Mathematical and Computer Modelling, vol. 51, no. 1-2, pp. 1–12, 2010.
20. C. Fu and Y. Wang, "An interval difference based evidential reasoning approach with unknown attribute weights and utilities of assessment grades," Computers and Industrial Engineering, vol. 81, pp. 109–117, 2015.
21. E. Triantaphyllou and S. H. Mann, "An examination of the effectiveness of multi-dimensional decision-making methods: a decision-making paradox," International Journal of Decision Support Systems, vol. 5, no. 3, pp. 303–312, 1989
22. L. A. Zadeh, "Fuzzy sets," Information and Control, vol. 8, no. 3, pp. 338–353, 1965.
23. P. J. M. van Laarhoven and W. Pedrycz, "A fuzzy extension of Saaty's priority theory," Fuzzy Sets and Systems, vol. 11, no. 3, pp. 229–241, 1983.
24. C. Kahraman, S. C. Onar, and B. Oztaysi, "Fuzzy multicriteria decision-making: a literature review,"International Journal of Computational Intelligence Systems, vol. 8, no. 4, pp. 637–666, 2015.
25. E. K. Zavadskas, J. Antucheviciene, S. H. R. Hajiagha, and S. S. Hashemi, "Extension of weighted aggregated sum product assessment with interval-valued intuitionistic fuzzy numbers (WASPAS-IVIF),"Applied Soft Computing, vol. 24, pp. 1013–1021, 2014.
26. S. H. Razavi Hajiagha, S. S. Hashemi, and E. K. Zavadskas, "A complex proportional assessment method for group decision making in an interval-valued intuitionistic fuzzy environment," Technological and Economic Development of Economy, vol. 19, no. 1, pp. 22–37, 2013.
27. H.-C. Liu, X.-J. Fan, P. Li, and Y.-Z. Chen, "Evaluating the risk of failure modes with extended MULTIMOORA method under fuzzy environment," Engineering Applications of Artificial Intelligence, vol. 34, pp. 168–177, 2014.
28. H.-C. Liu, J.-X. You, C. Lu, and Y.-Z. Chen, "Evaluating health-care waste treatment technologies using a hybrid multi-criteria decision making model," Renewable and Sustainable Energy Reviews, vol. 41, pp. 932–942, 2015.
29. H.-C. Liu, J.-X. You, Ch. Lu, and M.-M. Shan, "Application of interval 2-tuple linguistic MULTIMOORA method for health-care waste treatment technology evaluation and selection," Waste Management, vol. 34, no. 11, pp. 2355–2364, 2014.
30. T. Balezentis, S. Z. Zeng, and A. Balezentis, "MULTIMOORA-IFN: a MCDM method based on intuitionistic fuzzy number for performance management," Economic Computation and Economic Cybernetics Studies and Research, vol. 48, no. 4, pp. 85–102, 2014.

31. Z.-H. Li, "An extension of the MULTIMOORA method for multiple criteria group decision making based upon hesitant fuzzy sets," Journal of Applied Mathematics, vol. 2014, Article ID 527836, 16 pages, 2014.
32. Z. Turskis and E. K. Zavadskas, "A new fuzzy additive ratio assessment method (ARAS-F). Case study: the analysis of fuzzy Multiple Criteria in order to select the logistic centers location," Transport, vol. 25, no. 4, pp. 423–432, 2010.
33. M. Zamani, A. Rabbani, A. Yazdani-Chamzini, and Z. Turskis, "An integrated model for extending brand based on fuzzy ARAS and ANP methods," Journal of Business Economics and Management, vol. 15, no. 3, pp. 403–423, 2014.
34. A. S. Ghadikolaei, S. K. Esbouei, and J. Antucheviciene, "Applying fuzzy MCDM for financial performance evaluation of Iranian companies," Technological and Economic Development of Economy, vol. 20, no. 2, pp. 274–291, 2014.
35. N. Rikhtegar, N. Mansouri, A. A. Oroumieh, A. Yazdani-Chamzini, E. K. Zavadskas, and S. Kildienė, "Environmental impact assessment based on group decision-making methods in mining projects,"Economic Research-Ekonomska Istraživanja, vol. 27, no. 1, pp. 378–392, 2014.
36. M. Kabak, E. Köse, O. Kirilmaz, and S. Burmaoğlu, "A fuzzy multi-criteria decision making approach to assess building energy performance," Energy and Buildings, vol. 72, pp. 382–389, 2014.
37. M. Kucukvar, S. Gumus, G. Egilmez, and O. Tatari, "Ranking the sustainability performance of pavements: an intuitionistic fuzzy decision making method," Automation in Construction, vol. 40, pp. 33–43, 2014.
38. A. H. Azadnia, M. Z. M. Saman, and K. Y. Wong, "Sustainable supplier selection and order lot-sizing: an integrated multi-objective decision-making process," International Journal of Production Research, vol. 53, no. 2, pp. 383–408, 2015.
39. S. M. Hashemian, M. Behzadian, R. Samizadeh, and J. Ignatius, "A fuzzy hybrid group decision support system approach for the supplier evaluation process," The International Journal of Advanced Manufacturing Technology, vol. 73, no. 5–8, pp. 1105–1117, 2014.
40. M. K. Ghorabaee, M. Amiri, J. S. Sadaghiani, and G. H. Goodarzi, "Multiple criteria group decision-making for supplier selection based on COPRAS method with interval type-2 fuzzy sets," The International Journal of Advanced Manufacturing Technology, vol. 75, no. 5–8, pp. 1115–1130, 2014.
41. S. Zeng, T. Balezentis, J. Chen, and G. Luo, "A projection method for multiple attribute group decision making with intuitionistic fuzzy information," Informatica, vol. 24, no. 3, pp. 485–503, 2013.
42. S. M. Mousavi, B. Vahdani, R. Tavakkoli-Moghaddam, and N. Tajik, "Soft computing based on a fuzzy grey group compromise solution approach with an application to the selection problem of material handling equipment," International Journal of Computer Integrated Manufacturing, vol. 27, no. 6, pp. 547–569, 2014.
43. A. Hadi-Vencheh and A. Mohamadghasemi, "A new hybrid fuzzy multi-criteria decision making model for solving the material handling equipment selection problem," International Journal of Computer Integrated Manufacturing, vol. 28, no. 5, pp. 534–550, 2015.
44. S. Kiris, "Multi-criteria inventory classification by using a fuzzy analytic network process (ANP) approach," Informatica, vol. 24, no. 2, pp. 199–217, 2013.
45. B. Bairagi, B. Dey, B. Sarkar, and S. Sanyal, "Selection of robot for automated foundry operations using fuzzy multi-criteria decision making approaches," International Journal of Management Science and Engineering Management, vol. 9, no. 3, pp. 221–232, 2014.
46. Ü. Kurt, "The fuzzy TOPSIS and generalized Choquet fuzzy integral algorithm for nuclear power plant site selection—a case study from Turkey," Journal of Nuclear Science and Technology, vol. 51, no. 10, pp. 1241–1255, 2014.
47. E. K. Zavadskas, Z. Turskis, and V. Bagočius, "Multi-criteria selection of a deep-water port in the Eastern Baltic Sea," Applied Soft Computing Journal, vol. 26, pp. 180–192, 2014.
48. W. Zhang, K. Sun, C. Lei, Y. Zhang, H. Li, and B. F. Spencer, "Fuzzy analytic hierarchy process synthetic evaluation models for the health monitoring of shield tunnels," Computer-Aided Civil and Infrastructure Engineering, vol. 29, no. 9, pp. 676–688, 2014.
49. S. Avikal, R. Jain, and P. K. Mishra, "A Kano model, AHP and M-TOPSIS method-based technique for disassembly line balancing under fuzzy environment," Applied Soft Computing Journal, vol. 25, pp. 519–529, 2014.
50. S. Avikal, P. K. Mishra, and R. Jain, "A fuzzy AHP and PROMETHEE method-based heuristic for disassembly line balancing problems," International Journal of Production Research, vol. 52, no. 5, pp. 1306–1317, 2014.
51. C. H. Wang and H. S. Wu, "A novel framework to evaluate programmable logic controllers: a fuzzy MCDM perspective," Journal of Intelligent Manufacturing, 10 pages, 2014.
52. T.-M. Yeh, F.-Y. Pai, and C.-W. Liao, "Using a hybrid MCDM methodology to identify critical factors in new product development," Neural Computing and Applications, vol. 24, no. 3-4, pp. 957–971, 2014.
53. L. Anojkumar, M. Ilangkumaran, and V. Sasirekha, "Comparative analysis of MCDM methods for pipe material selection in sugar industry," Expert Systems with Applications, vol. 41, no. 6, pp. 2964–2980, 2014.
54. C. Tan, W. H. Ip, and X. Chen, "Stochastic multiple criteria decision making with aspiration level based on prospect stochastic dominance," Knowledge-Based Systems, vol. 70, pp. 231–241, 2014.
55. T. L. Saaty, "A scaling method for priorities in hierarchical structures," Journal of Mathematical Psychology, vol. 15, no. 3, pp. 234–281, 1977.

56. L. G. Vargas, "Reciprocal matrices with random coefficients," Mathematical Modelling, vol. 3, no. 1, pp. 69–81, 1982.

57. A. Saltelli, K. Chan, and E. M. Scott, Sensitivity Analysis, Wiley Series in Probability and Statistics, John Wiley & Sons, New York, NY, USA, 2004.

58. Y. Zhang, Z.-P. Fan, and Y. Liu, "A method based on stochastic dominance degrees for stochastic multiple criteria decision making," Computers and Industrial Engineering, vol. 58, no. 4, pp. 544–552, 2010.

59. J. Ren, Y. Gao, and C. Bian, "Multiple criteria decision making based on discrete linguistic stochastic variables," Mathematical Problems in Engineering, vol. 2013, Article ID 546508, 11 pages, 2013.

60. M. D. McKay, R. J. Beckman, and W. J. Conover, "A comparison of three methods for selecting values of input variables in the analysis of output from a computer code," Technometrics, vol. 21, no. 2, pp. 239–245, 1979.

61. R. L. Iman and W. J. Conover, "Small sample sensitivity analysis techniques for computer models with an application to risk assessment," Communications in Statistics—Theory and Methods, vol. 9, no. 17, pp. 1749–1842, 1980.

62. B. Möller and U. Reuter, Uncertainty Forecasting in Engineering, Springer, Berlin, Germany, 2007.

63. A. Saltelli, M. Ratto, T. Andres et al., Global Sensitivity Analysis: the Primer, John Wiley & Sons, Chichester, UK, 2008.

64. A. Saltelli, S. Tarantola, F. Campolongo, and M. Ratto, Sensitivity Analysis in Practice: A Guide to Assessing Scientific Models, John Wiley & Sons, Chichester, UK, 2004.

65. A. Ligmann-Zielinska, "Spatially-explicit sensitivity analysis of an agent-based model of land use change," International Journal of Geographical Information Science, vol. 27, no. 9, pp. 1764–1781, 2013.

66. A. Ligmann-Zielinska and P. Jankowski, "Spatially-explicit integrated uncertainty and sensitivity analysis of criteria weights in multicriteria land suitability evaluation," Environmental Modelling & Software, vol. 57, pp. 235–247, 2014.

67. C.-R. Wu, C.-T. Lin, and H.-C. Chen, "Optimal selection of location for Taiwanese hospitals to ensure a competitive advantage by using the analytic hierarchy process and sensitivity analysis," Building and Environment, vol. 42, no. 3, pp. 1431–1444, 2007.

68. A. Awasthi, S. S. Chauhan, and S. K. Goyal, "A multi-criteria decision making approach for location planning for urban distribution centers under uncertainty," Mathematical and Computer Modelling, vol. 53, no. 1-2, pp. 98–109, 2011.

69. J. Malczewski and C. Rinner, Multicriteria Decision Analysis in Geographic Information Science, Springer, New York, NY, USA, 2015.

70. A. Shanian and O. Savadogo, "A methodological concept for material selection of highly sensitive components based on multiple criteria decision analysis," Expert Systems with Applications, vol. 36, no. 2, pp. 1362–1370, 2009.

71. G. Büyüközkan and G. Çifçi, "A novel fuzzy multi-criteria decision framework for sustainable supplier selection with incomplete information," Computers in Industry, vol. 62, no. 2, pp. 164–174, 2011.

72. E. Triantaphyllou and A. Sánchez, "A sensitivity analysis approach for some deterministic multi-criteria decision-making methods," Decision Sciences, vol. 28, no. 1, pp. 151–194, 1997.

73. E. Triantaphyllou, Multi-Criteria Decision Making: A Comparative Study, Kluwer Academic Publishers, Dordrecht, The Netherlands, 2000.

74. M. Bevilacqua and M. Braglia, "The analytic hierarchy process applied to maintenance strategy selection," Reliability Engineering and System Safety, vol. 70, no. 1, pp. 71–83, 2000.

75. C.-W. Chang, C.-R. Wu, C.-T. Lin, and H.-C. Chen, "An application of AHP and sensitivity analysis for selecting the best slicing machine," Computers & Industrial Engineering, vol. 52, no. 2, pp. 296–307, 2007.

76. D. Choudhary and R. Shankar, "An STEEP-fuzzy AHP-TOPSIS framework for evaluation and selection of thermal power plant location: a case study from India," Energy, vol. 42, no. 1, pp. 510–521, 2012.

77. A. Awasthi, S. S. Chauhan, H. Omrani, and A. Panahi, "A hybrid approach based on SERVQUAL and fuzzy TOPSIS for evaluating transportation service quality," Computers & Industrial Engineering, vol. 61, no. 3, pp. 637–646, 2011.

78. A. Awasthi, S. S. Chauhan, and H. Omrani, "Application of fuzzy TOPSIS in evaluating sustainable transportation systems," Expert Systems with Applications, vol. 38, no. 10, pp. 12270–12280, 2011.

79. A. Awasthi and S. S. Chauhan, "A hybrid approach integrating Affinity Diagram, AHP and fuzzy TOPSIS for sustainable city logistics planning," Applied Mathematical Modelling, vol. 36, no. 2, pp. 573–584, 2012.

80. M. Saisana, A. Saltelli, and S. Tarantola, "Uncertainty and sensitivity analysis techniques as tools for the quality assessment of composite indicators," Journal of the Royal Statistical Society Series A: Statistics in Society, vol. 168, no. 2, pp. 307–323, 2005.

81. D. Ben-Arieh, "Sensitivity of multi-criteria decision making to linguistic quantifiers and aggregation means," Computers and Industrial Engineering, vol. 48, no. 2, pp. 289–309, 2005.

82. G. Büyüközkan and G. Çifçi, "A combined fuzzy AHP and fuzzy TOPSIS based strategic analysis of electronic service quality in healthcare industry," Expert Systems with Applications, vol. 39, no. 3, pp. 2341–2354, 2012.

83. M. Zarghami, "Soft computing of the Borda count by fuzzy linguistic quantifiers," Applied Soft Computing Journal, vol. 11, no. 1, pp. 1067–1073, 2011.

84. M. Zarghami and F. Szidarovszky, "On the relation between compromise programming and ordered weighted averaging operator," Information Sciences,

vol. 180, no. 11, pp. 2239–2248, 2010.

85. E. R. Vaidogas, E. K. Zavadskas, and Z. Turskis, "Reliability measures in multicriteria decision making as applied to engineering projects," International Journal of Management and Decision Making, vol. 8, no. 5-6, pp. 497–518, 2007.

86. B. J. Garrick, Quantifying and Controlling Catastrophic Risks, Elsevier, Amsterdam, The Netherlands, 2008.

87. L. F. Gay and S. K. Sinha, "Resilience of civil infrastructure systems: literature review for improved asset management," International Journal of Critical Infrastructures, vol. 9, no. 4, pp. 330–350, 2013.

88. B. J. Jennings, E. D. Vugrin, and D. K. Belasich, "Resilience certification for commercial buildings: a study of stakeholder perspectives," Environment Systems and Decisions, vol. 33, no. 2, pp. 184–194, 2013.

89. G. Lizarralde, K. Chmutina, L. Bosher, and A. Dainty, "Sustainability and resilience in the built environment: the challenges of establishing a turquoise agenda in the UK," Sustainable Cities and Society, vol. 15, pp. 96–104, 2015.

90. M. Rausand, Reliability of Safety-Critical Systems: Theory and Applications, Wiley, Chichester, UK, 2014.

91. E. R. Vaidogas and E. K. Zavadskas, "Introducing reliability measures into multi-criteria decision making," International Journal of Management and Decision Making, vol. 8, no. 5-6, pp. 475–496, 2007

92. N. D. Singpurwalla, Reliability and Risk. A Bayesian Perspective, Wiley, Chichester, UK, 2006.

93. E. R. Vaidogas, "On applying sparse and uncertain information to estimating the probability of failure due to rare abnormal situations," Information Technology and Control, vol. 38, no. 2, pp. 135–146, 2009.

94. J. C. Helton and W. L. Oberkampf, "Alternative representations of epistemic uncertainty," Reliability Engineering & System Safety, vol. 85, no. 1–3, pp. 1–10, 2004.

95. C. J. Roy and W. L. Oberkampf, "A comprehensive framework for verification, validation, and uncertainty quantification in scientific computing," Computer Methods in Applied Mechanics and Engineering, vol. 200, no. 25–28, pp. 2131–2144, 2011.

96. M. C. M. Troffaes, G. Walter, and D. Kelly, "A robust Bayesian approach to modeling epistemic uncertainty in common-cause failure models," Reliability Engineering and System Safety, vol. 125, no. 1, pp. 13–21, 2014.

97. E. K. Zavadskas and E. R. Vaidogas, "Multiattribute selection from alternative designs of infrastructure components for accidental situations," Computer-Aided Civil and Infrastructure Engineering, vol. 24, no. 5, pp. 346–358, 2009.

98. S. Mannan, Ed., Lees's Loss Prevention in the Process Industries, Elsevier, Amsterdam, The Netherlands, 2005.

99. E. Versluis, M. van Asselt, T. Fox, and A. Hommels, "The EU Seveso regime in practice: from uncertainty blindness to uncertainty tolerance," Journal of Hazardous Materials, vol. 184, no. 1–3, pp. 627–631, 2010.

100. L. T. Ostrom and C. A. Wilhelmsen, Risk Assessment. Tools, Techniques, and Their Applications, Wiley, Hoboken, NJ, USA, 2012.

101. E. R. Vaidogas and J. Šakėnaitė, "Protecting built property against fire disasters: multi-attribute decision making with respect to fire risk," International Journal of Strategic Property Management, vol. 14, no. 4, pp. 391–407, 2010.

102. J.-L. Zhou, Z.-H. Bai, and Z.-Y. Sun, "A hybrid approach for safety assessment in high-risk hydropower-construction-project work systems," Safety Science, vol. 64, pp. 163–172, 2014.

103. M. D. Catrinu and D. E. Nordgård, "Integrating risk analysis and multi-criteria decision support under uncertainty in electricity distribution system asset management," Reliability Engineering & System Safety, vol. 96, no. 6, pp. 663–670, 2011.

104. A. Nieto-Morote and F. Ruz-Vila, "A fuzzy approach to construction project risk assessment,"International Journal of Project Management, vol. 29, no. 2, pp. 220–231, 2011.

105. A. Nieto-Morote and F. Ruz-Vila, "A fuzzy multi-criteria decision-making model for construction contractor prequalification," Automation in Construction, vol. 25, pp. 8–19, 2012.

106. Y. Xiang, C. Liu, K. Zhang, and Q. Wu, "Risk analysis and management of submerged floating tunnel and its application," Procedia Engineering, vol. 4, pp. 107–116, 2010.

107. Y.-M. Wang, J. Liu, and T. M. S. Elhag, "An integrated AHP-DEA methodology for bridge risk assessment," Computers & Industrial Engineering, vol. 54, no. 3, pp. 513–525, 2008.

108. M. S. El-Abbasy, T. Zayed, M. Ahmed, H. Alzraiee, and M. Abouhamad, "Contractor selection model for highway projects using integrated simulation and analytic network process," Journal of Construction Engineering and Management, vol. 139, no. 7, pp. 755–767, 2013.

109. L. Hui, W. Yongqing, S. Shimei, and S. Baotie, "Study on safety assessment of fire hazard for the construction site," Procedia Engineering, vol. 43, pp. 369–373, 2012.

110. M. A. Alqassim and N. N. Daeid, "Fires and related incidents in Dubai, United Arab Emirates (2006–2013)," Case Studies in Fire Safety, vol. 2, pp. 28–36, 2014.

111. D. J. Rasbash, G. Ramachandran, B. Kandola, J. M. Watts, and M. Law, Evaluation of Fire Safety, John Wiley & Sons, Chichester, UK, 2004.

112. C. M. Zhao, S. M. Lo, J. A. Lu, and Z. Fang, "A simulation approach for ranking of fire safety attributes of existing buildings," Fire Safety Journal, vol. 39, no. 7, pp. 557–579, 2004.

113. J. Wong, H. Li, and J. Lai, "Evaluating the system intelligence of the intelligent building systems—part 1: development of key intelligent indicators and conceptual analytical framework," Automation in Construction, vol. 17, no. 3, pp. 284–302, 2008.

114. J. Wong, H. Li, and J. Lai, "Evaluating the system

intelligence of the intelligent building systems. Part 2: construction and validation of analytical models," Automation in Construction, vol. 17, no. 3, pp. 303–321, 2008.

115. E. R. Vaidogas and J. Šakėnaitė, "Multi-attribute decision-making in economics of fire protection,"Engineering Economics, vol. 22, no. 3, pp. 262–270, 2011.

116. E. R. Vaidogas and L. Linkutė, "Sitting the barrier aimed at protecting roadside property from accidental fires and explosions on road: a pre-optimisation stage," The Baltic Journal of Road and Bridge Engineering, vol. 7, no. 4, pp. 277–287, 2012.

117. E. K. Zavadskas and J. Antucheviciene, "Multiple criteria evaluation of rural building's regeneration alternatives," Building and Environment, vol. 42, no. 1, pp. 436–451, 2007.

118. E. Krygiel and B. Nies, Green BIM: Successful Sustainable Design with Building Information Modeling, Wiley, Indianapolis, Ind, USA, 2008.

119. AIA, Guide, Instructions and Commentary to the 2013 AIA Digital Practice Documents, The American Institute of Architects, Washington, DC, USA, 2013.

120. B. A. Wayland, Security for Business Professionals. How to Plan, Implement, and Manage your Company's Security Program, Elsevier, Amsterdam, The Netherlands, 2014.

121. D. Drengenberg and G. Corley, "Evolution of building code requirements in a post 9/11 world," CTBUH Journal, no. 3, pp. 32–35, 2011.

CHAPTER 2

THE INTERVAL-VALUED INTUITIONISTIC FUZZY MULTIMOORA METHOD FOR GROUP DECISION MAKING IN ENGINEERING

Edmundas Kazimieras Zavadskas, Jurgita Antucheviciene, Seyed Hossein Razavi Hajiagha, and Shide Sadat Hashemi

INTRODUCTION

Multicriteria decision making (MCDM) is a growing field of operations research both from theoretical and implementation perspectives. The importance of MCDM can be drawn from Zeleny [1]: "it has become more and more difficult to see the world around us in a uni-dimensional way and to use only a single criterion when judging what we see. We always compare, rank, and order the objects of our choice with respect to multiple criteria of choice." The MCDM field is further divided into two classes, including multiobjective decision making (MODM) and multiattribute decision making (MADM). These classes are, respectively, associated with planning and selection classification of Simon [2] for decision making problems. The main topic of this paper covers MADM.

A MADM problem can be formally characterized as the task to evaluate, compare, and rank a set of finite alternatives, options, or choices with regard to a set of finite attributes. According to Yu [3], a MADM problem is composed of the set of substitutive alternatives (1), the set of evaluation criteria (2), the outcome (or decision) matrix with regard to the alternatives scored based on the evaluation criteria (3), and the preference structure of decision making about the criterion significances or weights (4). The information about the third and fourth parts of a decision making problem is determined exactly. However, a decision maker always deals with approximate or partial information [4]. Therefore, exactness is an unrealistic assumption. Different frameworks have been proposed to handle uncertainty in practice. Liu and Lin [5] classified these frameworks into probability and statistics (1) and grey system theory and fuzzy set theory (2). Fuzzy set theory introduced by Zadeh [6] is widely applied to decision making problems. In other words, the fuzzy set theory is a generalization of ordinal sets, where a membership degree is assigned to each element of a set. Each type of uncertainty has its own characteristics and is appropriate for special cases. While probability is concerned with the occurrence of well-defined events, fuzzy sets deal with gradual concepts and describe their boundaries [7]. In fact, the fuzzy set theory is appropriate for recognition-based uncertainty that is common in decision making.

Determining a single membership degree is a difficult task for decision makers and, therefore, Grattan-Guinness [8] believes that presentation of a linguistic expression in the form of a fuzzy set is not sufficient. In 1986, Atanassov [9] introduced the notion of intuitionistic fuzzy sets (IFSs) as an extension of ordinal fuzzy sets. In addition to a membership degree of each element, IFS assigns a degree of nonmembership to each element. Later, Atanassov and Gargov [10] extended the interval-valued intuitionistic fuzzy sets (IVIFSs), where membership and nonmembership degrees are stated as closed intervals.

When it became obvious that the type 1 fuzzy sets were not always sufficient for MADM under uncertain environment, type 2 fuzzy sets, involving interval-valued as well as intuitionistic fuzzy sets, were proposed. As a result, the MULTIMOORA method was updated by using generalized interval-valued trapezoidal fuzzy numbers [11] or intuitionistic fuzzy numbers [12]. The method has been successfully applied to making economic, technological, or management decisions.

Citation: Edmundas Kazimieras Zavadskas, Jurgita Antucheviciene, Seyed Hossein Razavi Hajiagha, and Shide Sadat Hashemi, "The Interval-Valued Intuitionistic Fuzzy MULTIMOORA Method for Group Decision Making in Engineering," Mathematical Problems in Engineering, vol. 2015, Article ID 560690, 13 pages, 2015. doi:10.1155/2015/560690.

The applications of various approaches in 2006–2013, including crisp and extended methods, were summarized by T. Baležentis and A. Baležentis [13]. The most recent applications cover the extended versions of the method, in particular, the fuzzy MULTIMOORA [14, 15] or MULTIMOORA based on the interval 2-tuple linguistic variables [16]. Also, Li [17] extended MULTIMOORA with hesitant fuzzy numbers, where membership degree of elements is defined over a set of different values.

The current research aims to extend the MULTIMOORA method, using the interval-valued intuitionistic fuzzy sets. Taking into account the advantages of IVIFS, the interval-valued intuitionistic fuzzy MULTIMOORA (IVIF-MULTIMOORA) method for group decision making is presented in the paper. For this reason, three parts of MULTIMOORA were extended under IVIF information and the appropriate algorithm was proposed.

The paper is organized as follows. In Section 2, the literature review is presented and then the crisp version of MULTIMOORA is briefly overviewed in Section 3. Section 4 includes a brief description of the interval-valued intuitionistic fuzzy sets. The proposed IVIF-MULTIMOORA algorithm is presented in Section 5. The solution of two numerical examples of civil engineering problems is given in Section 6, while the results obtained are compared with the data yielded by some other methods of IVIF decision making. Finally, the discussion and the conclusions are presented in Sections 7 and 8.

LITERATURE REVIEW

Taking into account the flexibility and general character of IVIFS, researchers developed decision making methods under the IVIFS environment. Some researchers developed the well-known MADM methods in the IVIF form. Some authors [18–20] proposed different frameworks of TOPSIS method under the IVIF environment. Li [21, 22] developed some mathematical programming-based method to solve the MADM problems with both ratings of the alternatives on attributes and weights of attributes expressed with the help of IVIFS. Park et al. [23] presented an IVIF version of the VIKOR method. Li [24] proposed a closeness coefficient based on the nonlinear programming method for solving IVIF MADM problems. Chen et al. [25] solved the MADM problems based on the interval-valued intuitionistic fuzzy weighted average operator and newly defined fuzzy ranking method for intuitionistic fuzzy values. Yu et al. [26] introduced some IVIF aggregation operators and applied them to solving various decision making problems. Zhang and Yu [27] presented an optimization model to determine the attribute weights in an IVIF decision making problem and, then, used these weights in an extended TOPSIS to rank the alternatives. Ye [28] introduced a cosine similarity measure and a weighted cosine similarity measure to MADM problems. Meng et al. [29] also proposed two new aggregation operators, that is, the arithmetical interval-valued intuitionistic fuzzy generalized λ--Shapley Choquet operator and the geometric interval-valued intuitionistic fuzzy generalized λ-Shapley Choquet operator, and investigated their application to solving MADM problems Meng et al. [30] used Shapley function in extending a generalized IVIF hybrid Shapley averaging operator and proposed its application to solving MADM problems. Razavi Hajiagha et al. [31] developed Complex Proportional Assessment (COPRAS) method with IVIF data. Chai et al. [32] proposed a rule-based decision model when decision information in a group decision making problem is provided as IVIF values. Wan and Dong [33] defined the possibility degree of comparison between two IVIF numbers and introduced two ordered averaging operators based on the Karnik-Mendel algorithm to solve MADM problems with IVIF information. Chen [34] developed a method based on the traditional linear assignment method for solving decision making problems in the IVIF context. Xu and Shen [35] presented the IVIF outranking choice method to solve MADM problems, while Zavadskas et al. [36] extended the IVIF weighted aggregated sum product assessment (WASPAS) method. Geetha et al. [37] proposed a new ranking method of IVIF numbers and extended its application in decision making problems. Zhang et al. [38] proposed a new definition of IVIF entropy an entropy-based MADM method. Later, Wei, and Zhang [39] proposed applying an entropy measure for IFSs and IVIFSs to assess the experts' weights for multicriteria fuzzy group decision making. Tong and Yu [40] introduced a novel approach for ranking the alternatives based on the IVIF cross entropy and TOPSIS in the IVIF environment. Wu and Chiclana [41, 42] introduced a risk attitudinal ranking method for IVIF numbers and applied this ranking method in a MADM problem. They imposed a risk attitude parameter over IVIF numbers ordinal score and accuracy functions [43] and used this method for solving MADM problems.

As mentioned before the current paper has focused on the extension of the MULTIMOORA (Multiobjective Optimization by Ratio Analysis plus Full Multiplicative Form) method. This method belongs to the group of complete aggregation methods, based on the reference point technique. The crisp MOORA method (Multiobjective Optimization by Ratio Analysis) was presented by Brauers and Zavadskas in 2006 [44]. This method was further supplemented by the full multiplicative form. The MULTIMOORA was suggested by Brauers and Zavadskas in 2010 [45]. The method is based on the theory of dominance [46] and enables us to summarize MOORA, consisting of the ratio system and the reference point approaches, as well as the full multiplicative form. The robustness of the method has been proved [47] and it has been successfully applied to evaluate decisions by using a single approach or its combinations with other MADM methods, such as TOPSIS [48] and WASPAS [49, 50].

Several developments of MULTIMOORA for the uncertain environment have been presented. The fuzzy MULTIMOORA was suggested by Brauers et al. [52]. To overcome some drawbacks of the fuzzy set theory, the two-tuple linguistic representation method for computing with words can be applied [53]. Accordingly, the considered method has been extended for group decision making based on two tuples (MULTIMOORA-2T-G) [54, 55]. The fuzzy MULTIMOORA, based on triangular fuzzy numbers and designed for group decision making (MULTIMOORA-FG), was introduced [56, 57]. As a further modification of the method, MOORA with grey numbers was developed [58, 59].

MULTIMOORA

The method summarizes three approaches, that is, MOORA, consisting of the ratio system and the reference point, and the full multiplicative form.

Regardless of the approach used, initial decision criteria are transformed by applying vector normalization:

$$\bar{x}_{ij} = \frac{x_{ij}}{\sqrt{\sum_{i=1}^{m} x_{ij}^2}}, \quad (1)$$

where x_{ij} is the initial criterion value, that is, the response of the alternative i to objective j; $i = 1, 2, \ldots, m$; m is the number of alternatives; $j = 1, 2, \ldots, n$; n is the number of decision criteria (objectives); $\bar{x}_{ij}$ is a dimensionless value of a decision criterion

The first part of the approach is based on the ratio system [44]. To calculate the relative significance, $\bar{y}_i$, of each alternative i with respect to all objectives j, the weighted normalized criteria values should be added in the case of maximization, while, in the case of minimization, the weighted normalized criteria values should be subtracted as follows:

$$\bar{y}_i = \sum_{j=1}^{j=g} \bar{x}_{ij} w_j - \sum_{j=g+1}^{j=n} \bar{x}_{ij} w_j, \quad (2)$$

where $j = 1, 2, \ldots, g$ are maximized decision criteria; $j = g + 1, g + 2, \ldots, n$ are minimized decision criteria; w_j is the weight (or relative significance) of a criterion.

The values of variables $\bar{y}_i$, show the preference of the alternatives according to the ratio system approach.

The second part of the approach is based on the maximal objective reference point [44]. First, the desirable ideal alternative should to be established. The virtual ideal alternative consists of the best values of the considered criteria r_j. It is formed by selecting the best criteria values from every decision alternative based on their optimization direction, that is, the maximal values from the criteria set $j = 1, 2, \ldots, g$ and the minimal values from the criteria set $j = g + 1, g + 2, \ldots, n$.

All criteria values were transformed by applying vector normalization (1). Having the dimensionless values of the criteria $\bar{x}_{ij}$ and relative significances (weights) of the criteria w^j, decision alternatives are ranked based on the Min-Max metric of Tchebycheff [60]:

$$\underset{(i)}{\text{Min}} \left\{ \underset{(j)}{\max} \left| r_j - \bar{x}_{ij} w_j \right| \right\}. \quad (3)$$

The second part of the approach is based on the full multiplicative form, as presented by Brauers and Zavadskas [44]. The full multiplicative form for calculating the utility of the alternatives Ui is applied as follows:

$$U_i = \frac{A_i}{B_i}, \quad (4)$$

where A_i and B_i are calculated separately for maximized decision criteria $j = 1, 2, \ldots, g$ and minimized decision criteria $j = g + 1, g + 2, \ldots, n$, respectively. A_i and B_i are calculated as follows:

$$A_i = \prod_{j=1}^{g} w_j \bar{x}_{ij},$$

$$B_i = \prod_{j=g+1}^{n} w_j \bar{x}_{ij}. \quad (5)$$

The MULTIMOORA is based on the theory of dominance [46] and summarizes the MOORA, involving the above described ratio system and the reference point approaches, as well as the full multiplicative form [45].

Robustness of the method has been verified proving that accuracy of aggregated approaches is larger comparing to accuracy of single ones [47]. The method has been widely applied to evaluate alternatives and to select the best decisions in engineering technology or management problems by using a single approach or its combinations with other MADM methods [13, 48–50]. However, the farther the more complex decisions should be made and decision makers usually face a larger amount of approximate or partial information. Accordingly, extensions of conventional MADM methods under uncertain environment for group decision making are essential. The further developed IVIF-MULTIMOORA method could be applied in innovation themes in engineering, such as sustainable building life-cycle modelling and evaluating and selecting new business models to finance, build, and manage public buildings and infrastructures.

INTERVAL-VALUED INTUITIONISTIC FUZZY SETS

Zadeh [6] generalized the characteristic function of classic sets into membership function and introduced fuzzy sets, where a membership degree was assigned to each element of a fuzzy set. Later, Atanassov [9] proposed the idea of intuitionistic fuzzy sets (IFSs), when a nonmembership degree is assigned to each element of the set alongside its membership. The interval-valued intuitionistic fuzzy sets (IVIFSs) present an extended form of IFSs.

Let [0, 1] be the set of all closed subintervals of the interval [0, 1] and let $X(\neq \Phi)$ be a given set. An IVIFS in X was defined as $\tilde{A} = \{\langle x, \mu_{\tilde{A}}(x), \nu_{\tilde{A}}(x)\rangle \mid x \in X\}$, where $\mu_{\tilde{A}} : X \to D[0,1]$ and $\nu_{\tilde{A}} : X \to D[0,1]$ with condition $0 < \sup_x \mu_{\tilde{A}}(x) + \sup_x \nu_{\tilde{A}}(x) \leq 1$. The intervals $\mu_{\tilde{A}}(x)$ and $\nu_{\tilde{A}}(x)$ denote the degree of membership and nonmembership of the element x of the set A. Thus, for each $x \in X, \mu_{\tilde{A}}(x)$ are closed intervals whose lower and upper end points are denoted by $\mu_{AL}(x), \mu_{AU}(x), \nu_{AL}(x)$, and $\nu_{AU}(x)$.

The IVIFS was denoted by

$$A = \{\langle x, [\mu_{AL}(x), \mu_{AU}(x)], [\nu_{AL}(x), \nu_{AU}(x)]\rangle \mid x \in X\}, \quad (6)$$

where $0 < \mu_{AU}(x) + \nu_{AU}(x) \leq 1, \mu_{AL}(x), \nu_{AL}(x) \geq 0$. For convenience, the IVIFS value was denoted by $\tilde{A} = ([a,b],[c,d])$ and referred to as an interval-valued intuitionistic fuzzy number (IVIFN).

The algebraic operations were extended over IVIFSs. Let $\tilde{A}_1 = ([a_1,b_1],[c_1,d_1])$ and $\tilde{A}_2 = ([a_2,b_2],[c_2,d_2])$ be any two IVIFNs;

then [43]

$$\widetilde{A}_1 + \widetilde{A}_2 = ([a_1 + a_2 - a_1a_2, b_1 + b_2 - b_1b_2], [c_1c_2, d_1d_2]), \tag{7}$$

$$\widetilde{A}_1 \cdot \widetilde{A}_2 = ([a_1a_2, b_1b_2], [c_1 + c_2 - c_1c_2, d_1 + d_2 - d_1d_2]), \tag{8}$$

$$= \left(\left[1-(1-a_1)^{\lambda}, 1-(1-b_1)^{\lambda}\right], \left[c_1^{\lambda}, d_1^{\lambda}\right]\right), \quad \lambda \geq 0. \tag{9}$$

The division and subtraction operations were defined as follows, using the extension principle [61]:

$$\widetilde{A}_1 \div \widetilde{A}_2 = \{[\min(a_1, a_2), \min(b_1, b_2)], [\max(c_1, c_2), \max(d_1, d_2)]\}, \tag{10}$$

$$\widetilde{A}_1 - \widetilde{A}_2 = \{[\min(a_1, a_2), \min(b_1, b_2)], [\max(c_1, c_2), \max(d_1, d_2)]\}. \tag{11}$$

Since the final rating of the alternatives is usually determined as IVIFNs, it requires their comparison. This comparison was made, using the score and accuracy functions [43]. Let $\widetilde{A} = ([a,b],[c,d])$ be an IVIFN. Then,

$$s(\widetilde{A}) = \frac{1}{2}(a - c + b - d) \tag{12}$$

was called the score function of $\widetilde{A}$, where $s(\widetilde{A}) \in [-1, 1]$, while

$$h(\widetilde{A}) = \frac{1}{2}(a + c + b + d) \tag{13}$$

was referred to as the accuracy function of A, where $h(\widetilde{A}) \in [0, 1]$.

For $\tilde{A}_1$ and $\tilde{A}_2$ as two IVIFNs, it follows that [43]

(1) if $s(\tilde{A}_1) \prec s(\tilde{A}_2)$, then $\tilde{A}_1$ is smaller than $\tilde{A}_2$, $\tilde{A}_1 \prec \tilde{A}_2$;

(2) if $s(\tilde{A}_1) = s(\tilde{A}_2)$, then

(a) if $h(\widetilde{A}_1) = h(\widetilde{A}_2)$, then $\widetilde{A}_1 = \widetilde{A}_2$,

(b) if $h(\widetilde{A}_1) \prec h(\widetilde{A}_2)$, then $\widetilde{A}_1$ is smaller than $\widetilde{A}_2$, $\widetilde{A}_1 \prec \widetilde{A}_2$.

Another requirement of group decision making is associated with the aggregation of different judgments of decision makers into a single estimate. In this case, aggregation operators on IVIFNs can be used. Let $\widetilde{A}_j = ([a_j, b_j], [c_j, d_j]),\ j = 1, 2, \ldots, n,$, be a collection of IVIFNs. Then, the generalized interval intuitionistic fuzzy weighted average $\mathrm{GIIFWA}_w(\widetilde{A}_1, \widetilde{A}_2, \ldots, \widetilde{A}_n)$ was defined as

$$\mathrm{GIIFWA}_w(A_1, A_2, \ldots, A_n) = \left(w_1\widetilde{A}_1^{\lambda} + w_2\widetilde{A}_2^{\lambda} + \cdots + w_n\widetilde{A}_n^{\lambda}\right)^{1/\lambda}, \tag{14}$$

where $\lambda > 0$, and $w = (w_1, w_2, \ldots, w_n)^T$ is the weight vector with $w_j \geq 0,\ j = 1, 2, \ldots, n$, and $\sum_{j=1}^{n} w_j = 1$. It can be shown that GIIFWA is also an IVIFN and can be calculated as follows [62]:

$$\mathrm{GIIFWA}_w(\widetilde{A}_1, \widetilde{A}_2, \ldots, \widetilde{A}_n) = \left(\left[\left(1 - \prod_{j=1}^{n}(1 - a_j^{\lambda})^{w_j}\right)^{1/\lambda}, \left(1 - \prod_{j=1}^{n}(1 - b_j^{\lambda})^{w_j}\right)^{1/\lambda}\right], \left[1 - \left(1 - \prod_{j=1}^{n}\left(1 - (1 - c_j)^{\lambda}\right)^{w_j}\right)^{1/\lambda}, 1 - \left(1 - \prod_{j=1}^{n}\left(1 - (1 - d_j)^{\lambda}\right)^{w_j}\right)^{1/\lambda}\right]\right). \tag{15}$$

If $\lambda = 1$,, then GIIFWA is turned into the interval intuitionistic fuzzy weighted average (IIFWA).

IVIF-MULTIMOORA FOR GROUP DECISION MAKING

IVIF-MULTIMOORA

Assume that a group of K decision makers (experts) wants to appraise a set of m alternatives A_1, A_2,..., based on a set of n criteria C_1, C_2,...,C_n. Also, the weight vector w_1, w_2,..., with $w_j \geq 0, j = 1, 2,...,n$, and $\sum_{j=1}^{n} w_j = 1$ is determined that shows the relative importance of different criteria on final decision. Due to incomplete and ill-defined data, the required information in the considered problem is expressed by IVIFNs.

At the first step, each decision maker expresses his/her decision matrix $\widetilde{X}^k,\ k = 1, 2, \ldots, K$, as follows:

where $\widetilde{x}_{ij}^{k}$ represents the IVIF performance of the alternative A_i over the criterion C_j from the viewpoint of kth decision maker. The aim of a decision making group is to rank the alternatives

To solve this MAGDM problem, an IVIF version of the MULTIMOORA method, called IVIF-MULTIMOORA, was extended and described in this section. Initially, the aggregation was required to transform the set of K decision matrices to an aggregated decision matrix. This aggregated decision matrix had the form of $\widetilde{X} = [\widetilde{x}_{ij}]$. If an equal weight was assumed for decision makers, then

$$\widetilde{x}_{ij} = \left(\left[\mu_{ij}^{l}, \mu_{ij}^{u}\right], \left[v_{ij}^{l}, v_{ij}^{u}\right]\right) = \mathrm{IIFWA}\left(\widetilde{x}_{ij}^{1}, \widetilde{x}_{ij}^{2}, \ldots, \widetilde{x}_{ij}^{K}\right) \tag{17}$$

The aggregation operation was performed using the GIIFWA operator and considering the weight vector $V = (V_1, V_2,..., V_K)$ for different decision makers.

Since the aggregated decision matrix $\widetilde{X} = [\widetilde{x}_{ij}]$ was available, the IVIF-MULTIMOORA method was proposed to solve the MAGDM problem. As mentioned in Section 3, the MULTIMOORA method involves three parts: the ratio system (1), the reference point approach (2), and the full multiplicative form (3).

The Part of MOORA Based on the IVIF-Ratio System

The first step in the ratio system is the normalization of the decision matrix. However, since IVIFNs are commensurable numbers, normalization is not required. The set of criteria was decomposed into two subsets of the benefit criteria (where more is better) and cost criteria (where less is better). Then, the $\tilde{y}_i$, $i = 1, 2, \ldots, m$, score of alternatives was computed by adding its weighted benefit criteria and subtracting the weighted cost criteria. If the benefit criteria were ordered as $C_1, C_2, \ldots$, and cost criteria as $C_{g+1}, C_{g+2}, \ldots, C_n$, then

$$\tilde{y}_i = \sum_{j=1}^{g} w_j \tilde{x}_{ij} - \sum_{j=g+1}^{n} w_j \tilde{x}_{ij}. \tag{18}$$

This score could be computed by applying the IVIFNs algebraic operators and the extension principle given in Section 4. However, a slight modification could simplify the computations. Let $W^+ = \sum_{j=1}^{g} w_j$ and $W^- = \sum_{j=g+1}^{n} w_j$. Then, $w'_j = w_j/W^+$ for each $j \in \{1, 2, \ldots, g\}$, and $w''_j = w_j/W^-$ for each $j \in \{g+1, g+2, \ldots, n\}$. It was clear that $w'_j \geq$, and $0,\ j = 1, 2, \ldots, g$, and $\sum_{j=1}^{g} w'_j = 1$. Similarly $w''_j \geq 0,\ j =$, $1, 2, \ldots, g$, and $\sum_{j=1}^{g} w''_j = 1$.

Then

$$\tilde{y}'_i = \frac{\tilde{y}_i}{W^+ W^-}. \tag{19}$$

Multiplying both sides of (18) by positive value of $1/(W^+W^-)$, (18) was transformed as follows:

$$\tilde{y}'_i = \frac{1}{W^-} \sum_{j=1}^{g} w'_j \tilde{x}_{ij} - \frac{1}{W^+} \sum_{j=g+1}^{n} w''_j \tilde{x}_{ij}. \tag{20}$$

Based on the IIFWA operator definition, (18) was represented as

$$\tilde{y}'_i = \frac{1}{W^-} \mathrm{IIFWA}_{W^+} \left(\tilde{x}_{i1}, \tilde{x}_{i2}, \ldots, \tilde{x}_{ig}\right) - \frac{1}{W^+} \mathrm{IIFWA}_{W^-} \left(\tilde{x}_{ig+1}, \tilde{x}_{ig+2}, \ldots, \tilde{x}_{in}\right). \tag{21}$$

The multiplication of $1/(W^-)$ and $1/(W^+)$ was performed using (9). In addition, the subtraction of two IVIFNs in (21) was performed by applying (16). $\tilde{y}'_i$, $i = 1, 2 \ldots m$, is relative significance of the alternative i. The comparison of $\tilde{y}'_i$, $i = 1, 2 \ldots m$, values based on the score and the accuracy function resulted in the ranking of the alternatives based on the ratio system.

The IVIF-Reference Point Portion of the MOORA Method

According to the second part of the MOORA, the maximal objective reference point approach was used. The desirable ideal alternative in the IVIF environment was the IVIF vector with the coordinates $\tilde{r}_j$, $j = 1, 2, \ldots, n$, which was formed by selecting the data from every considered decision alternative and taking into account the optimization direction of every particular criterion. To find the reference point based on the idea of the TOPSIS method, the weighted decision matrix $\tilde{N} = [\tilde{n}_{ij}] = W \cdot \tilde{X}$ was constructed first, where

$$\tilde{n}_{ij} = \left(\left[\eta^l_{ij}, \eta^u_{ij}\right], \left[\delta^l_{ij}, \delta^u_{ij}\right]\right) = w_j \tilde{x}_{ij}. \tag{22}$$

The scalar multiplication was performed by applying (9). Then, for benefit criteria, $C_1, C_2, \ldots$, the reference point was constructed as follows:

$$\tilde{r}_j = \left(\left[\eta^{*l}_{ij}, \eta^{*u}_{ij}\right], \left[\delta^{*l}_{ij}, \delta^{*u}_{ij}\right]\right) = \left(\left[\max_{1 \le i \le m} \left(\eta^l_{ij}\right), \max_{1 \le i \le m} \left(\eta^u_{ij}\right)\right], \left[\min_{1 \le i \le m} \left(\delta^l_{ij}\right), \min_{1 \le i \le m} \left(\delta^u_{ij}\right)\right]\right), \tag{23}$$

and for cost criteria, $C_{g+1}, C_{g+2}, \ldots, C_n$,

$$\tilde{r}_j = \left(\left[\eta^{*l}_{ij}, \eta^{*u}_{ij}\right], \left[\delta^{*l}_{ij}, \delta^{*u}_{ij}\right]\right) = \left(\left[\min_{1 \le i \le m} \left(\mu^l_{ij}\right), \min_{1 \le i \le m} \left(\mu^u_{ij}\right)\right], \left[\max_{1 \le i \le m} \left(v^l_{ij}\right), \max_{1 \le i \le m} \left(v^u_{ij}\right)\right]\right). \tag{24}$$

Then, the Min-Max metric of Tchebycheff [26] was used for ranking the alternatives:

$$\operatorname*{Min}_{1 \le i \le m} \left\{ \max_{1 \le j \le n} \left|\tilde{r}_j - \tilde{x}_{ij} w_j\right| \right\}. \tag{25}$$

Based on (22)–(24), (25) could be simplified. The following steps were taken to determine the rankings of the alternatives, using the reference point approach.

Step 1. Compute for each alternative A_i, $i = 1, 2, \ldots$, the score function in the weighted decision matrix:

$$S\left(\tilde{n}_{ij}\right) = \eta^l_{ij} - \delta^l_{ij} + \eta^u_{ij} - \delta^u_{ij}. \tag{26}$$

Step 2. Compute the score function of the reference points:

$$S\left(\tilde{r}_j\right) = \eta^{*l}_{ij} - \delta^{*l}_{ij} + \eta^{*u}_{ij} - \delta^{*u}_{ij}. \tag{27}$$

Step 3. Find the maximum deviation from the reference points:

$$d_i = \max_{1 \le j \le n} \left|\tilde{r}_j - \tilde{x}_{ij} w_j\right| = \max_{1 \le j \le n} \left|S\left(\tilde{r}_j\right) - S\left(\tilde{n}_{ij}\right)\right|. \tag{28}$$

Step 4. Rank the alternatives in the ascending order of d_i, $i = 1, 2, \ldots$.

The IVIF-Full Multiplicative Form

Finally, the full multiplicative form was applied as follows:

$$\tilde{U}_i = \frac{\tilde{A}_i}{\tilde{B}_i}, \tag{29}$$

where $\tilde{U}_i$ denotes the overall utility of the alternative i and

$$\tilde{A}_i = \prod_{j=1}^{g} w_j \tilde{x}_{ij},$$

$$\tilde{B}_i = \prod_{j=g+1}^{n} w_j \tilde{x}_{ij}. \tag{30}$$

Multiplication of a set of IVIFNs was performed by sequentially applying (8). To give a compact form of this multiplication, it was supposed that $\widetilde{x}_{ij},\ j = 1, 2, \ldots, g$, was determined as an IVIFN as indicated in (17). First $\widetilde{n}_{ij} = w_j \widetilde{x}_{ij},\ j = 1, 2, \ldots, g$, was formed by applying (9) in the same manner as previously performed in (22). Then,

$$
\begin{aligned}
\widetilde{A}_i &= \prod_{j=1}^{g} \widetilde{n}_{ij} = \left(\left[\underline{\eta}^l_{ij}, \overline{\eta}^l_{ij} \right], \left[\underline{\delta}^l_{ij}, \overline{\delta}^u_{ij} \right] \right) \\
&= \left(\left[\prod_{j=1}^{g} \eta^l_{ij}, \prod_{j=1}^{g} \eta^u_{ij} \right], \left[\sum_{j=1}^{g} \delta^l_{ij} - \sum_{\substack{j_1 < j_2 \\ j_1, j_2 \in \{1,2,\ldots,g\}}} \delta^l_{ij_1} \cdot \delta^l_{ij_2} \right. \right. \\
&\quad + \cdots + (-1)^{k+1} \sum_{\substack{j_1 < j_2 < \cdots < j_k \\ j_1, j_2, \ldots, j_k \in \{1,2,\ldots,g\}}} \left(\delta^l_{ij_1} \cdot \delta^l_{ij_2} \cdots \delta^l_{ij_k} \right) + \cdots \\
&\quad + (-1)^{g+1} \left(\delta^l_{i1} \cdot \delta^l_{i2} \cdots \delta^l_{ig} \right),\ \sum_{j=1}^{g} \delta^u_{ij} \\
&\quad - \sum_{\substack{j_1 < j_2 \\ j_1, j_2 \in \{1,2,\ldots,g\}}} \delta^u_{ij_1} \cdot \delta^u_{ij_2} + \cdots \\
&\quad + (-1)^{k+1} \sum_{\substack{j_1 < j_2 < \cdots < j_k \\ j_1, j_2, \ldots, j_k \in \{1,2,\ldots,g\}}} \left(\delta^u_{ij_1} \cdot \delta^u_{ij_2} \cdots \delta^u_{ij_k} \right) + \cdots \\
&\quad \left. \left. + (-1)^{g+1} \left(\delta^u_{i1} \cdot \delta^u_{i2} \cdots \delta^u_{ig} \right) \right] \right).
\end{aligned}
\tag{31}
$$

Then $\widetilde{B}_i,\ i = 1, 2, \ldots, m$, was also calculated in a similar way. Then $\widetilde{U}_i,\ i = 1, 2, \ldots, m$, was determined by (10) for division. The ranking of the alternatives was specified by calculating the score functions $S(\widetilde{U}_i),\ i = 1, 2, \ldots, m$, and ranking them in a descending order.

Applying three distinct parts of MULTIMOORA, three sets of rankings of the alternatives were determined. The obtained rankings formed a ranking pool, based on which the final ranking of the alternatives could be determined. To find the final ranking of the alternatives, the theory of dominance [46] could be used. Also, the techniques, such as the Borda count method [63] or Copeland pairwise aggregation method [64], could be applied to determine the final ranking of the alternatives.

THE IVIF-MULTIMOORA ALGORITHM

MULTIMOORA is the extension of the MOORA method and the full multiplicative form of multiple-objective. The distinct parts of the MULTIMOORA method are presented in Sections 5.1 to 5.3. The algorithmic scheme of MULTIMOORA is presented in the current section. The IVIF-MULTIMOORA algorithm includes five different stages: (1) initialization, (2) IVIF-ratio system, (3) IVIF-reference point approach, (4) IVIF-full multiplicative form, and (5) final ranking. These stages are presented in Figure 1.

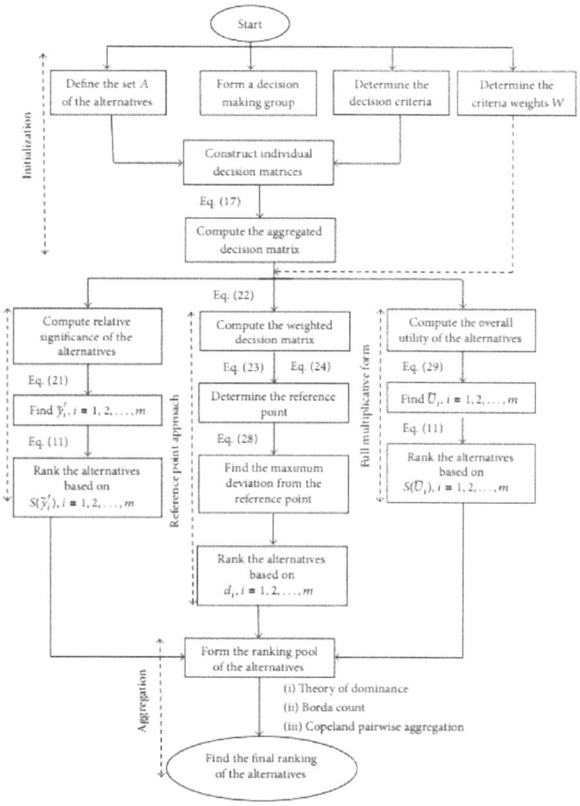

Figure 1: The IVIF-MULTIMOORA algorithm.

A NUMERICAL EXAMPLE

In this section, two applications of the proposed method are presented and the results are compared to the rankings yielded by other multiple criteria decision making methods.

The First Example

Zavadskas et al. [36] solved a problem of rational revitalization of derelict and mismanaged buildings in Lithuanian rural areas. Three alternatives included the reconstruction of rural buildings and their adaptation to production (or commercial) activities (A_1), as well as their improvement and use for farming (A_2) or dismantling and recycling the demolition waste materials (A_3). The following criteria were taken into consideration: the average soil fertility in the area x_1 (points), quality of life of the local population x_2 (points), population activity index x_3 (%), GDP proportion with respect to the average GDP of the country x_4 (%), material investment in the area x5 (LTL per resident), foreign investments in the area x^6 (LTL × 10^3 per resident), building redevelopment costs x^7 (LTL × 10^6), increase in the local population's income x_8 (LTL × 10^6 per year), increase in sales in the area x^9 (%), increase in employment in the area x_{10} (%), state income from business and property taxes x_{11} (LTL × 106 per year), business outlook x_{12}, difficulties in changing the original purpose of the site x_{13}, the degree of contamination x_{14}, and the attractiveness of the countryside (i.e., image and landscape) x_{15}.

A decision matrix is presented in Table 1.

Table 1: Decision matrix of rational revitalization of derelict and mismanaged buildings.

Criteria	Optimization direction	Weight	A_1	A_2	A_3
$\tilde{x}_1$	Max	0.0600	([0.70, 0.80], [0.10, 0.15])	([0.45, 0.60], [0.20, 0.30])	([0.30, 0.40], [0.45, 0.50])
$\tilde{x}_2$	Min	0.0727	([0.60, 0.70], [0.15, 0.20])	([0.45, 0.55], [0.20, 0.30])	([0.25, 0.40], [0.45, 0.55])
$\tilde{x}_3$	Max	0.0747	([0.75, 0.80], [0.10, 0.15])	([0.45, 0.55], [0.25, 0.35])	([0.30, 0.40], [0.40, 0.50])
$\tilde{x}_4$	Max	0.0627	([0.80, 0.90], [0, 0.05])	([0.50, 0.65], [0.15, 0.25])	([0.35, 0.50], [0.30, 0.40])
$\tilde{x}_5$	Max	0.0673	([0.70, 0.85], [0.05, 0.10])	([0.55, 0.65], [0.20, 0.25])	([0.35, 0.45], [0.30, 0.40])
$\tilde{x}_6$	Max	0.0627	([0.75, 0.85], [0.10, 0.15])	([0.55, 0.65], [0.20, 0.25])	([0.35, 0.45], [0.30, 0.40])
$\tilde{x}_7$	Min	0.0667	([0.80, 0.85], [0.10, 0.15])	([0.45, 0.55], [0.25, 0.35])	([0.15, 0.35], [0.40, 0.50])
$\tilde{x}_8$	Max	0.0667	([0.85, 0.90], [0, 0.05])	([0.20, 0.35], [0.35, 0.50])	([0.10, 0.15], [0.75, 0.80])
$\tilde{x}_9$	Max	0.0667	([0.80, 0.90], [0, 0.05])	([0.10, 0.25], [0.55, 0.65])	([0.05, 0.10], [0.80, 0.85])
$\tilde{x}_{10}$	Max	0.0667	([0.45, 0.60], [0.25, 0.35])	([0.10, 0.25], [0.55, 0.65])	([0.05, 0.10], [0.80, 0.85])
$\tilde{x}_{11}$	Max	0.0667	([0.75, 0.80], [0.10, 0.15])	([0.15, 0.30], [0.30, 0.50])	([0.05, 0.10], [0.75, 0.85])
$\tilde{x}_{12}$	Max	0.0667	([0.75, 0.85], [0.05, 0.10])	([0.25, 0.40], [0.20, 0.40])	([0.70, 0.80], [0.10, 0.15])
$\tilde{x}_{13}$	Min	0.0667	([0.70, 0.85], [0.05, 0.10])	([0.10, 0.15], [0.75, 0.80])	([0.50, 0.55], [0.35, 0.40])
$\tilde{x}_{14}$	Min	0.0667	([0.70, 0.80], [0.10, 0.15])	([0.50, 0.55], [0.30, 0.40])	([0.10, 0.15], [0.70, 0.80])
$\tilde{x}_{15}$	Max	0.0667	([0.60, 0.75], [0.15, 0.20])	([0.45, 0.60], [0.20, 0.30])	([0.25, 0.45], [0.25, 0.40])

This problem was solved with the proposed IVIF-MULTIMOORA method as follows.

Ratio System

In the ratio system ranking, the relative importance (significance) of the alternatives was determined by (21). For this purpose, initially, the benefit criteria, including x_1, x_2, x_3, x_4, x_5, x_6, x_8, x_9, x_{10}, x_{11}, x_{12}, and x_{15}, were separated from the cost criteria, including x_7, x_{13}, and x_{14}. Then, the aggregations of the benefit criteria weights, W^+, and cost criteria weights, W^-, were computed. These weights were then used in (21), and the values of $\tilde{y}'_i$, $i = 1, 2, 3$, were calculated as follows:

$$\tilde{y}'_1 = ([0.285, 0.363], [0.531, 0.602]),$$
$$\tilde{y}'_2 = ([0.110, 0.136], [0.787, 0.833]),$$
$$\tilde{y}'_3 = ([0.077, 0.110], [0.824, 0.858]). \quad (32)$$

Ranking these values according to their score functions led to the ranking of the alternatives as $A_1 \succ A_2 \succ A_3$.

Reference Point Approach

Ranking of the alternatives in this approach was initialized by determining a reference point from the weighted decision matrix. Then, the deviation of each alternative from the reference point was computed by (28). These deviations were as follows: $d_1 = 0.468$, $d_2 = 1.332$, and $d_3 = 1.391$. By ranking the alternatives in the descending order of deviations, the final ranking was determined as $A_1 \succ A_2 \succ A_3$.

Full Multiplicative Form

In applying the third part of the method, the full multiplicative approach was used to rank the alternatives. Using (30), the multiplicative forms of benefit and cost criteria were obtained. These multiplicative forms were constructed by sequential multiplication of a set of IVIFNs. The overall utility of the alternatives was determined as follows:

$$\tilde{U}_1 = ([6.010E-14, 2.028E-12], [9.999999825E-1, 9.9999999998E-1]),$$
$$\tilde{U}_2 = ([4.127E-20, 2.310E-17], [9.99999999999909E-1, 9.99999999999998E-1]),$$
$$\tilde{U}_3 = ([0,0], [1,1]). \quad (33)$$

Ranking these values according to their score functions led to ranking of the alternatives as $A_1 \succ A_2 \succ A_3$.

The three sets of rankings are summarized in Table 2. The final ranking was obtained by using the dominance theory in the last column of the table.

Table 2: IVIF-MULTIMOORA ranking of the alternatives.

Alternatives	IVIF-MOORA ratio system	IVIF-MOORA reference point	IVIF-full multiplicative form	IVIF-MULTIMOORA
A_1	1	1	1	1
A_2	2	2	2	2
A_3	3	3	3	3

Table 3 also presents the rankings obtained by some different methods, including IFOWA [62], TOPSIS-IVIF [19], COPRAS-IVIF [31], and WASPAS-IVIF [36].

Table 3: Ranking of the alternatives by applying different methods.

Alternatives	IFOWA	TOPSIS-IVIF	COPRAS-IVIF	WASPAS-IVIF	IVIF-MULTIMOORA
A_1	1	1	1	1	1
A_2	2	2	3	2	2
A_3	3	3	2	3	3

Kendall's coefficient of concordance in Table 3 is 0.840, which shows a high degree of concordance. Moreover, there is complete coincidence among the IVIF-MULTIMOORA method and the other three methods. These facts imply that there is great homogeneity between IVIF-MULTIMOORA and other methods.

The Second Example

Consider the problem of appraising four different investment alternatives $\{A_1, A_2, A_3, A_4\}$ based on four criteria of risk (C_1), growth (C_2), sociopolitical issues (C_3), and environmental impacts (C_4). The vector of criteria significance is (0.13, 0.17, 0.39, 0.31). Also, C_1 and C_3 were considered as cost criteria, while C_2 and C_4 were benefit criteria. The problem was initially formulated by Wang et al. [51] as the decision matrix which is given below (Table 4).

Table 4: IVIF decision matrix for investment alternatives.

	C_1	C_2	C_3	C_4
A_1	([0.42, 0.48], [0.4, 0.5])	([0.6, 0.7], [0.05, 0.025])	([0.4, 0.5], [0.2, 0.5])	([0.55, 0.75], [0.15, 0.25])
A_2	([0.4, 0.5], [0.4, 0.5])	([0.5, 0.8], [0.1, 0.2])	([0.3, 0.6], [0.3, 0.4])	([0.6, 0.7], [0.1, 0.3])
A_3	([0.3, 0.5], [0.4, 0.5])	([0.1, 0.3], [0.2, 0.4])	([0.7, 0.8], [0.1, 0.2])	([0.5, 0.7], [0.1, 0.2])
A_4	([0.2, 0.4], [0.4, 0.5])	([0.6, 0.7], [0.2, 0.3])	([0.5, 0.6], [0.2, 0.3])	([0.7, 0.8], [0.1, 0.2])

The above problem was solved by using the proposed IVIF-MULTIMOORA method. The results obtained by using the ratio system, the reference point approach, and the full multiplicative form are presented in Table 5. Moreover, the final ranking obtained based on the dominance theory is presented in the last column of the table.

Table 5: IVIF-MULTIMOORA ranking of the investment alternatives.

Alternatives	IVIF-MOORA ratio system	IVIF-MOORA reference point	IVIF-full multiplicative form	IVIF-MULTIMOORA
A_1	3	2	3	3
A_2	4	1	4	4
A_3	1	4	1	1
A_4	2	3	2	2

The results obtained by solving the above problem by different methods, including the method of Wang et al. [51], IFOWA [62], TOPSIS-IVIF [19], COPRAS-IVIF [31], and WASPAS-IVIF [36], are presented in Table 6.

Table 6: Ranking of the investment alternatives by applying different methods.

Alternatives	Wang et al.	IFOWA	TOPSIS-IVIF	COPRAS-IVIF	WASPAS-IVIF	IVIF-MULTIMOORA
A_1	3	4	2	3	3	3
A_2	4	3	4	4	4	4
A_3	2	1	1	2	1	1
A_4	1	2	3	1	2	2

In this case, Kendall's coefficient of concordance is 0.766, which shows a high degree of concordance, compared to the critical value of 0.421 [65]. Spearman's rank correlation is used to demonstrate the similarity between different rankings. Table 7 shows Spearman's rank correlation between the considered methods. The percentages specified in parenthesis in Table 7 show the matched ranks obtained by using the considered five methods with ranks of Wang et al. [51] method as the percentages of the matched alternatives.

Table 7: Ranking similarity of the investment alternatives by applying different methods.

Methods	Wang et al.	IFOWA	TOPSIS-IVIF	CORAS-IVIF	WASPAS-IVIF	IVIF-MULTIMOORA
Wang et al. [51]	—	0.6 (0%)	0.4 (25%)	1 (100%)	0.8 (50%)	0.8 (50%)
IFOWA		—	0.4 (25%)	0.6 (0%)	0.8 (50%)	0.8 (50%)
TOPSIS-IVIF			—	0.4 (25%)	0.8 (50%)	0.8 (50%)
CORAS-IVIF				—	0.8 (50%)	0.8 (50%)
WASPAS-IVIF					—	1 (100%)
IVIF-MULTIMOORA						—

Based on the above findings, it can be concluded that, except for COPRAS-IVIF, which completely matches the method of Wang et al. [51], the IVIF-MULTIMOORA, along with WASPAS-IVIF, has the highest similarity (reaching 80%) to all other methods. Moreover, the percentage of the matched rankings with those of Wang, Li, and Wang is 50%, which is the second high value of congruence. These results have shown great similarity of IVIF-MULTIMOORA to other methods, which can be considered to be one of its advantages.

DISCUSSION

Vincke [66] believes that the main difficulty in solving multiple criteria decision making problems lies in the fact that, usually, there is no optimal solution to such a problem, which could dominate other solutions in all the criteria. As a response to this dilemma, various methods have been proposed for solving MCDM problems. The selection of the most appropriate and the most robustly and effectively useable method is a continuous object of scientific discussions for decades [67, 68].

This means that the experts usually look for Pareto optimal solution. Therefore, while an optimal solution is obtained by several methods, this emphasizes that proposed method will achieve the near to optimum answer. In this regard, MULTIMOORA method has a strong advantage over the other methods as it summarizes three approaches. The robustness of the aggregated approach has been substantiated in the previous research [47].

On the other hand, the above-mentioned difficulty may be considered from two additional perspectives. First, the uncertainty of decision maker in his judgments should be taken into consideration. To deal with this challenge, several frameworks have been introduced, starting from application of Zadeh fuzzy numbers [6] and followed by numerous extensions of decision making methods using type 1 and type 2 fuzzy sets, involving interval-valued as well as intuitionistic fuzzy sets [69]. The extensions of the considered MULTIMOORA method till 2013 were summarized in a review paper [13]. The most recent extensions not covered by [13] should also be mentioned [14–17]. Seeking the best imagination of uncertainty, another up-to-date extension, namely, the interval-valued intuitionistic fuzzy MULTIMOORA method is presented in the current research.

Second, decisions are usually made by several decision makers. This makes it necessary to consider a suitable procedure to aggregate the opinions. Therefore, this problem is handled as a group decision making problem.

As it is not easy to prove that a newly proposed method is applicable, it is useful to test it in solving several multiple criteria problems. To make sure that the method is better or at least it is not worse than the other existing methods, it is appropriate to apply several related approaches to compare their ranking results for the same problem [70]. Accordingly, two illustrative examples have been presented to fulfill the task. It is encouraging that the results have shown great similarity of IVIF-MULTIMOORA to other methods. The fact can be considered to be one of advantages of the novel approach.

CONCLUSIONS AND FUTURE WORK

In the current paper, the interval-valued intuitionistic fuzzy MULTIMOORA method is proposed for solving a group decision making problem under uncertainty. The IVIF is a generalized form of fuzzy sets which considers the nonmembership of objects in a set beside their membership and with respect to hesitancy of decision makers.

The MULTIMOORA method is a decision making procedure composed of three parts, including the ratio system, the reference point approach, and full multiplicative form. The rankings yielded by this method are finally aggregated, and a unique ranking is determined.

The proposed algorithm benefits from the abilities of IVIFSs to be used in imagination of uncertainty and MULTIMOORA method to consider three different viewpoints in analyzing decision alternatives. The application of the proposed method is illustrated by two numerical examples and the obtained results are compared to the results yielded by other methods of decision making with IVIF information. The comparison, using Kendall coefficient of concordance and Spearman's rank correlation, has shown a high consistency between the proposed method and the other approaches that verifies the proposed method can be applied in case of similar problems in ambiguous and uncertain situation. This advantage strengthens the applicability of IVIF-MULTIMOORA as a method for group decision making under uncertainty when solving engineering technology and management complex problems.

Future directions of the research will be focused on searching modifications of the method, the best reflecting uncertain environment, and producing the most reliable results. At the moment two novel and up-to-date possible extensions of the MULTIMOORA method could be suggested. One of the approaches could be based on the neutrosophic sets [71, 72] that are a generalization of the fuzzy and intuitionistic fuzzy sets. The other possible approach could be to apply D numbers that were recently proposed to handle uncertain information [73, 74].

CONFLICT OF INTERESTS

The authors declare that there is no conflict of interests regarding the publication of this paper.

REFERENCES

1. M. Zeleny, Multiple Criteria Decision Making, McGraw-Hill, New York, NY, USA, 1982.
2. H. A. Simon, Models of Discovery and Other Topics in the Methods of Science, D. Reidel, Boston, Mass, USA, 1977.
3. P. L. Yu, Forming Winning Strategies: An Integrated Theory of Habitual Domains, Springer, Heidelberg, Germany, 1990.
4. M. C. Yovits, Advances in Computers, Academic Press, Orlando, Fla, USA, 1984.
5. S. F. Liu and Y. Lin, Grey Systems Theory and Application, Springer, Heidelberg, Germany, 2011.
6. L. A. Zadeh, "Fuzzy sets," Information and Computation, vol. 8, no. 3, pp. 338–353, 1965.
7. W. Pedrycz and F. Gomide, An Introduction to Fuzzy Sets: Analysis and Design, MIT Press, Cambridge, UK, 1998.
8. Grattan-Guinness, "Fuzzy membership mapped onto intervals and many-valued quantities,"Mathematical Logic Quarterly, vol. 22, no. 2, pp. 149–160, 1976.
9. K. T. Atanassov, "Intuitionistic fuzzy sets," Fuzzy Sets and Systems, vol. 20, no. 1, pp. 87–96, 1986.
10. K. Atanassov and G. Gargov, "Interval valued intuitionistic fuzzy sets," Fuzzy Sets and Systems, vol. 31, no. 3, pp. 343–349, 1989.
11. T. Baležentis and S. Zeng, "Group multi-criteria decision making based upon interval-valued fuzzy numbers: an extension of the MULTIMOORA method," Expert Systems with Applications, vol. 40, no. 2, pp. 543–550, 2013.
12. T. Baležentis, S. Z. Zeng, and A. Baležentis, "MULTIMOORA-IFN: a MCDM method based on intuitionistic fuzzy number for performance management," Economic Computation and Economic Cybernetics Studies and Research, vol. 48, no. 4, pp. 85–102, 2014.
13. T. Baležentis and A. Baležentis, "A survey on development and applications of the multi-criteria decision making method MULTIMOORA," Journal of Multi-Criteria Decision Analysis, vol. 21, no. 3-4, pp. 209–222, 2014.
14. H. Liu, X. Fan, P. Li, and Y. Chen, "Evaluating the risk of failure modes with extended MULTIMOORA method under fuzzy environment," Engineering Applications of Artificial Intelligence, vol. 34, pp. 168–177, 2014.
15. H. Liu, J. X. You, Ch. Lu, and Y. Z. Chen, "Evaluating health-care waste treatment technologies using a hybrid multi-criteria decision making model," Renewable and Sustainable Energy Reviews, vol. 41, pp. 932–942, 2015.
16. H. Liu, J. X. You, C. Lu, and M. M. Shan, "Application of interval 2-tuple linguistic MULTIMOORA method for health-care waste treatment technology evaluation and selection," Waste Management, vol. 34, no. 11, pp. 2355–2364, 2014.
17. Z. H. Li, "An extension of the MULTIMOORA method for multiple criteria group decision making based upon hesitant fuzzy sets," Journal of Applied Mathematics, vol. 2014, Article ID 527836, 16 pages, 2014.
18. K. Verma, R. Verma, and N. C. Mahanti, "Facility location selection: an interval valued intuitionistic fuzzy TOPSIS approach," Journal of Modern Mathematics and Statistics, vol. 4, no. 2, pp. 68–72, 2010.
19. F. Ye, "An extended TOPSIS method with interval-valued intuitionistic fuzzy numbers for virtual enterprise partner selection," Expert Systems with Applications, vol. 37, no. 10, pp. 7050–7055, 2010.
20. J. H. Park, I. Y. Park, Y. C. Kwun, and X. Tan, "Extension of the TOPSIS method for decision making problems under interval-valued intuitionistic fuzzy environment," Applied Mathematical Modelling, vol. 35, no. 5, pp. 2544–2556, 2011.
21. D.-F. Li, "TOPSIS-based nonlinear-programming methodology for multiattribute decision making with interval-valued intuitionistic fuzzy sets," IEEE Transactions on Fuzzy Systems, vol. 18, no. 2, pp. 299–311, 2010.
22. D.-F. Li, "Linear programming method for MADM with interval-valued intuitionistic fuzzy sets," Expert Systems with Applications, vol. 37, no. 8, pp. 5939–5945, 2010.
23. J. H. Park, H. J. Cho, and Y. C. Kwun, "Extension of the VIKOR method for group decision making with interval-valued intuitionistic fuzzy information," Fuzzy Optimization and Decision Making, vol. 10, no. 3, pp. 233–253, 2011.
24. D.-F. Li, "Closeness coefficient based nonlinear programming method for interval-valued intuitionistic fuzzy multiattribute decision making with incomplete

preference information," Applied Soft Computing Journal, vol. 11, no. 4, pp. 3402–3418, 2011.

25. S.-M. Chen, L.-W. Lee, H.-C. Liu, and S.-W. Yang, "Multiattribute decision making based on interval-valued intuitionistic fuzzy values," Expert Systems with Applications, vol. 39, no. 12, pp. 10343–10351, 2012.
26. D. Yu, Y. Wu, and T. Lu, "Interval-valued intuitionistic fuzzy prioritized operators and their application in group decision making," Knowledge-Based Systems, vol. 30, pp. 57–66, 2012.
27. H. Zhang and L. Yu, "MADM method based on cross-entropy and extended TOPSIS with interval-valued intuitionistic fuzzy sets," Knowledge-Based Systems, vol. 30, pp. 115–120, 2012.
28. J. Ye, "Interval-valued intuitionistic fuzzy cosine similarity measures for multiple attribute decision-making," International Journal of General Systems, vol. 42, no. 8, pp. 883–891, 2013.
29. F. Meng, Q. Zhang, and H. Cheng, "Approaches to multiple-criteria group decision making based on interval-valued intuitionistic fuzzy Choquet integral with respect to the generalized λ-Shapley index,"Knowledge-Based Systems, vol. 37, pp. 237–249, 2013.
30. F. Meng, C. Tan, and Q. Zhang, "The induced generalized interval-valued intuitionistic fuzzy hybrid Shapley averaging operator and its application in decision making," Knowledge-Based Systems, vol. 42, pp. 9–19, 2013.
31. S. H. Razavi Hajiagha, S. S. Hashemi, and E. K. Zavadskas, "A complex proportional assessment method for group decision making in an interval-valued intuitionistic fuzzy environment," Technological and Economic Development of Economy, vol. 19, no. 1, pp. 22–37, 2013.
32. J. Chai, J. N. K. Liu, and Z. Xu, "A rule-based group decision model for warehouse evaluation under interval-valued Intuitionistic fuzzy environments," Expert Systems with Applications, vol. 40, no. 6, pp. 1959–1970, 2013.
33. S. Wan and J. Dong, "A possibility degree method for interval-valued intuitionistic fuzzy multi-attribute group decision making," Journal of Computer and System Sciences, vol. 80, no. 1, pp. 237–256, 2014.
34. T.-Y. Chen, "The extended linear assignment method for multiple criteria decision analysis based on interval-valued intuitionistic fuzzy sets," Applied Mathematical Modelling, vol. 38, no. 7-8, pp. 2101–2117, 2014.
35. J. Xu and F. Shen, "A new outranking choice method for group decision making under Atanassov's interval-valued intuitionistic fuzzy environment," Knowledge-Based Systems, vol. 70, pp. 177–188, 2014.
36. E. K. Zavadskas, J. Antucheviciene, S. H. R. Hajiagha, and S. S. Hashemi, "Extension of weighted aggregated sum product assessment with interval-valued intuitionistic fuzzy numbers (WASPAS-IVIF),"Applied Soft Computing, vol. 24, pp. 1013–1021, 2014.
37. S. Geetha, V. L. G. Nayagam, and R. Ponalagusamy, "A complete ranking of incomplete interval information," Expert Systems with Applications, vol. 41, no. 4, pp. 1947–1954, 2014.
38. Y. Zhang, P. Li, Y. Wang, P. Ma, and X. Su, "Multiattribute decision making based on entropy under interval-valued intuitionistic fuzzy environment," Mathematical Problems in Engineering, vol. 2013, Article ID 526871, 8 pages, 2013.
39. C. Wei and Y. Zhang, "Entropy measures for interval-valued intuitionistic fuzzy sets and their application in group decision-making," Mathematical Problems in Engineering, vol. 2015, Article ID 563745, 13 pages, 2015.
40. X. Tong and L. Yu, "A novel MADM approach based on fuzzy cross entropy with interval-valued intuitionistic fuzzy sets," Mathematical Problems in Engineering, vol. 2015, Article ID 965040, 9 pages, 2015.
41. J. Wu and F. Chiclana, "Non-dominance and attitudinal prioritisation methods for intuitionistic and interval-valued intuitionistic fuzzy preference relations," Expert Systems with Applications, vol. 39, no. 18, pp. 13409–13416, 2012.
42. J. Wu and F. Chiclana, "A risk attitudinal ranking method for interval-valued intuitionistic fuzzy numbers based on novel attitudinal expected score and accuracy functions," Applied Soft Computing, vol. 22, pp. 272–286, 2014.
43. Z.-S. Xu, "Methods for aggregating interval-valued intuitionistic fuzzy information and their application to decision making," Control and Decision, vol. 22, no. 2, pp. 215–219, 2007.
44. W. K. M. Brauers and E. K. Zavadskas, "The MOORA method and its application to privatization in a transition economy," Control and Cybernetics, vol. 35, no. 2, pp. 445–469, 2006.
45. W. K. M. Brauers and E. K. Zavadskas, "Project management by MULTIMOORA as an instrument for transition economies," Technological and Economic Development of Economy, vol. 16, no. 1, pp. 5–24, 2010.
46. W. K. M. Brauers and E. K. Zavadskas, "Multimoora optimization used to decide on a bank loan to buy property," Technological and Economic Development of Economy, vol. 17, no. 1, pp. 174–188, 2011.
47. W. K. Brauers and E. K. Zavadskas, "Robustness of MULTIMOORA: a method for multi-objective optimization," Informatica, vol. 23, no. 1, pp. 1–25, 2012.
48. D. Streimikiene, T. Balezentis, I. Krisciukaitien, and A. Balezentis, "Prioritizing sustainable electricity production technologies: MCDM approach," Renewable and Sustainable Energy Reviews, vol. 16, no. 5, pp. 3302–3311, 2012.
49. E. K. Zavadskas, J. Antucheviciene, J. Saparauskas, and Z. Turskis, "MCDM methods WASPAS and MULTIMOORA: verification of robustness of methods when assessing alternative solutions," Economic Computation and Economic Cybernetics Studies and Research, vol. 47, no. 2, pp. 5–20, 2013.
50. E. K. Zavadskas, J. Antucheviciene, J. Saparauskas, and Z. Turskis, "Multi-criteria assessment of facades

alternatives: peculiarities of ranking methodology," Procedia Engineering, vol. 57, pp. 107–112, 2013.

51. Z. Wang, K. W. Li, and W. Wang, "An approach to multiattribute decision making with interval-valued intuitionistic fuzzy assessments and incomplete weights," Information Sciences, vol. 179, no. 17, pp. 3026–3040, 2009.

52. W. K. M. Brauers, A. Baležentis, and T. Baležentis, "Multimoora for the EU member States updated with fuzzy number theory," Technological and Economic Development of Economy, vol. 17, no. 2, pp. 259–290, 2011.

53. F. Herrera and L. Martinez, "A 2-tuple fuzzy linguistic represent model for computing with words," IEEE Transactions on Fuzzy Systems, vol. 8, no. 6, pp. 746–752, 2000.

54. Baležentis and T. Baležentis, "An innovative multi-criteria supplier selection based on two-tuple MULTIMOORA and hybrid data," Economic Computation and Economic Cybernetics Studies and Research, vol. 45, no. 2, pp. 37–56, 2011.

55. Baležentis and T. Baležentis, "A novel method for group multi-attribute decision making with two-tuple linguistic computing: supplier evaluation under uncertainty," Economic Computation and Economic Cybernetics Studies and Research, vol. 45, no. 4, pp. 5–30, 2011.

56. Baležentis, T. Baležentis, and W. K. M. Brauers, "Personnel selection based on computing with words and fuzzy MULTIMOORA," Expert Systems with Applications, vol. 39, no. 9, pp. 7961–7967, 2012.

57. Baležentis, T. Baležentis, and W. K. M. Brauers, "MULTIMOORA-FG: a multi-objective decision making method for linguistic reasoning with an application to personnel selection," Informatica, vol. 23, no. 2, pp. 173–190, 2012.

58. D. Stanujkic, N. Magdalinovic, S. Stojanovic, and R. Jovanovic, "Extension of ratio system part of MOORA method for solving decision-making problems with interval data," Informatica, vol. 23, no. 1, pp. 141–154, 2012.

59. D. Stanujkic, N. Magdalinovic, R. Jovanovic, and S. Stojanovic, "An objective multi-criteria approach to optimization using MOORA method and interval grey numbers," Technological and Economic Development of Economy, vol. 18, no. 2, pp. 331–363, 2012.

60. S. Karlin and W. J. Studden, Tchebycheff Systems: With Applications in Analysis and Statistics, Interscience Publishers, New York, NY, USA, 1966.

61. D.-F. Li, "Extension principles for interval-valued intuitionistic fuzzy sets and algebraic operations,"Fuzzy Optimization and Decision Making, vol. 10, no. 1, pp. 45–58, 2011.

62. H. Zhao, Z. Xu, M. Ni, and S. Liu, "Generalized aggregation operators for intuitionistic fuzzy sets,"International Journal of Intelligent Systems, vol. 25, no. 1, pp. 1–30, 2010.

63. L. Dym, Principles of Mathematical Modeling, Elsevier Academic Press, Boston, Mass, USA, 2nd edition, 2001.

64. J. C. Pomerol and S. Barbara-Romero, Multicriterion Decision in Management: Principles and Practice, Kluwer Academic Publishers, Boston, Mass, USA, 2000.

65. J. Sheskin, Handbook of Parametric and Nonparametric Statistical Procedures, Chapman and Hall, Boca Raton, Fla, USA, 5th edition, 2011.

66. P. Vincke, Multicriteria Decision-Aid, John Wiley & Sons, Chichester, UK, 1992.

67. V. M. Ozernoy, "Choosing the best multiple criteria decision-making method," INFOR, vol. 30, no. 2, pp. 159–171, 1992.

68. M. Aruldoss, T. M. Lakshmi, and V. P. Venkatesan, "A survey on multi criteria decision making methods and its applications," American Journal of Information Systems, vol. 1, no. 1, pp. 31–43, 2013.

69. Mardani, A. Jusoh, and E. K. Zavadskas, "Fuzzy multiple criteria decision-making techniques and applications—two decades review from 1994 to 2014," Expert Systems with Applications, vol. 42, no. 8, pp. 4126–4148, 2015.

70. P. Wang, Y. Li, Y.-H. Wang, and Z.-Q. Zhu, "A new method based on TOPSIS and response surface method for MCDM problems with interval numbers," Mathematical Problems in Engineering, vol. 2015, Article ID 938535, 11 pages, 2015.

71. F. Smarandache, "A geometric interpretation of the neutrosophic set—a generalization of the intuitionistic fuzzy set," in Neutrosophic Theory and Its Applications, F. Smarandache, Ed., Europa Nova, Brussels, Belgium, 2014.

72. J. J. Peng, J. Q. Wang, H. Y. Zhang, and X. H. Chen, "An outranking approach for multi-criteria decision-making problems with simplified neutrosophic sets," Applied Soft Computing, vol. 25, pp. 336–346, 2014.

73. Y. Deng, "D numbers: theory and applications," Journal of Information and Computational Science, vol. 9, no. 9, pp. 2421–2428, 2012.

74. X. Deng, X. Lu, F. T. Chan, R. Sadiq, S. Mahadevan, and Y. Deng, "D-CFPR: D numbers extended consistent fuzzy preference relations," Knowledge-Based Systems, vol. 73, pp. 61–68, 2015.

CHAPTER 3

AN INTERVAL-PARAMETER FUZZY LINEAR PROGRAMMING WITH STOCHASTIC VERTICES MODEL FOR WATER RESOURCES MANAGEMENT UNDER UNCERTAINTY

Yan Han, Yuefei Huang, Shaofeng Jia, and Jiahong Liu

INTRODUCTION

Water shortage has become serious issue in process of urbanization as well as socioeconomic development gradually, especially in metropolis, where water resources are limited. Therefore, effective allocation of water resources to various users is important. There are many uncertain factors in practical management and decision making of complex water resources system, which might result in significant difficulties in optimizing water resources allocation. Conventional deterministic optimization methods have difficulties in reflecting these uncertainties [1]. Many researchers have tried to tackle these uncertain problems through fuzzy programming, interval programming, and stochastic programming [2–10]. In many real-world problems, several types of uncertainties may exist together in a complex system. Therefore, hybrid uncertainty methods have been desired for solving the problem with several types of uncertainties. Based on inexact chance-constrained programming (ICCP) method, Huang [11] proposed a hybrid inexact-stochastic water management model, which improves upon the existing inexact and stochastic programming approaches by allowing both distribution information in right hand and uncertainties in left hand or coefficients of objective. Hybrid uncertainties, including interval and stochastic distribution information in parameters and coefficients, can be directly communicated into the optimization process through representing the uncertain parameters or coefficients as fuzzy sets and random variables [3, 9, 10, 12–15]. In these hybrid uncertainty approaches, each coefficient or parameter has only one kind of uncertainty, and the stochastic distribution information is treated with the discrete way.

Due to the complexity of the real world, highly uncertain information may exist. Such boundaries of interval parameters of the optimization model are also uncertain. Nie et al. [16] proposed an interval parameter fuzzy-robust programming (IFRP) model through introducing the concept of fuzzy boundary interval. The parameters of the IFRP model were represented as interval numbers with fuzzy uncertain boundary, and the uncertainties were directly communicated into the optimization process and resulting solution. In this way, the robustness of the optimization process and solution can be enhanced. Considering the interval uncertain of boundaries of interval parameters, Liu and Huang [17] proposed a dual interval two-stage restricted-recourse programming (DITRP) method for flood-diversion planning. Liu et al. [18] established a dual-interval linear programming (DILP) model by introducing ILP approach into the existing interval-parameter linear programming framework. Some of coefficients and parameters in the DILP model were represented as interval-parameter with interval vertices. The DILP approach improved the ILP method by allowing dual uncertainties (presented as dual intervals) to be incorporated into the optimization process. For the fuzzy feature of boundaries of interval parameters, Li

Citation: Yan Han, Yuefei Huang, Shaofeng Jia, and Jiahong Liu, "An Interval-Parameter Fuzzy Linear Programming with Stochastic Vertices Model for Water Resources Management under Uncertainty," Mathematical Problems in Engineering, vol. 2013, Article ID 942343, 12 pages, 2013. doi:10.1155/2013/942343.

et al. [19] proposed a dual-interval vertex (DIV) method by incorporating the vertex method within an interval-parameter programming framework, and a fuzzy vertex analysis approach was proposed for solving the DIV model. The DIV approach can tackle uncertainties expressed as dual intervals that exist in both objective function and left-hand and right-hand sides of the constraints. However, these above methods hardly deal with the dual uncertainties including stochastic distribution attributes. Considering the stochastic attribute of boundaries of interval parameters in objective functions and constraints, Han et al. [20] proposed an interval linear programming with stochastic vertices (ILPSV) method to tackle dual uncertainties, which were presented as interval parameter with stochastic vertices problem. The fuzzy attribute of objective and constraints is not for concern in the ILPSV model. Therefore, one potential approach for better accounting for integrated uncertainties of parameters of model is to incorporate the stochastic distribution within a general fuzzy linear programming framework. This leads to an interval-parameter fuzzy linear programming with stochastic vertices method under dual uncertainty.

The objective of this paper is to propose an interval-parameter fuzzy linear programming model with stochastic vertices by coupling inexact fuzzy linear programming (IFLP) and stochastic vertices method. Highly uncertain information for the lower and upper bounds of interval parameters that exist in optimization model due to the complexity of the real world can be effectively handled through allowing the stochastic boundary of interval parameter to be incorporated into the optimization processes. In addition, the dual uncertainty concept (being stochastic boundaries of interval) is presented when the available information is highly uncertain for boundaries of interval parameter of objective functions and constraints. A hybrid intelligent algorithm based on Liu [21, 22] has been proposed for solving the developed model. The developed IFLPSV model is then applied to allocation of multisource water to multiple users in Beijing city of China in 2020, where water resources shortage is a challenging issue.

METHODOLOGY

Dual Uncertain Linear Programming Model

In many practical problems, the lower and upper bounds of some interval parameters in a water resources management system can rarely be acquired as deterministic. Instead, they can only be expressed by interval, fuzzy, or stochastic numbers. For a system with such dual uncertainty, an interval-parameter linear programming with stochastic vertices is generated as follows:

$$\max \quad \bar{\bar{f}} = \sum_{j=1}^{n} \left(\bar{c}_j + \bar{\bar{d}}_j \right) \bar{x}_j \tag{1a}$$

$$\text{Subject to} \quad \sum_{j=1}^{n} \bar{a}_{rj} \bar{x}_j \le \bar{b}_r, \quad (r = 1, 2, \ldots, s), \tag{1b}$$

$$\sum_{j=1}^{n} \bar{\bar{a}}_{tj} \bar{x}_j \le \bar{\bar{b}}_t, \quad (t = s+1, s+2, \ldots, m), \tag{1c}$$

$$\bar{x}_j \ge 0, \quad \forall j, \tag{1d}$$

where $\bar{c}_j$, $\bar{a}_{rj}$, and $\bar{b}_r$ are the intervals with deterministic lower and upper bounds, respectively; $\bar{\bar{a}}_{tj}$, $\bar{\bar{b}}_t$, and $\bar{\bar{d}}_j$ are the uncertain intervals with stochastic lower and upper vertices, respectively; j is the index of decision variables; n is the total number of decision variables; r is the index of single interval constraints; s is the number of single uncertain constraints; t is the index of dual uncertain constraints; and m is the total number of constraints. The models (1a), (1b), (1c), and (1d) can then be reformulated as follows:

$$\max \quad \bar{\bar{f}} = \sum_{j=1}^{n} \bar{c}_j \bar{x}_j + \sum_{j=1}^{n} \left[\bar{d}_j^L, \bar{d}_j^U \right] \bar{x}_j \tag{2a}$$

$$\text{Subject to} \quad \sum_{j=1}^{n} \bar{a}_{rj} \bar{x}_j \le \bar{b}_r, \quad (r = 1, 2, \ldots, s), \tag{2b}$$

$$\sum_{j=1}^{n} \left[\bar{a}_{tj}^L, \bar{a}_{tj}^U \right] \bar{x}_j \le \left[\bar{b}_t^L, \bar{b}_t^U \right], \tag{2c}$$

$$(t = s+1, s+2, \ldots, m),$$

$$\bar{x}_j \ge 0, \quad \forall j, \tag{2d}$$

Where $\bar{a}_{tj}^L$ and $\bar{a}_{tj}^U$, $\bar{b}_t^L$, and $\bar{b}_t^U$, $\bar{d}_j^L$, and $\bar{d}_j^U$ are the stochastic lower and upper bounds $\bar{\bar{a}}_{tj}$, $\bar{\bar{b}}_t$, and $\bar{\bar{d}}_j$ ($\bar{a}_{tj}^L \le \bar{a}_{tj}^U$, $\bar{b}_t^L \le \bar{b}_t^U$, and $\bar{d}_j^L \le \bar{d}_j^U$), , respectively. The lower and upper bounds of interval number in ILP are certain as shown in Figure 1(a). The interval number of dual uncertainty linear programming is shown in Figure 1(b), and the vertices may possess a certain distribution function, such as formal distribution. The -axis in Figure 1does not represent any physical meaning since the figure is used to illustrate an interval number or an interval number with randomly distributed lower and upper bounds. The models (2a), (2b), (2c), and (2d) are an interval parameter linear programming with stochastic vertices, and it possesses the randomly distributed attribute of stochastic vertices and interval attribute of interval parameter; therefore, it is a dual uncertain optimization model. The traditional solution algorithm for interval-parameter linear programming (ILP) is not applicable to solve the models (2a), (2b), (2c) and (2d) due to stochastic variables existence.

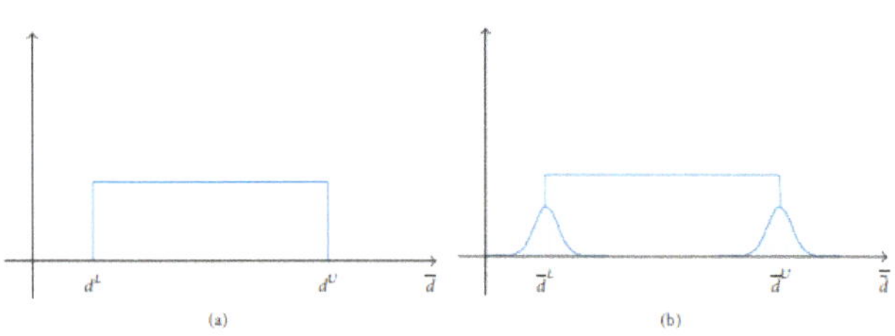

Figure 1: Interval number with certain and uncertain vertices.

Considering the fuzzy features of interval number of linear programming model under dual uncertainty, a potential approach for handling such complexities in the dual uncertain linear programming framework is to introduce IFLP technique into models (2a), (2b), (2c), and (2d). This leads to an interval-parameter fuzzy linear programming with stochastic vertices (IFLPSV) model as follows:

$$\max \quad \bar{\lambda} \tag{1a}$$

$$\text{subject to} \quad \sum_{j=1}^{k_1} \left[c_j^L + \bar{d}_j^L\right] x_j^L + \sum_{j=k_1+1}^{n} \left[c_j^L + \bar{d}_j^L\right] x_j^U \geq \bar{f}_{\text{opt}}^L + \bar{\lambda}\left(\bar{f}_{\text{opt}}^U - \bar{f}_{\text{opt}}^L\right), \tag{1b}$$

$$\sum_{j=1}^{k_1} \left|a_{rj}\right|^U \operatorname{Sign}\left(a_{rj}^U\right) x_j^L + \sum_{j=k_1+1}^{n} \left|a_{rj}\right|^L \operatorname{Sign}\left(a_{rj}^L\right) x_j^U \leq b_r^U - \bar{\lambda}\left(b_r^U - b_r^L\right), \quad (r = 1, 2, \ldots, s), \tag{1c}$$

$$\sum_{j=1}^{k_1} \left|\bar{a}_{tj}\right|^U \operatorname{Sign}\left(\bar{a}_{tj}^U\right) x_j^L + \sum_{j=k_1+1}^{n} \left|\bar{a}_{tj}\right|^L \operatorname{Sign}\left(\bar{a}_{tj}^L\right) x_j^U \leq \bar{b}_t^U - \bar{\lambda}\left(\bar{b}_t^U - \bar{b}_t^L\right), \quad (t = s+1, s+2, \ldots, m), \tag{1d}$$

$$x_j^L \geq 0, \quad (j = 1, 2, \ldots, k_1), \tag{1e}$$

$$x_j^U \geq 0, \quad (j = k_1 + 1, k_1 + 2, \ldots, n), \tag{1f}$$

where $\bar{\lambda}$ is the fuzzy membership grade, x_j^L $(j = 1, 2, \ldots, k_1)$ are variables with positive coefficients in the objective function, and x_j^U $(j = k_1 + 1, k_1 + 2, \ldots, n)$ are variables with negative coefficients in the objective function. $\bar{f}_{\text{opt}}^U$ and $\bar{f}_{\text{opt}}^L$ are upper and lower bounds of objective values from models (2a), (2b), (2c), and (2d).

Solving the IFLPSV

According to Huang et al. [23, 24], the above models (3a), (3b), (3c), (3d), (3e), and (3f) can be transformed into two submodels, which correspond to the upper and lower bounds of the desired objective function value. With a two-step interactive algorithm, one submodel corresponding to λ^U (lower bound of the membership grade) can be first formulated as

$$\max \quad \lambda^U \tag{4a}$$

$$\text{subject to} \quad \sum_{j=1}^{k_1} \left[c_j^U + \bar{d}_j^U\right] x_j^U + \sum_{j=k_1+1}^{n} \left[c_j^U + \bar{d}_j^U\right] x_j^U \geq \bar{f}_{\text{opt}}^L + \lambda^U\left(\bar{f}_{\text{opt}}^U - \bar{f}_{\text{opt}}^L\right), \tag{4b}$$

$$\sum_{j=1}^{k_1} \left|a_{rj}\right|^L \operatorname{Sign}\left(a_{rj}^L\right) x_j^U + \sum_{j=k_1+1}^{n} \left|a_{rj}\right|^U \operatorname{Sign}\left(a_{rj}^U\right) x_j^L \leq b_r^U - \lambda^U\left(b_r^U - b_r^L\right), \quad (r = 1, 2, \ldots, s), \tag{4c}$$

$$\sum_{j=1}^{k_1} \left|\bar{a}_{tj}\right|^L \operatorname{Sign}\left(\bar{a}_{tj}^L\right) x_j^U + \sum_{j=k_1+1}^{n} \left|\bar{a}_{tj}\right|^U \operatorname{Sign}\left(\bar{a}_{tj}^U\right) x_j^L \leq \bar{b}_t^U - \lambda^U\left(\bar{b}_t^U - \bar{b}_t^L\right), \tag{4d}$$

$$(t = s+1, s+2, \ldots, m), \quad x_j^U \geq 0, \quad (j = 1, 2, \ldots, k_1), \tag{4e}$$

$$x_j^L \geq 0, \quad (j = k_1 + 1, k_1 + 2, \ldots, n), \quad 0 \leq \lambda^U \leq 1. \tag{4f}$$

With the solutions of models (4a), (4b), (4c), (4d), (4e), (4f), and (4g), another submodel corresponding to λ^L (lower bound of the membership grade) can be formulated as

$$\max \quad \lambda^L \tag{5a}$$

$$\text{subject to} \quad \sum_{j=1}^{k_1} \left[c_j^L + \bar{d}_j^L\right] x_j^L + \sum_{j=k_1+1}^{n} \left[c_j^L + \bar{d}_j^L\right] x_j^L \geq \bar{f}_{\text{opt}}^L + \lambda^L\left(\bar{f}_{\text{opt}}^U - \bar{f}_{\text{opt}}^L\right), \tag{5b}$$

$$\sum_{j=1}^{k_1} \left|a_{rj}\right|^U \mathrm{Sign}\left(a_{rj}^U\right) x_j^L + \sum_{j=k_1+1}^{n} \left|a_{rj}\right|^L \mathrm{Sign}\left(a_{rj}^L\right) x_j^U \le b_r^U - \lambda^U \left(b_r^U - b_r^L\right), \quad (r = 1, 2, \ldots, s), \tag{5c}$$

$$\sum_{j=1}^{k_1} \left|\overline{a}_{tj}\right|^U \mathrm{Sign}\left(\overline{a}_{tj}^U\right) x_j^L + \sum_{j=k_1+1}^{n} \left|\overline{a}_{tj}\right|^L \mathrm{Sign}\left(\overline{a}_{tj}^L\right) x_j^U \le \overline{b}_t^U - \lambda^U \left(\overline{b}_t^U - \overline{b}_t^L\right), \quad (t = s+1, s+2, \ldots, m), \tag{5d}$$

$$0 \le x_j^L \le x_{j\,\mathrm{opt}}^U, \quad (j = 1, 2, \ldots, k_1), \tag{5e}$$

$$x_j^U \ge x_{j\,\mathrm{opt}}^L, \quad (j = k_1 + 1, k_1 + 2, \ldots, n), \tag{5f}$$

$$0 \le \lambda^L \le \lambda_{\mathrm{opt}}^U, \tag{5g}$$

where $x_{j\,\mathrm{opt}}^U$ $(j = 1, 2, \ldots, k_1)$, $x_{j\,\mathrm{opt}}^L$ $(j = k_1 + 1, k_1 + 2, \ldots, n)$, and λ_{opt}^U are solutions of models (4a), (4b), (4c), (4d), (4e), (4f), and (4g). Due to the stochastic variables existence in the models (4a), (4b), (4c), (4d), (4e), (4f), (4g), (5a), (5b), (5c), (5d), (5e), (5f), and (5g), the traditional solving algorithms of ILP (Liu et al. [18]) and DIV (Li et al. [19]) methods are impracticable.

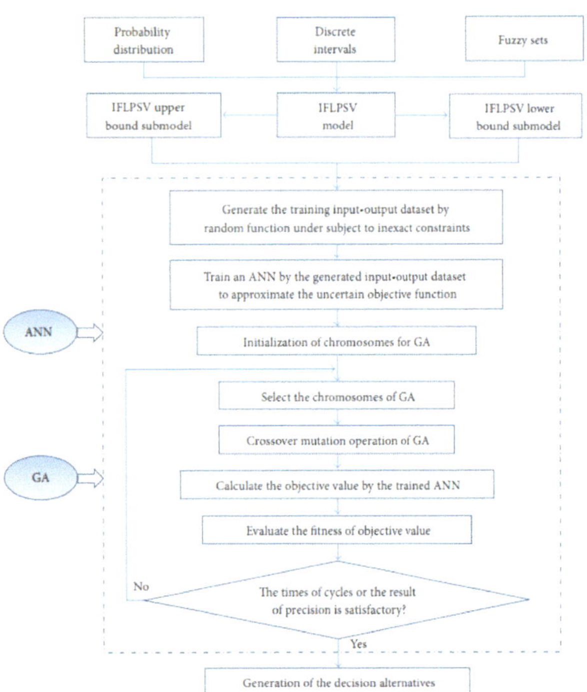

Figure 2: Flow chart of solving the IFLPSV model by hybrid intelligent algorithm.

The hybrid intelligent algorithm, which incorporates ANN and GA, can be used to deal with the stochastic variables in objective function, left-hand and right-hand sides of constraints. The stochastic variables of objective function and constraints need not to be initially set up, and they are automatically determined through many times iteration of the GA and ANN. Therefore, it is proposed as an intelligent algorithm. The procedure of solving the IFLPSV model by hybrid intelligent algorithm is shown in Figure 2.

ILLUSTRATIVE EXAMPLE

The following illustrative example can be formulated to demonstrate the applicability of the proposed method:

$$\max \quad \overline{f} = [\xi_1, \xi_2]\,\overline{x}_1 - [5.5, 6.0]\,\overline{x}_2 \tag{6a}$$

$$\text{Subject to} \quad [8, 10]\,\overline{x}_1 - [12, 14]\,\overline{x}_2 \le [\eta_1, \eta_2], \tag{6b}$$

$$[\gamma_1, \gamma_2]\,\overline{x}_1 + [3, 4]\,\overline{x}_2 \le [6.0, 6.5], \tag{6c}$$

$$\overline{x}_1, \overline{x}_2 \ge 0, \tag{6d}$$

where, ξ_1, ξ_2, γ_1, γ_2, η_1, and η_2 are the stochastic variables with normal distributions of $N(26.5, 0.02)$, $N(29.5, 0.02)$, $N(2.4, 0.01)$, $N(2.8, 0.01)$, $N(3.8, 0.01)$, and $N(4.2, 0.01)$, respectively. According to two-step method, the models (6a), (6b), (6c), and (6d) can be converted into two submodels. A running of the hybrid intelligent algorithm (200 iteration simulations, 500 data in ANN, and 50 generations in GA) was undertaken to solve the two sub-models. Through two-step method and hybrid artificial algorithm, the interval solutions of models (6a), (6b), (6c), and (6d) are $\overline{x}_1 = [1.3104, 1.6401]$, $\overline{x}_2 = [0.6376, 0.7748]$, and $\overline{f} = [29.5070, 45.0840]$(as shown in Table 1, no units for all results since it is only an illustrative example).

Table 1: Comparison of results of ILP, DIV, ILPSV, and IFLPSV method.

Methods	ILP	DIV	ILPSV	IFLPSV
Objective value	[[29.438, 32.150], [42.172, 45.784]]	[[30.099, 31.455], [43.053, 44.895]]	[29.5070, 45.0840]	[30.9548, 42.6913]

Based on formulas (3a), (3b), (3c), (3d), (3e), and (3f) and Huang et al.'s [24], the models (6a), (6b), (6c), and (6d) can be converted to an IFLP problem as follows:

$$\max \quad \overline{\lambda} \tag{7a}$$

$$\text{Subject to} \quad [\xi_1, \xi_2]\,\overline{x}_1 - [5.5, 6.0]\,\overline{x}_2 \ge f_{\mathrm{opt1}}^L + \overline{\lambda}\left[f_{\mathrm{opt1}}^U - f_{\mathrm{opt1}}^L\right], \tag{7b}$$

$$[8, 10]\,\overline{x}_1 - [12, 14]\,\overline{x}_2 \le \eta_2 - \overline{\lambda}\left[\eta_2 - \eta_1\right], \tag{7c}$$

$$[\gamma_1, \gamma_2]\,\overline{x}_1 + [3, 4]\,\overline{x}_2 \le 6.5 - \overline{\lambda}\,[6.5 - 6.0], \tag{7d}$$

$$0 \le \overline{\lambda} \le 1, \tag{7e}$$

where f_{opt1}^U and f_{opt1}^L are upper and lower bounds of objective value of models (6a), (6b), (6c), and (6d), $f_{\mathrm{opt1}}^U = 45.0840$, and $f_{\mathrm{opt1}}^L = 29.5070$. The above models (7a), (7b), (7c), (7d), and (7e) can be solved through a two-step method where a submodel corresponding to λ^U is first formulated and solved. This is based on the fact that the λ^Ucorresponding to f^U and the system objective are to be maximized. In the second step, the other sub-model corresponding to λ^L can then be formulated supported by the solution of the first sub-model. The first sub-model can be formulated as follows:

$$\max \quad \lambda^U \tag{8a}$$

$$\text{Subject to} \quad -\xi_2 x_1^U + 5.5x_2^L + 15.7928\lambda^U \le -29.2912, \tag{8b}$$

$$8x_1^U - 14x_2^L + (\eta_2 - \eta_1)\lambda^U \le \eta_2, \tag{8c}$$

$$\gamma_1 x_1^U + 4x_2^L + 0.5\lambda^U \le 6.5, \tag{8d}$$

$$0 \le \lambda^U \le 1. \tag{8e}$$

With the hybrid intelligent algorithm, the solutions of models (8a), (8b), (8c), (8d), and (8e) are $x_1^U = 1.5333$ and $x_2^L = 0.6014$, and the corresponding upper bound of membership grade value is $\lambda^U = 0.7539$ (as Figure 3).

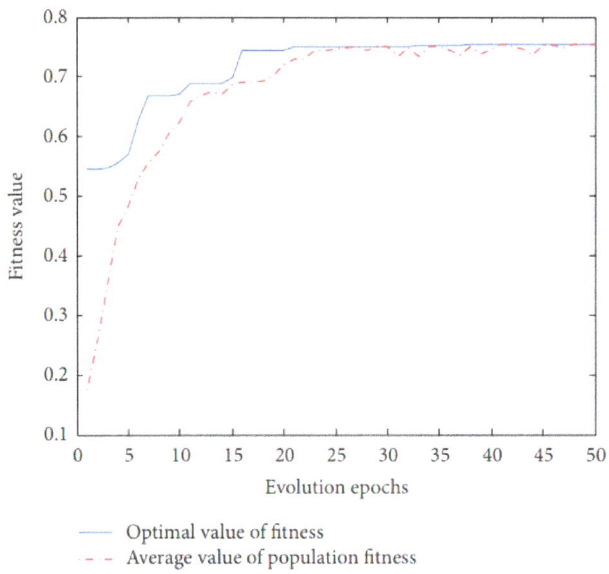

Figure 3: Evolution process of upper bound of membership function by genetic algorithm.

Then the sub-model corresponding to λ^L can be formulated as follows:

$$\max \quad \lambda^L \tag{9a}$$

$$\text{Subject to} \quad -\xi_1 x_1^L + 6.0x_2^U + 15.7928\lambda^L \le -29.2912, \tag{9b}$$

$$10x_1^L - 12x_2^U + (\eta_2 - \eta_1)\lambda^L \le \eta_2, \tag{9c}$$

$$\gamma_2 x_1^L + 4x_2^U + 0.5\lambda^L \le 6.5, \tag{9d}$$

$$x_1^L \le x_{1\,opt}^U, \tag{9e}$$

$$x_2^U \ge x_{2\,opt}^L, \tag{9f}$$

$$0 \le \lambda^L \le \lambda_{opt}^U. \tag{9g}$$

The solutions of models (9a), (9b), (9c), (9d), (9e), (9f), and (9g) are $x_1^L = 1.4078$ and $x_2^U = 0.6080$, and corresponding lower bound of membership grade value is $\lambda L = 0.2164$ (as Figure 4).

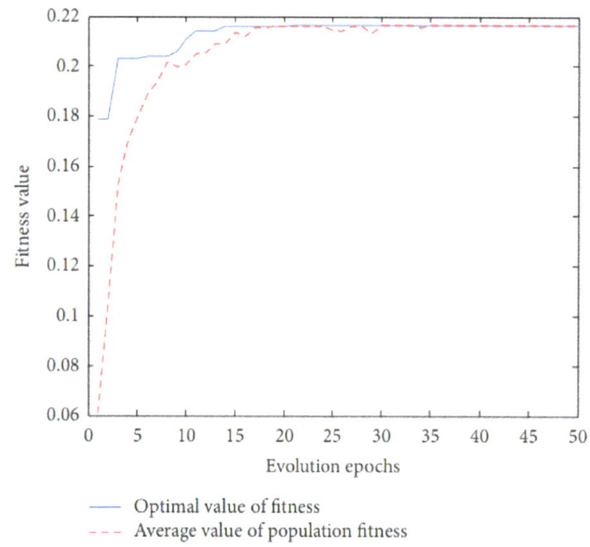

Figure 4: Evolution process of lower bound of membership function by genetic algorithm.

Therefore, the solutions of problem models (7a), (7b), (7c), (7d), and (7e) are $\bar{x}_1 = [1.4078, 1.5333]$ and $\bar{x}_2 = [0.6014, 0.6080]$, and interval objective value is $\bar{f} = [30.9548, 42.6913]$. In the work by Li et al. [19], dual uncertain parameters were interval values, and the DIV method was employed. The results of illustrative examples (6a), (6b), (6c), and (6d) by ILP, DIV, ILPSV, and IFLPSV method are shown in Table 1.

The ILP method can tackle uncertainties expressed as interval values with known lower and upper bounds, but the distribution functions of uncertain parameters are unknown. The result of objective value from ILP is [[29.438, 32.150], [42.172, 45.784]]. The solutions might compose fuzzy information due to the parameter's large boundary range. The DIV can tackle the uncertainties presented as interval parameter with fuzzy vertices by incorporating the vertex method within an interval parameter programming. The result of objective value from DIV is [[30.099, 31.455], [43.053, 44.895]]. Some feasible solutions might be missed by DIV method due to using the discrete vertex way, and the DIV method is unable to deal with the vertex presented as certain stochastic distribution function problem. The upper bound (42.6913) of objective value from IFLPSV is smaller than the interval [43.053, 44.895] of upper bound of DIV method.The reason is that normal distribution of upper and lower bounds of the interval parameter has more information than that of DIV model. The ILPSV method can tackle the interval-parameter linear programming with stochastic vertices. The result of objective value from IFLPSV is [30.9548, 42.6913]. The upper and lower bounds of the objective value from IFLPSV have less uncertainty than ILPSV since parameters with membership function have less uncertainty than those with interval numbers. Furthermore, the upper and lower bounds of the objective values from IFLPSV model all lie in the interval of the upper and lower bounds from ILP model. The reason is that coefficients of IFLPSV model with fuzzy membership function and stochastic vertices have less uncertainty degree than interval parameters of ILP with interval bounds. Therefore, the uncertainty degree of the solution from the IFLPSV is also less. The developed IFLPSV model is an integrated optimization model under dual uncertainty, which considers fuzzy, interval, and stochastic attributes of parameters and coefficients of optimization model.

APPLICATION

Overview of the Study Area

Beijing city, the capital of China, is located in the northern part of the North China Plain as shown in Figure 5. The annual precipitation is about 590 mm in Beijing, with 85% of the annual precipitation falling during the period from July to September. The main surface water source is from Guanting reservoir and Miyun reservoir in Beijing city. The amount of exploitable water resources ranges from 3.0 to 4.12 billion m^3 per year. However, under changing climate and overexploitation of water resources, the total of exploitable water resources is 2.308 billion m^3 in 2010 [26], including surface water resources 0.722 billion m^3 and groundwater 1.586 billion m^3. The water consumption users include industry, agriculture, domestic, and environment in Beijing city. The amount of water consumption of Beijing city was 3.52 billion m^3 in 2010. Water shortage problem is severe, and there has appeared four times water crisis in urbanization process of Beijing city. The per capita water resource in Beijing city is less than 300 cubic meters, which is one-thirtieth of the world average. Beijing has become one of the most water-scarce megacities in the world. Water resource has become a key factor which limits urbanization process and socioeconomic development in Beijing city.

Figure 5: Location illustration of main water resources in the study area.

In order to meet water requirements of Northwest and North China, the South-to-North Water Diversion Projects have been developed early. The Middle Route of South-to-North Water Diversion Project will transfer water from the Danjiangkou Reservoir on the Hanjiang river, a tributary of the Yangtze river, to Beijing city through opening channel. The Middle Route Project will be completed in 2014. The total amount of transferred water to Beijing city ranges from 1.0 to 1.4 billion m^3 per year, which will be another important part of water resources for Beijing city in the future.

IFLPSV Optimization Model for Water Resources Allocation in Beijing City

Reasonable allocation of urban water resources is usually a multiobjective problem. In this study, the objective functions include net economic benefit maximization and greenbelt irrigation area maximization as shown in formulas (10a) and (10b). The objective functions are subject to ten kinds of constraints, including water delivery capacity limit from river or lake as formula (10c), Guanting reservoir water supply capacity limit as formula (10d), Miyun reservoir water supply capacity limit as formula (10e), groundwater supply capacity limit as formula (10f), reused water supply capacity limit as formula (10g), South-to-North Water Transfer supply capacity limit as formula (10h), the lowest requirement limit of domestic water consumption as formula (10i), the lowest requirement limit of industry water consumption as formula (10j), the lowest requirement limit of agriculture water consumption as formula (10k), the lowest requirement limit of greenbelt water irrigation as formula (10l), nonnegativity constraint is as formula (10m):

$$\overline{f}_1 = \sum_{i=1}^{m}\sum_{j=1}^{n} \overline{b}_{ij}\overline{x}_{ij} \text{ (net benefit of economy)}, \tag{10a}$$

$$\overline{f}_2 = \sum_{i=1}^{m} \frac{1}{\overline{p}}\overline{x}_i \text{ (greenbelt irrigation area)}, \tag{10b}$$

Subject to

$$\sum_{j=1}^{n} \overline{x}_j \le \overline{Q}_{\text{deliv}} \text{ (water deliver capacity constraint from river or lake)}, \tag{10c}$$

$$\sum_{j=1}^{n} \overline{x}_j \le \overline{Q}_{\text{reser1}} \text{ (Guanting reservoir water supply capacity constraint)}, \tag{10d}$$

$$\sum_{j=1}^{n} \overline{x}_j \le \overline{Q}_{\text{reser2}} \text{ (Miyun reservoir water supply capacity constraint)}, \tag{10e}$$

$$\sum_{j=1}^{n} \overline{x}_j \le \overline{Q}_{\text{ground}} \text{ (groundwater supply capacity constraint)}, \tag{10f}$$

$$\sum_{j=1}^{n} \overline{x}_j \le \overline{Q}_{\text{reuse}} \text{ (reused water supply capacity constraint)}, \tag{10g}$$

$$\sum_{j=1}^{n} \overline{x}_{ij} \leq \overline{Q}_{\text{trans}} \text{ (South-to-North Water Transfer supply capacity constraint)}, \quad (10h)$$

$$\sum_{i=1}^{m} \overline{x}_{i} \geq \overline{Q}_{\text{domes}} \text{ (the lowest requirement of domestic water consumption)}, \quad (10i)$$

$$\sum_{i=1}^{m} \overline{x}_{i} \geq \overline{Q}_{\text{indus}} \text{ (the lowest requirement of industry water consumption)}, \quad (10j)$$

$$\sum_{i=1}^{m} \overline{x}_{i} \geq \overline{Q}_{\text{agri}} \text{ (the lowest requirement of agriculture water consumption)}, \quad (10k)$$

$$\sum_{i=1}^{m} \overline{x}_{ij} \geq \overline{Q}_{\text{env}} \text{ (the lowest requirement of greenbelt water irrigation)}, \quad (10l)$$

$$\overline{x}_{ij} \geq 0 \text{ (non-negativity constraint)}, \quad (10m)$$

where $\overline{f}_1, \overline{f}_2$ are the interval objective function of the net economic benefit maximization and greenbelt irrigation area maximization, respectively, $\overline{x}_{ij}$ (i = 1, 2, 3, . . . , m; j = 1, 2, 3, . . . , n) denotes the interval amount of water source i supply to j water user, m=6 is the total number of water source (lift or deliver water from river and lake, Guanting reservoir, Miyun reservoir, groundwater, reused water, and South-to-North Water Transfer); n=4 is the total number of water users (domestic, industry, agriculture and environment), $\overline{b}_{ij}$ is the benefit coefficient of the water source i supply to j user, which belongs an interval number, p is the interval coefficient of greenbelt irrigation water consumption per acre, $\overline{Q}_{\text{deliv}}$ is the maximization water deliver capacity from rivers and lakes, which is an interval number, $\overline{Q}_{\text{reser1}}$ is the water supply capacity of the Guanting reservoir, which is an interval number, $\overline{Q}_{\text{reser2}}$ is the water supply capacity of the Miyun reservoir, which is an interval number, $\overline{Q}_{\text{ground}}$ is the groundwater supply capacity, which is an interval number, $\overline{Q}_{\text{reuse}}$ is the reused water supply capacity, which is an interval range, $\overline{Q}_{\text{trans}}$ is the available water amount from the South-to-North Water Diversion Project, which is an interval number, $\overline{Q}_{\text{domes}}$ is the lowest amount of domestic water requirement, which belongs an interval range, the reused water for domestic is small, it is ignored, $\overline{Q}_{\text{indus}}$ is the lowest amount of industry water requirement, which is an interval number, $\overline{Q}_{\text{agri}}$ is the lowest amount of agriculture water requirement, depending on weather and rainfall. The value of agriculture water requirement is an interval number with normal distribution boundaries, (12.05×10^8, 10000) m^3 and N(12.01×10^8, 10000) m^3 respectively; $\overline{Q}_{\text{env}}$ is the lowest amount of greenbelt irrigation water requirement, which is an interval number. In this study, the greenbelt irrigation water requirement includes river water supplement and urban roadway watering.

The developed model is a multiobjective optimization model, including two conflict objectives. In this study, the multiobjective weighted method is adopted to solve the multiobjective model. The two objective functions are transferred to a single objective function as follows:

$$F(X) = \sum_{i=1}^{m} w_i f_i(x), \quad (11)$$

where m=2 and w_i is the weight coefficient of the i objective function $f_i(x)$, $(\sum_i^{mi} w_i = 1)$. In this study, $w_1 = w_2 = 0.5$, and it means that economic benefit and environment protection are considered to be of the same importance. Table 2 shows the coefficients and parameters in models (10a), (10b), (10c), (10d), (10e), (10f), (10g), (10h), (10i), (10j), (10k), (10l), and (10m).

Table 2: Coefficients and parameters of model.

Water source	Benefit coefficients (RMB/m^3)			Greenbelt irrigation ration (m^3/acre)
	Domestic	Industry	Agriculture	environment
Deliver water from river or lake	[4.0, 4.5]	[455, 465]	[0.07, 0.08]	[1.5, 2.0]
Guanting reservoir	[4.0, 4.5]	[460, 470]	[0.06, 0.08]	[1.5, 2.0]
Miyun reservoir	[4.0, 4.5]	[460, 470]	[0.06, 0.08]	[1.5, 2.0]
Groundwater	[4.0, 4.5]	[460, 470]	[0.05, 0.07]	[1.5, 2.0]
Reused water	[0.0, 0.0]	[460, 470]	[0.00, 0.00]	[1.5, 2.0]
South-to-North Water Transfer	[4.0, 4.5]	[460, 470]	[0.06, 0.08]	[1.5, 2.0]

RMB: monetary symbol in China.

The capacity of water source supply under different flow frequencies is shown in Table 3.

Table 3: Capacity of water source supply under three scenarios.

Flow frequency	Deliver water from rivers or lakes	Guanting reservoir	Miyun reservoir	Ground water	Reused water	South-to-North Water Transfer
Scenario 1 (50%)	[1.38, 1.40]	[7.51, 7.55]	[9.15, 9.20]	[26.3, 26.5]	[8.0, 8.2]	[9.5, 10.0]
Scenario 2 (75%)	[1.05, 1.08]	[4.45, 4.49]	[6.20, 6.24]	[26.3, 26.5]	[8.0, 8.2]	[11.5, 12.0]
Scenario 3 (90%)	[0.70, 0.72]	[1.90, 1.94]	[3.60, 3.64]	[26.3, 26.5]	[8.0, 8.2]	[13.5, 14.0]

Data source from [25].
(units 10^8 m^3).

Scenario 1 denotes the normal flow year, in which frequency of flow is 50%. The total amount of water resource is [6.184, 6.285] billion cubic meters in scenario 1. The amount of available water from local rivers and lakes is [1.38, 1.40] × 10^8 m^3 . The available water from the Guanting and Miyun reservoirs is [7.51, 7.55] × 10^8 m^3 and [9.15, 9.20] × 10^8 m^3 , respectively. The amount of water from the South-to-North Middle Rout Project is [9.5, 10.0] × 10^8 m^3 in scenario 1. Scenario 2 denotes the dry year, in which frequency of flow is 75%. The total amount of water resource is [5.751, 5.851] billion cubic meters in scenario 2, which decline 0.434 billion cubic meters comparing with scenario 1. The amount of available water from local rivers and lakes is [1.05, 1.08] × 10^8 m^3 in scenario 2, which declines to [0.32, 0.33] × 10^8 m^3 compared with scenario 1. The water amounts from the Guanting and Miyun reservoirs are [4.45, 4.49] × 10^8 m^3 and [6.20, 6.24] × 10^8 m^3 , respectively, in scenario 2, which corresponding declines to [3.06, 3.06] × 10^8 m^3 and [3.04, 3.05] × 10^8 m^3 compared with scenario 1. The amount of water from the South-to-North Middle Route Project is [11.5, 12.0] × 10^8 m^3 in scenario 2, in which increase 0.2 billion cubic meters comparing with scenario 1. Scenario 3 denotes the extraordinary dry year, which frequency of flow is 90%. The total amount of water resource is [5.40, 5.50] billion cubic meters in scenario 3. The amount of available water from local rivers and lakes is [0.70, 0.72] × 10^8 m^3 in scenario 3, in which declines to 0.68×108 m3 compared with scenario 1.The amounts of water from the Guanting and Miyun reservoirs are only [1.90, 1.94] × 10^8 m^3 and [3.60, 3.64] × 10^8 m^3 , respectively, in scenario 3, in which declines to 5.61 × 108 m3 and [5.55, 5.56] × 10^8 m^3 compared with scenario 1. The amount of water from South-to-North Middle Route Project is [13.5, 14.0] × 10^8 m^3 in scenario 3, which increase 0.4 billion cubic meters compared with scenario 1. The amount of exploitable groundwater and reused water keep the same for all three scenarios.

Results and Discussion

Table 4 shows the results of water allocation to water users under three scenarios. The total amount of allocated water is [6184, 6285] million m^3 in normal flow year (50%). The total amount of allocated water is [5751, 5851] million m^3 in dry year (75%), less [433, 434] million m^3 than that in scenario 1. The total amount of allocated water is only [5400, 5500] million m^3 in scenario 3, less [784, 785] million m3 than that in scenario 1.

Table 4: Results of water resources allocation among water users under three scenarios.

Scenario no.	Domestic	Industry	Agriculture	Environment	Total water consumption
Scenario 1 (50%)	[1634, 1634]	[2146, 2146]	[1205, 1220]	[1199, 1285]	[6184, 6285]
Scenario 2 (75%)	[1634, 1634]	[1712, 1712]	[1205, 1220]	[1200, 1285]	[5751, 5851]
Scenario 3 (90%)	[1632, 1634]	[1361, 1361]	[1205, 1220]	[1201, 1285]	[5400, 5500]

(units million m^3).

In Table 4 and Figure 6, the industry and domestic are main water utilization sectors in Beijing city. For example, in scenario 1, the amount of water allocation to industry is [2146, 2146] million cubic meters, which accounts for [34.1%, 34.7%] of total allocated water. The amount of allocated water to domestic is [1634, 1634] million cubic meters, which accounts for [26.0%, 26.4%] of total allocated water. In scenario 2, the amount of water allocation to industry is declined to [1712, 1712] million cubic meters, which accounts for [29.3%, 29.8%] of total allocated water. The amount of allocated water to domestic is [1634, 1634] million cubic meters, which accounts for [27.9%, 28.4%] of total allocated water. In scenario 3, the industrial water consumption is declined to [1361, 1361] million cubic meters, which only accounts for [24.7%, 25.2%] of total allocated water. The amount of allocated water to domestic is [1632, 1634] million cubic meters, which accounts for [29.7%, 30.2%] of total allocated water. Although the amount of domestic water consumption is almost not changed, its proportion is more than industrial in scenario 3. The proportion of domestic water consumption increases from [26.0%, 26.4%] in scenario 1 to [29.7%, 30.2%] in scenario 3. The results show that domestic has higher priority than industry on water consumption under water shortage situation. The amount of water consumption of agriculture and environment reach its least requirement in Beijing city. Therefore, the amounts of agricultural and environmental water consumption are almost the same under three scenarios. In the dry year or extraordinary dry year, the water crisis might be relieved by reducing industrial water consumption.

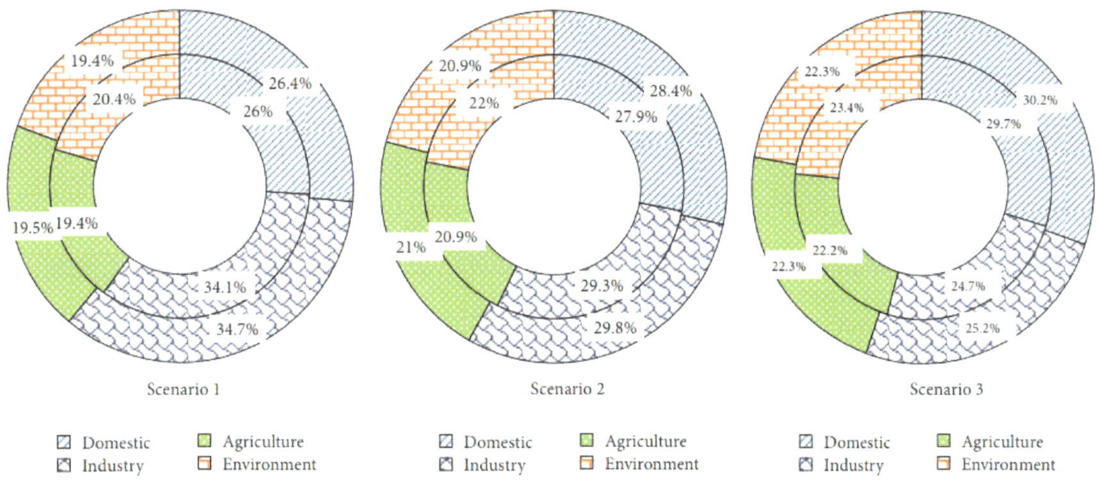

Figure 6: Rate of water resources allocation among water users under three scenarios.

Figure 7 shows the allocation results of different water source supplies to different water users in Beijing city under three scenarios. Domestic water consumption of Beijing city is mainly from groundwater, reservoirs, and South-to-North Middle Route Project in 2020. Although the total amount of the domestic water consumption is the same under three scenarios, the amounts of water in Guanting and Miyun reservoirs (reservoir1 and reservoir 2 in Figure 2) for the domestic sector evidently decreased with rainfall reduction. Moreover, groundwater is the primary and reliable water source for the domestic sector. The allocated groundwater for domestic sector is raised from [738.2, 738.2] million m^3 in scenario 1 to [764.3, 764.3] million m^3 in scenario 2 and to [808.2, 808.2] million m3 in scenario 3. Meanwhile, the water amount for the domestic sector from South-toNorth Middle Route Project is raised from [325.7, 325.7] million m^3 in scenario 1 to [401.8, 401.8] million m^3 in scenario 2 and to [495.7, 495.7] million m^3 in scenario 3. The South-to-North Middle Route Project gradually becomes the second water source for the domestic sector when water supply capacity of the Guanting and Miyun reservoirs decreases in dry and extraordinary dry year. Industry sector mainly uses groundwater, reused water, reservoirs water, and South-to-North Middle Route Project water. The allocated water amount from these water sources for industry are all declined with total available water reducing. For example, the groundwater for industry is significantly declined, from [742.2, 742.2] million m^3 in scenario 1 to [667.1, 667.1] billion m^3 in scenario 2 and to [566.8, 566.8] million m3 in scenario 3. Although the total amount of reused water supply is same, the amount of reused water for industry is declined from [496.1, 496.1] million m^3 in scenario 1 to [451.1, 451.1] million m^3 in scenario 2 and to [399.2, 399.2] million m^3 in scenario 3. The agricultural water consumption is mainly from rivers and lakes, reservoirs, groundwater, and South-to-North Middle Route Project. Because the available water from rivers, lakes, and reservoirs is declined with rainfall reduction in dry year and extraordinary dry year, the exploitation amount of groundwater and South-to-North Middle Route Project water is raised to compensate water shortage of the agriculture sector. The allocated groundwater for agriculture is raised from [586.7, 599.7] million m^3 in scenario 1 to [620.8, 633.8] million m^3 in scenario 2 and to [679.9, 686.7] million m3 in scenario 3. The amount of South-to-North Middle Route Project water for agriculture is also raised from [187.2, 187.2] million m^3 in scenario 1 to [271.3, 271.3] million m^3 in scenario 2 and to [374.2, 374.2] million m3 in scenario 3. The environment mainly uses reservoirs water, groundwater, reused water, and South-toNorth Middle Route Project water. The water amount of Guanting and Miyun reservoirs for environment is declined with rainfall reduction. Contrarily, the groundwater, reused water, and transfer water for environment are all raised. For example, the reused water amount for environment is remarkably raised from [303.9, 323.9] million m^3 in scenario 1 to [348.9, 368.9] million m^3 in scenario 2 and to [400.8, 420.8] million m^3 in scenario 3. The results show that transfer water from South-to-North Middle Route Project has significant impact on water supply structure in Beijing city. It will mitigate the water shortage issue of social and economy development in Beijing city in some extent, particularly in dry year and extraordinary dry year.

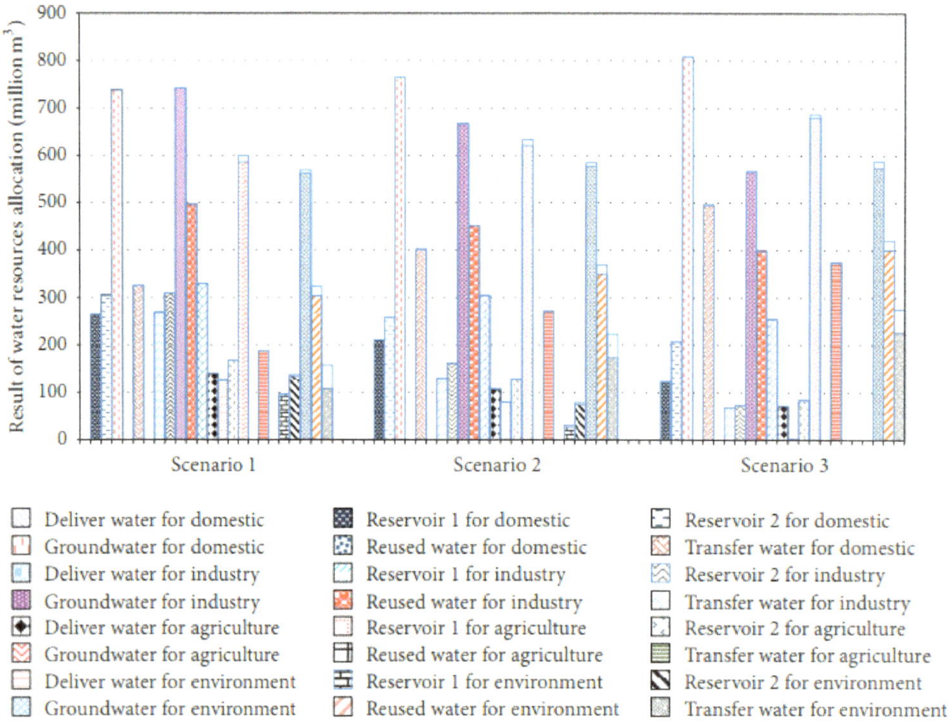

Figure 7: Result of water source supply among sectors under three scenarios.

CONCLUSION

An interval-parameter fuzzy linear programming with stochastic vertices (IFLPSV) has been developed for water resources allocation under dual uncertainty. The developed IFLPSV model improves upon the existing IFLP method by allowing the uncertain boundaries of interval parameter to be incorporated into the optimization processes. A hybrid intelligent algorithm based on genetic algorithm and artificial neural network was used to solve the developed model. The developed IFLPSV considers fuzzy, interval, and stochastic attributes of parameters and coefficients of objective functions and constraints. The application results indicated that it is effective for regional water resources allocation and planning under dual uncertainty. IFLPSV may provide more satisfactory solutions for an optimization problem under dual uncertainty. The developed IFLPSV model has been applied to multisource water allocation among multiple users in Beijing city in 2020, where water resources shortage is a challenging issue. The results indicate that transfering water from South-to-North Middle Route Project has an important impact on water supply structure in Beijing city, particularly in dry year and extraordinary dry year. The developed model and solution algorithm can be extended to other water resources system and environment management, where bounds of interval parameters or coefficients are dual uncertainty.

In this study, only two objectives including economic and greenbelt irrigation area maximization have been considered when carrying out water allocation in Beijing city. In the future, other objectives, such as social and environmental objective, might also be involved into water resources allocation in Beijing city. Moreover, interactive method for solving multi-objective optimization model is supposed to be incorporated into the methodology, so that the decision makers can make compromise among conflict objectives. There are many uncertain factors in practical management and decision making of water resources system. The interval programming is an effective method to tackle uncertainties expressed as interval values with known lower and upper bounds. However, due to high complexity of water resources system, sometimes it is hard to determine the exact values of both the lower and upper bounds. They might be with characteristics of interval, fuzzy, or random. Therefore, in this paper, the dual uncertainties are defined as interval or fuzzy numbers with lower and upper bounds with interval, fuzzy, or random characteristics. Although the bounds of interval parameters might be randomly distributed, the elements within the interval are regarded as certain numbers. In future research, physical meaning of the stochastic attribute of upper and lower bounds of interval parameters is suggested to in-depth investigation. Further discussion should be undertaken to answer whats the difference between interval number with randomly distributed bounds when the elements within the interval are simply considered as certain numbers and interval parameter as simply a random parameter with its own distribution.

ACKNOWLEDGMENTS

This research has been supported by the Open Research Fund of State Key Laboratory of Simulation and Regulation of Water Cycle in River Basin (China Institute of Water Resources and Hydropower Research). Grant no. IWHR-SKL-201214, and the National Natural Science Foundation of China (no. 91125018). The authors are grateful to the editor and the anonymous reviewers for their insightful and helpful comments and suggestions.

REFERENCES

1. Y. P. Li, G. H. Huang, G. Q. Wang, and Y. F. Huang, "FSWM: a hybrid fuzzy-stochastic water-management model for agricultural sustainability under uncertainty," Agricultural Water Management, vol. 96, no. 12, pp. 1807–1818, 2009.
2. Abrishamchi, M. A. Marino, and A. Afshar, "Reservoir planning for irrigation district," Journal of Water Resources Planning and Management, vol. 117, no. 1, pp. 74–85, 1991.
3. J. Kindler, "Rationalizing water requirements with aid of fuzzy allocation model," Journal of Water Resources Planning & Management, vol. 118, no. 3, pp. 308–323, 1992.
4. J. M. Wagner, U. Shamir, and D. H. Marks, "Containing groundwater contamination: planning models using stochastic programming with recourse," European Journal of Operational Research, vol. 77, no. 1, pp. 1–26, 1994.
5. O. Beaumont, "Solving interval linear systems with linear programming techniques," Linear Algebra and Its Applications, vol. 281, no. 1–3, pp. 293–309, 1998.
6. P. G. Jairaj and S. Vedula, "Multi-reservoir system optimization using fuzzy mathematical programming,"Water Resources Management, vol. 14, no. 6, pp. 457–472, 2000.
7. J. Q. Ma, S. Y. Chen, and L. Qiu, "A multi-objective fuzzy optimization model for cropping structure and water resources and its method," Agriculture Science & Technology, vol. 5, no. 1, pp. 5–10, 2004.
8. Maqsood, G. H. Huang, and J. S. Yeomans, "An interval-parameter fuzzy two-stage stochastic program for water resources management under uncertainty," European Journal of Operational Research, vol. 167, no. 1, pp. 208–225, 2005.
9. Y. P. Li, G. H. Huang, and S. L. Nie, "An interval-parameter multi-stage stochastic programming model for water resources management under uncertainty," Advances in Water Resources, vol. 29, no. 5, pp. 776–789, 2006.
10. Y. P. Li, G. H. Huang, Z. F. Yang, and S. L. Nie, "IFMP: interval-fuzzy multistage programming for water resources management under uncertainty," Resources, Conservation and Recycling, vol. 52, no. 5, pp. 800–812, 2008.
11. G. H. Huang, "A hybrid inexact-stochastic water management model," European Journal of Operational Research, vol. 107, no. 1, pp. 137–158, 1998.
12. G. H. Huang and D. P. Loucks, "An inexact two-stage stochastic programming model for water resources management under uncertainty," Civil Engineering and Environmental Systems, vol. 17, no. 2, pp. 95–118, 2000.
13. Q. G. Lin, G. H. Huang, B. Bass, and X. S. Qin, "IFTEM: an interval-fuzzy two-stage stochastic optimization model for regional energy systems planning under uncertainty," Energy Policy, vol. 37, no. 3, pp. 868–878, 2009.
14. H. W. Lu, G. H. Huang, G. M. Zeng, I. Maqsood, and L. He, "An inexact two-stage fuzzy-stochastic programming model for water resources management," Water Resources Management, vol. 22, no. 8, pp. 991–1016, 2008.
15. H. W. Lu, G. H. Huang, and L. He, "A semi-infinite analysis-based inexact two-stage stochastic fuzzy linear programming approach for water resources management," Engineering Optimization, vol. 41, no. 1, pp. 73–85, 2009.
16. X. H. Nie, G. H. Huang, Y. P. Li, and L. Liu, "IFRP: a hybrid interval-parameter fuzzy robust programming approach for waste management planning under uncertainty," Journal of Environmental Management, vol. 84, no. 1, pp. 1–11, 2007.
17. Z. F. Liu and G. H. Huang, "Dual-interval two-stage optimization for flood management and risk analyses," Water Resources Management, vol. 23, no. 11, pp. 2141–2162, 2009.
18. Z. F. Liu, G. H. Huang, X. H. Nie, and L. He, "Dual-interval linear programming model and its application to solid waste management planning," Environmental Engineering Science, vol. 26, no. 6, pp. 1033–1045, 2009.
19. Y. P. Li, G. H. Huang, P. Guo, Z. F. Yang, and S. L. Nie, "A dual-interval vertex analysis method and its application to environmental decision making under uncertainty," European Journal of Operational Research, vol. 200, no. 2, pp. 536–550, 2010.
20. Y. Han, Y. F. Huang, and G. Q. Wang, "Interval-parameter linear optimization model with stochastic vertices for land and water resources allocation under dual uncertainty," Environmental Engineering Science, vol. 28, no. 3, pp. 197–205, 2011.
21. B. D. Liu, Theory and Practice of Uncertain Programming, Physica, Heidelberg, Germany, 2002.
22. B. D. Liu, Theory and Practice of Uncertain Programming, Springer, Berlin, Germany, 3rd edition, 2009.
23. G. H. Huang, B. W. Baetz, and G. G. Patry, "A grey linear programming approach for municipal solid waste management planning under uncertainty," Civil Engineering Systems, vol. 9, no. 4, pp. 319–335, 1992.
24. G. H. Huang, B. W. Baetz, and G. G. Patry, "A grey fuzzy linear programming approach for municipal solid waste management planning under uncertainty," Civil Engineering Systems, vol. 10, no. 2, pp. 123–146, 1993.
25. B.-Y. Wei and J. Wang, "Analysis of water resources supply and demand in the Beijing city," South-to-North Water Transfers and Water Science & Technology, vol. 7, no. 2, pp. 41–42, 2009 (Chinese).
26. Beijing hydrology and water resources station, Beijing water resources Bulletin, Beijing water authority, 2010.

CHAPTER 4

ATTRACTOR OF BEAM EQUATION WITH STRUCTURAL DAMPING UNDER NONLINEAR BOUNDARY CONDITIONS

Danxia Wang, Jianwen Zhang, Yinzhu Wang, and Sufang Zhang

INTRODUCTION

In this paper, we will consider a kind of more general Kirchhoff-type beam equation. The physical origin of the problem lies in the theory of vibrations of an extensible beam of length L; moreover, during vibration, the elements of a beam not only perform a translatory motion but also rotate.

A mathematical model for this problem is an initial boundary value problem for the nonlinear Kirchhoff-type beam equation

$$\begin{aligned} & u_{tt} - u_{xxtt} + u_{xxxx} - \sigma\left(\int_0^L (u_x)^2\,dx\right)u_{xx} \\ & \quad - \phi\left(\int_0^L (u_x)^2\,dx\right)u_{xxt} \\ & \quad = q(x), \quad (x,t) \in (0,L) \times R^+. \end{aligned} \tag{1}$$

We assume the nonlinear boundary conditions

$$u(0,t) = u_x(L,t) = u_{xx}(0,t) = 0, \tag{2}$$

$$u_{xxx}(L,t) = f(u(L,t)) + g(u_t(L,t)) \tag{3}$$

and the initial conditions

$$u(x,0) = u^0(x), \quad u_t(x,0) = u^1(x), \quad x \in (0,L). \tag{4}$$

Here the unknown function (x,t) is the elevation of the surface of beam, $u^0(x)$ and $u^1(x)$ are the given initial value functions, and the subscript t and x denote derivative with respect to t and x, respectively. $-u_{xxtt}$ expresses the rotational inertia, and nonlinear terms $\sigma(\int_0^L (u_x)^2 dx)u_{xx}$ and $\phi(\int_0^L (u_x)^2 dx)u_{xxt}$ represent the extensibility effects and the structural damping, respectively. (x) is a static load. Moreover the assumptions on nonlinear functions $(\cdot)$, $(\cdot)$, $(\cdot)$, and $(\cdot)$ and the external force function $q(x)$ will be specified later.

In (1), when the structural damping term and the rotational inertia term are absent, (1) is a model for vibrations of tensible beam. This was proposed by Woinowsky-Krieger [1] in the form

$$u_{tt} + u_{xxxx} - \left(\alpha + \beta\int_0^L u_x^2\,dx\right)u_{xx} = 0. \tag{5}$$

One of the first mathematical analyses for equation

$$u_{tt} + u_{xxxx} - M\left(\int_0^L (u_x)^2\,dx\right)u_{xx} = 0 \tag{6}$$

was done by Ball [2] which was later extended to an abstract setting by defining a linear operator A by Medeiros [3]. In [4], Patcheu obtained the decay of the energy for above equation when a nonlinear damping (u_t) was effective in Ω. In addition, the attractor on extensible beams with null boundary conditions was considered by several authors. We quote, for instance, [5–7], and so on. But the longtime behavior of the beam equation with nonlinear boundary conditions was paid little at-

Citation: Danxia Wang, Jianwen Zhang, Yinzhu Wang, and Sufang Zhang, "Attractor of Beam Equation with Structural Damping under Nonlinear Boundary Conditions," Mathematical Problems in Engineering, vol. 2015, Article ID 857920, 10 pages, 2015. doi:10.1155/2015/857920.

tention. We also refer the reader to a few works. One of the first studies in this direction was done by Pazoto and Menzala [8], where stabilization of a thermoelastic extensible beam was considered. Motivated by the result, Ma proved the existence of global solutions and the existence of a global attractor in [9] and [10], respectively, for the Kirchhoff-type beam equation

$$u_{tt} + u_{xxxx} - M\left(\int_0^L (u_x)^2\,dx\right)u_{xx} = 0, \tag{7}$$

with the absence of the structural damping and the rotational inertia, subjected to the nonlinear boundary conditions

$$\begin{aligned} &u(0,t) = u_x(0,t) = u_{xx}(L,t) = 0,\\ &u_{xxx}(L,t) - M\left(\int_0^L (u_x)^2\,dx\right)u_x(L,t)\\ &\quad = f(u(L,t)) + g(u_t(L,t)). \end{aligned} \tag{8}$$

In the following, we mentioned some results on longtime behavior of beam equation with the rotational inertia term. Under null boundary conditions, Geredeli and Lasiecka [11] considered the existence of a compact attractor of beam

$$\begin{aligned} &u_{tt} - \alpha\Delta u_{tt} + \Delta^2 u + d(x)\,g(u_t)\\ &\quad - \alpha\,\mathrm{div}\,(d(x))\,g(\nabla u_t) - F(u) = p(x) \end{aligned} \tag{9}$$

with a rotational inertia term. Under nonlinear boundary conditions

$$u = 0, \quad \Delta u = -g\left(\frac{\partial u_t}{\partial v}\right), \tag{10}$$

Ji and Lasiecka [12] considered the semilinear Kirchhoff equation

$$u_{tt} - \gamma\Delta u_{tt} + \Delta^2 u + f(u) = 0 \tag{11}$$

with rotational inertia, and they showed that the above problem is uniformly stabilized with uniform energy decay rates

In addition, we also mentioned some results on longtime behavior of the equation with the structural damping term. Chueshov [13] studied the global attractor with a structural damping of the form $(\|\nabla u\|^2)(-\Delta)^\theta ut$ with $1/2 \le \theta \le 1$.

Chueshov [14] and Yang et al. [15] considered the global attractor for the Kirchhoff-type equation

$$u_{tt} - \sigma\left(\|\nabla u\|^2\right)\Delta u_t - \phi\left(\|\nabla u\|^2\right)\Delta u + f(u) = h(x) \tag{12}$$

with structural damping under null boundary conditions, respectively.

On (1) under the following other nonlinear boundary conditions

$$\begin{aligned} &u(L,t) = u_x(0,t) = u_{xx}(L,t) = 0,\\ &u_{xxx}(0,t) = f(u(0,t)) + g(u_t(0,t)) \end{aligned} \tag{13}$$

we also can get the same on the existence of global solutions and the existence of global attractor.

Our fundamental assumptions on (·), (·), (·), (·), and $q(x)$ are given as follows.

Assumption 1. We assume that (·), (·) ∈ $C^1(R)$ are all nondecreasing and satisfy

$$\sigma(z)\,z \ge \hat{\sigma}(z) \ge 0, \quad \forall z \ge 0, \tag{14}$$

where $\hat{\sigma}(z) = \int_0^z \sigma(s)ds \ge 0$. Moreover

$$\sigma(0) = 0, \qquad \phi(0) = 0, \tag{15}$$

$$\sigma(s) \ge \phi(s) \ge \alpha + \beta s^\gamma \quad (\alpha, \beta > 0,\ \gamma \ge 1),\ \forall s \in R^+. \tag{16}$$

Assumption 2. The function (·): $R \to R$ is of class $C^1(R)$ and satisfies (0) = 0, and there exist constants k_1, k_2 and $r \ge 0$ such that

$$(g(u) - g(v))(u - v) \ge k_1|u - v|^2, \quad \forall u, v \in R, \tag{17}$$

$$|g(u) - g(v)| \le k_2\left(1 + |u|^r + |v|^r\right)|u - v|, \quad \forall u, v \in R. \tag{18}$$

Assumption 3. The function (·): $R \to R$ is of class $C^1(R)$ and satisfies (0) = 0, and there exist constants k and $\varrho \ge 0$ such that

$$|f(u) - f(v)| \le k\left(1 + |u|^\varrho + |v|^\varrho\right)|u - v|, \quad \forall u, v \in R, \tag{19}$$

$$-L_0 \le \hat{f}(u) \le f(u)u + L_1, \tag{20}$$

Where $\hat{f}(z) = \int_0^z f(s)ds \ge 0.$

Assumption 4. Consider $(x) \in L^2(0, L)$.

Under the above assumptions, we prove the existence of global solutions and the existence of a global attractor of extensible beam equation system (1)–(4). And the paper is organized as follows. In Section 2, we introduce some Sobolev spaces. In Section 3, we discuss the existence of global strong and weak solutions. In Section 4, we establish the result of the existence of a global attractor in $V \times U$.

PRELIMINARIES

Our analysis is based on the following Sobolev spaces. Let

$$\begin{aligned} &U = \{u \in H^1(0,L);\ u(0) = 0\},\\ &V = \{u \in H^2(0,L);\ u(0) = u_x(L) = 0\},\\ &W = \{u \in H^4(0,L) \cap V;\ u_{xx} = 0\}. \end{aligned} \tag{21}$$

Motivated by the boundary condition (3) we assume, for regular solutions, that data (u^0, u^1) satisfies the following compatibility condition:

$$u^0_{xxx}(L) = f\left(u^0(L)\right) + g\left(u^1(L)\right). \tag{22}$$

Then for regular solutions we consider the phase space

$$H_1 = \left\{\left(u^0, u^1\right) \in W \times W;\ \text{satisfies condition (22)}\right\}. \tag{23}$$

In the case of weak solutions we consider the phase space

$$H_0 = V \times U = \overline{H_1}, \quad (24)$$

Which guarantees that, for regular data, the nonlinear condition (22) holds. In H_0 we adopt the norm defined by

$$\|u, v\|_{H_0}^2 = \|(u, v)\|_{V\times U}^2 = \|u_{xx}\|^2 + \|v_x\|^2. \quad (25)$$

THE EXISTENCE OF GLOBAL SOLUTIONS

Firstly, using the classical Galerkin method, we can establish the existence and uniqueness of regular solution to problem (1)–(4). We state it as follows.

Theorem 1. Assume Assumptions 1–4 and (22)-(23) hold, for any initial data $(u^0, u^1) \in H_1$; then problem (1)–(4) has a unique regular solution (t) with

$$u \in L^{\infty}_{\text{loc}}(R^+, W) \cap C^0([0,\infty); V) \cap C^1([0,\infty); U). \quad (26)$$

Moreover

$$\|u_t\|^2 + \|u_{xt}\|^2 + \|u_{xx}\|^2 + \hat{\sigma}(\|u_x\|^2) \le M_1, \quad (27)$$

Where $M_1 > 0$ depends on the initial data and q, but not on $t>0$.

Proof. Let us consider the variational problem associated with (1)–(4): find $(t) \in W$ such that

$$\int_0^L u_{tt}\omega\, dx + \int_0^L u_{xtt}\omega_x dx + \int_0^L u_{xx}\omega_{xx} dx + \sigma(\|u_x\|^2)\int_0^L u_x\omega_x dx + \phi(\|u_x\|^2)\int_0^L u_{xt}\omega_x dx + f(u(L,t))\,\omega(L) + g(u_t(L,t))\,\omega(L) = \int_0^L q(x)\,\omega\, dx \quad (28)$$

For all $\omega \in V$. This is done with the Galerkin approximation methods which is standard. Here we denote the approximate solution by (t). We can get the theorem by proving the existence of approximation solution, the estimate of approximation solution, convergence, uniqueness, and $u \in C^0([0, \infty); V) \cap C^1([0, \infty); U)$. In the following we give the estimates of approximation solution, the proof of uniqueness of solution, and the proof of $u \in C^0([0, \infty); V) \cap C^1([0, \infty); U)$.

Estimate 1.In approximate equation of (28), putting $\omega = u_t^m(t)$ and considering $\hat{\sigma}(z) = \int_0^z \sigma(s)ds$, $\hat{f}(z) = \int_0^z f(s)ds$, using Schwarz inequality, and then integrating from 0 to $t<t_m$, we see that

$$\|u_t^m\|^2 + \|u_{xt}^m\|^2 + \|u_{xx}^m\|^2 + \hat{\sigma}(\|u_x^m\|^2) + \hat{f}(u^m(l,t)) + 2\int_0^t \phi(\|u_x^m\|^2)\|u_{xt}^m\|^2 ds + 2\int_0^t g(u_t^m(l,t))\,u_t^m(l,t)\,ds \le \int_0^t \|q(x)\|^2 ds + \int_0^t \|u_t^m\|^2 ds + \|u_t^m(0)\|^2 + \|u_{xt}^m(0)\|^2 + \|u_{xx}^m(0)\|^2 + \hat{\sigma}(\|u_x^m(0)\|^2) + \hat{f}(u^m(l,0)). \quad (29)$$

Taking into account the assumptions $\int_0^t g(u_t^m(l,t))u_t^m(l,t)ds \ge 0$, $\phi(\cdot) \ge 0$, and $\hat{f}, \hat{\sigma} \ge 0$ of f, g, σ, and ϕ, we see that there exists $M_1 > 0$ such that

$$\|u_t^m\|^2 + \|u_{xt}^m\|^2 + \|u_{xx}^m\|^2 + \hat{\sigma}(\|u_x^m\|^2) \le M_1 \quad (30)$$

for all $t \in [0, T]$ and for all $m \in N$.

Estimate 2. In approximate equation of (28), integrating by parts with $\omega = u_{tt}^m(0)$ and $t = 0$ and considering the compatibility condition (22) and then using Schwarz inequality and the mean value inequality, we see that there exists $M_2 > 0$ such that

$$\|u_{tt}^m(0)\|^2 + \|u_{xtt}^m(0)\|^2 \le M_2 \quad (31)$$

for all $t \in [0, T]$ and for all $m \in N$.

Estimate 3. Let us fix $t, \xi > 0$ such that $\xi < T-t$. Taking the difference of approximate equation of (28) with $t=t+\xi$ and $t=t$ and replacing ω by $u_t^m(t+\xi) - u_t^m(t)$, we can find a constant $M_3 > 0$, depending only on T, such that

$$\|u_{tt}^m\|^2 + \|u_{xtt}^m\|^2 + \|u_{xxt}^m\|^2 \le M_3, \quad \forall m \in N, \ \forall t \in [0, T]. \quad (32)$$

Uniqueness. Let u, V be two solutions of (1)–(4) with the same initial data. Then writing $z=u-$ V and taking the difference (28) with $u=u$ and $u =$ V and replacing ω by zt and then using mean value theorem and the Young inequalities combined with Estimates 1 and 3, we deduce that, for some constant $C>0$,

$$\frac{d}{dt}\left(\|z_t^m\|^2 + \|z_{xt}^m\|^2 + \|z_{xx}^m\|^2\right) \le C\left(\|z_t^m\|^2 + \|z_{xt}^m\|^2 + \|z_{xx}^m\|^2\right), \quad \forall t \in (0, T). \quad (33)$$

Then from Gronwall's lemma we see that $u =$ V.

The Proof of $u \in C^0([0,\infty); V) \cap C^1([0,\infty); U)$. Since u_{xx}, $u_{xxt} \in L^2(0,\infty; L2(0, L))$, we get $u(x, t) \in C^0([0,\infty); V)$. Similarly, $(x, t) \in C^1([0,\infty); U)$. The proof of Theorem 1 is completed.

Theorem 2. Assume the assumptions of Theorem 1 and (24) hold; if the initial data $(u^0, u^1) \in H_0$, then there exists a unique weak solution $u(t)$ of problem (1)–(4) which depends continuously on initial data with respect to the norm of $V\times U$.

Proof. Let us consider $\{u^0, u^1\} \in \overline{W \times W} = V \times L^2$, and since $W\times W$ is dense in $V\times U$, there exists $\{u_\mu^0, u_\mu^1\} \subset W \times W$, such that

$$u_\mu^0 \longrightarrow u^0, \quad \text{in } V;$$
$$u_\mu^1 \longrightarrow u^1, \quad \text{in } U. \quad (34)$$

We observe that, for each $\mu \in N$, there exists u_μ, smooth solution of the initial boundary value problem (1)–(4) which satisfies

$$u_{tt\mu} - u_{xxtt\mu} + u_{xxxx\mu} - \sigma\left(\int_0^L (u_{x\mu})^2 dx\right) u_{xx\mu}$$
$$- \phi\left(\int_0^L (u_{x\mu})^2 dx\right) u_{xxt\mu} = q(x), \quad (x,t) \in (0,L) \times R^+,$$
$$u_\mu(x,0) = u_\mu^0(x), \quad u_{t\mu}(x,0) = u_\mu^1(x), \quad x \in (0,L),$$
$$u_\mu(0,t) = u_{x\mu}(L,t) = u_{xx\mu}(0,t) = 0, \quad t \in (0,\infty),$$
$$u_{xxx\mu}(L,t) = f(u_\mu(L,t)) + g(u_{t\mu}(L,t)), \quad t \in (0,\infty). \tag{35}$$

Considering the arguments usedin the estimate of the existence of solution, we obtain

$$\|u_{t\mu}\|^2 + \|u_{xt\mu}\|^2 + \|u_{xx\mu}\|^2 \le C_0, \tag{36}$$

Where C_0 is a positive constant independent of $\mu \in N$.

Defining $Z_{\mu,} = u_\mu - u_\sigma$: $\mu, \sigma \in N$ where u_μ and u_σ are regular solutions of (35), following the steps already used in the uniqueness of regular solution for (1)–(4) and considering the convergence given in (34) ($u_\mu^0 \to u^0$ in V and $u_\mu^1 \to u^1$ in U), we deduce that there exists u such that

$$u_\mu \longrightarrow u \quad \text{strongly in } C([0,T];V),$$
$$u_{t\mu} \longrightarrow u_t \quad \text{strongly in } C([0,T];U). \tag{37}$$

From the above convergence, we can pass to the limit using standard arguments in order to obtain

$$u_{tt} - u_{xxtt} + u_{xxxx} - \sigma\left(\int_0^L (u_x)^2 dx\right) u_{xx}$$
$$- \phi\left(\int_0^L (u_x)^2 dx\right) u_{xxt}$$
$$= q(x), \quad (x,t) \in (0,L) \times R^+,$$
$$u(x,0) = u^0(x), \quad u_t(x,0) = u^1(x), \quad x \in (0,L). \tag{38}$$

Theorem 2 is proved.

Remark 3. Theorem 2 implies that problem (1)–(4) defines a nonlinear C_0-semigroup (t) on H_0. Indeed, let us set $(t)(u^0, u^1) = (u(t), u_t(t))$, where u is the unique solution corresponding to initial data $(u^0, u^1) \in H_0$. Moreover, the operator (t) defined in H_0 maps H_0 into itself and it enjoys the usual semigroup properties

$$S(t+\tau) = S(t) + S(\tau), \quad \forall t, \tau \in R,$$
$$S(0) = I. \tag{39}$$

And it is obvious that the map $\{(t), t \ge 0\}$ is continuous in space H_0.

THE EXISTENCE OF GLOBAL ATTRACTOR

In this section, we give the existence of a global attractor

A global attractor for a $C0$-semigroup (t) defined on a complete metric space H is a bounded closed subset $\mathbb{A} \subset H$ which is positive fully invariant, that is, $(t)\ \mathbb{A} = \mathbb{A}$, for all $t \ge 0$, and uniformly attracting, that is,

$$\text{dist}(S(t)B, \mathbb{A}) = \sup_{x \in S(t)B} \inf_{y \in \mathbb{A}} d(x,y) \longrightarrow 0 \quad \text{as } t \longrightarrow \infty, \tag{40}$$

For any bounded set $B \subset H$.

A bounded set $\mathbb{B} \subset H$ is an absorbing set for (t) if, for any bounded set $B \subset H$, there exists $t_B = (B) \ge 0$ such that

$$S(t)B \subset \mathbb{B}, \quad \forall t \ge t_B, \tag{41}$$

Which defines $(H, (t))$ as a dissipative dynamical system

Theorem 4. Assume the hypotheses of Theorem 2 and $\varrho = r = 0$, $k > 0$ is sufficiently small, and then the corresponding semigroup (t) of problem (1)–(4) has an absorbing set in H_0.

Proof. Now, we show that semigroup (t) has an absorbing set in H_0. Firstly, we can calculate the total energy functional

$$E(t) = \frac{1}{2}\left(\|u_t\|^2 + \|u_{xt}\|^2 + \|u_{xx}\|^2 + \hat{\sigma}\left(\|u_x\|^2\right)\right)$$
$$+ \hat{f}(u(L,t)) - \int_0^L qu(t)\,dx. \tag{42}$$

Let us fix an arbitrary bounded set $B \subset H_0$ and consider the solutions of problem (1)–(4) given by $((t), (t)) = (t)(u^0, u^1)$ with $(u^0, u^1) \in B$. Our analysis is based on the modified energy functional

$$\tilde{E}(t) = E(t) + L^4 \|q\|^2 + L_0. \tag{43}$$

It is easy to see that $(d/dt)E(t) = (d/dt)\tilde{E}(t)$. Indeed, since (0) = 0, we have

$$|u(L,t)|^2 = |u(L,t) - u(0,t)|^2$$
$$= \left|\int_0^L u_x dx\right|^2$$
$$\le \int_0^L dx \int_0^L u_x^2 dx = L\|u_x\|^2. \tag{44}$$

In a similar way, since $(0, t) = (L, t) = u_x(0, t) = 0$, the following inequalities hold

$$\|u\| \le \sqrt{L}\|u_x\|, \qquad \|u_x\| \le L\|u_{xx}\|,$$
$$\|u\| \le L^2\|u_{xx}\|. \tag{45}$$

Now let us define

$$\varphi(t) = \int_0^L (u_t - u_{xxt})\,u\,dx. \tag{46}$$

By multiplying (1) by u_t and integrating over $(0, L)$, we have

$$\frac{d}{dt}E(t) + \phi\left(\|u_x\|^2\right)\|u_{xt}\|^2 + g(u_t(l,t))\,u_t(l,t) = 0. \tag{47}$$

Then, multiplying (1) by u and integrating over $(0, L)$, we obtain

$$\frac{d}{dt}\varphi(t) - \|u_t\|^2 - \|u_{xt}\|^2 + \|u_{xx}\|^2 + u(L)\,u_{xxx}(L)$$
$$+ \sigma\left(\|u_x\|^2\right)\|u_x\|^2 + \phi\left(\|u_x\|^2\right)\int_0^L u_x u_{xt} dx = \int_0^L qu\,dx. \tag{48}$$

Taking into account the boundary condition (3), we get

$$\begin{aligned}\frac{d}{dt}\varphi(t) &- \|u_t\|^2 - \|u_{xt}\|^2 + \|u_{xx}\|^2 \\ &+ u(L)\left[f(u(L,t)) + g(u_t(L,t))\right] \\ &+ \sigma\left(\|u_x\|^2\right)\|u_x\|^2 + \phi\left(\|u_x\|^2\right)\int_0^L u_x u_{xt}\,dx \\ &= \int_0^L qu\,dx. \end{aligned} \tag{49}$$

Taking the sum of (47) with ε times (49) and using the Schwarz inequality, we have

$$\begin{aligned}\frac{d}{dt}E(t) &+ \varepsilon\frac{d}{dt}\varphi(t) + \varepsilon E(t) \\ &+ \frac{\varepsilon}{2}\|u_t\|^2 + \frac{\varepsilon}{2}\|u_{xt}\|^2 + \frac{\varepsilon}{2}\|u_{xx}\|^2 \\ &+ \varepsilon\sigma\left(\|u_x\|^2\right)\|u_x\|^2 - \frac{\varepsilon}{2}\hat{\sigma}\left(\|u_x\|^2\right) \\ &+ \phi\left(\|u_x\|^2\right)\|u_{xt}\|^2 + g(u_t(L,t))\,u_t(L,t) \\ &= 2\varepsilon\left(\|u_t\|^2 + \|u_{xt}\|^2\right) - \varepsilon\phi\left(\|u_x\|^2\right)\int_0^L u_x u_{xt}\,dx \\ &+ \varepsilon\hat{f}(u(L,t)) - \varepsilon f(u(L,t))\,u(L,t) \\ &- \varepsilon g(u_t(L,t))\,u(L,t). \end{aligned} \tag{50}$$

So, using (15), we obtain

$$g(u_t(L,t))\,u_t(L,t) \geq k_1\left|u_t(L,t)\right|^2 \tag{51}$$

and using (16) and (44), we obtain

$$\begin{aligned}\varepsilon g(u_t(L,t))\,u(L,t) &\leq 3\varepsilon k_2\left|u_t(L,t)\right|\left|u(L,t)\right| \\ &\leq 3L^{3/2}\varepsilon k_2\varepsilon\|u_{xx}\|\left|u_t(L,t)\right| \\ &\leq 9\varepsilon k_2^2 L^3\left|u_t(L,t)\right|^2 + \frac{\varepsilon}{4}\|u_{xx}\|^2. \end{aligned} \tag{52}$$

Taking into account (18) and (44)-(45), we have

$$\begin{aligned}&\left|\varepsilon\hat{f}(u(L,t)) - \varepsilon f(u(L,t))\,u(L,t)\right| \\ &\leq \left|\frac{\varepsilon}{2}f(u(L,t))\,u(L,t)\right| + \varepsilon L_1 \\ &\leq \frac{\varepsilon}{2}3k\|u\|_\infty^2 + \varepsilon L_1 \\ &\leq \frac{\varepsilon}{2}3kL^3\|u_{xx}\|^2 + \varepsilon L_1. \end{aligned} \tag{53}$$

Using the mean value inequality, we get

$$\begin{aligned}&\left|-\varepsilon\phi\left(\|u_x\|^2\right)\int_0^L u_x u_{xt}\,dx\right| \\ &\leq \frac{\varepsilon}{2}\phi\left(\|u_x\|^2\right)\left(\|u_x\|^2 + \|u_{xt}\|^2\right). \end{aligned} \tag{54}$$

Then inserting (51)–(54) into (50), we obtain

$$\begin{aligned}\frac{d}{dt}E(t) &+ \varepsilon\frac{d}{dt}\varphi(t) + \varepsilon E(t) + \frac{\varepsilon}{2}\|u_t\|^2 \\ &+ \frac{\varepsilon}{2}\|u_{xt}\|^2 + \frac{\varepsilon}{2}\|u_{xx}\|^2 \\ &+ \varepsilon\sigma\left(\|u_x\|^2\right)\|u_x\|^2 - \frac{\varepsilon}{2}\hat{\sigma}\left(\|u_x\|^2\right) \\ &+ \phi\left(\|u_x\|^2\right)\|u_{xt}\|^2 + k_1\left|u_t(L)\right|^2 \\ &\leq 2\varepsilon\left(\|u_t\|^2 + \|u_{xt}\|^2\right) + \frac{\varepsilon}{2}\phi\left(\|u_x\|^2\right)\|u_{xt}\|^2 \\ &+ \frac{\varepsilon}{2}\phi\left(\|u_x\|^2\right)\|u_x\|^2 \\ &+ \varepsilon 3kL^3\|u_{xx}\|^2 + \varepsilon L_1 + 9\varepsilon k_2^2 L^3\left|u_t(L,t)\right|^2 + \frac{\varepsilon}{4}\|u_{xx}\|^2. \end{aligned} \tag{55}$$

Taking $0 < \varepsilon \leq \min\{k_1/k_2^2 L^3, 1\} = \varepsilon_0$ small enough, we get

$$k_1\left|u_t(L,t)\right|^2 > 9\varepsilon k_2^2 L^3\left|u_t(L,t)\right|^2. \tag{56}$$

Since with $k>0$ sufficiently small we have $1 - 6kL^3 \geq 0$, we get

$$\frac{\varepsilon}{2}\|u_{xx}\|^2 \geq \frac{1}{2}\varepsilon 3kL^3\|u_{xx}\|^2 + \frac{\varepsilon}{4}\|u_{xx}\|^2. \tag{57}$$

Using $(s) \geq \alpha +$, ($\gamma \geq 1$, for all $s \in R$), we get

$$\begin{aligned}\frac{1}{2}\phi\left(\|u_x\|^2\right)\|u_{xt}\|^2 &\geq \frac{\alpha}{2}\|u_{xt}\|^2 + \frac{\beta}{2}\|u_x\|^{2\gamma}\|u_{xt}\|^2 \\ &\geq \frac{\alpha}{2}\|u_{xt}\|^2 \\ &\geq \frac{\alpha}{4\sqrt{L}}\|u_t\|^2 + \frac{\alpha}{4}\|u_{xt}\|^2. \end{aligned} \tag{58}$$

So setting $0<\varepsilon\leq \min\{\varepsilon_0, \alpha/8\sqrt{L}\}$ small enough,

$$\frac{\alpha}{4\sqrt{L}}\|u_t\|^2 + \frac{\alpha}{4}\|u_{xt}\|^2 \geq 2\varepsilon\left(\|u_t\|^2 + \|u_{xt}\|^2\right). \tag{59}$$

Considering that $\sigma(z)z - \hat{\sigma}(z) \geq 0$ and $\sigma(s) \geq \phi(s)$, for all $s \in R$, we obtain

$$\varepsilon\sigma\left(\|u_x\|^2\right)\|u_x\|^2 - \frac{\varepsilon}{2}\hat{\sigma}\left(\|u_x\|^2\right) - \frac{\varepsilon}{2}\phi\left(\|u_x\|^2\right)\|u_x\|^2 \geq 0. \tag{60}$$

Therefore with (56)–(60), (55) is transformed into

$$\frac{d}{dt}E(t) + \varepsilon\frac{d}{dt}\varphi(t) + \varepsilon E(t) \leq \varepsilon L_1. \tag{61}$$

Adding $\varepsilon L_0 + \varepsilon L^4\|h\|^2$ on both sides of inequality (61) and taking into account that $(d/dt)E(t) = (d/dt)\tilde{E}(t)$, we obtain

$$\frac{d}{dt}\left(\tilde{E}(t) + \varepsilon\varphi(t)\right) + \varepsilon\tilde{E}(t) \leq \varepsilon\left(L_0 + L_1 + L^4\|q\|^2\right). \tag{62}$$

Now, let us set

$$\tilde{E}_\varepsilon(t) = \tilde{E}(t) + \varepsilon\varphi(t). \tag{63}$$

Then since from (45) $-\int_0^L qu\,dx \geq -L^2\|q\|\|u_{xx}\| \geq -L^4\|q\|^2 - (1/4)\|u_{xx}\|^2$, also $\hat{f}(u(L,t)) \geq -L_0$ and $\hat{\sigma}(\|u_x\|^2) \geq 0$, we can get

$$\tilde{E}(t) \geq \frac{1}{4}\left(\|u_t\|^2 + \|u_{xt}\|^2 + \|u_{xx}\|^2\right)$$
$$= \frac{1}{4}\|(u(t), u_t(t))\|^2_{V\times U}. \quad (64)$$

Thus, $\tilde{E}(t)$ dominates $\|(u(t), u_t(t))\|_{V\times U}$. Also, from (45), we have

$$\begin{aligned}|\tilde{E}_\varepsilon(t) - \tilde{E}(t)| &= |\varepsilon\varphi(t)| \\ &= \left|\varepsilon\int_0^L (u_t - u_{xxt})\,u\,dx\right| \\ &\leq \frac{\varepsilon(L^2+L^4)}{2}\left(\|u_t\|^2 + \|u_{xt}\|^2 + \|u_{xx}\|^2\right) \\ &\leq 2\varepsilon(L^2+L^4)\tilde{E}(t) \end{aligned} \quad (65)$$

which implies that, for ε sufficiently small enough,

$$\left(1 - 2\varepsilon(L^2+L^4)\right)\tilde{E}(t) \leq \tilde{E}_\varepsilon(t) \leq \left(1 + 2\varepsilon(L^2+L^4)\right)\tilde{E}(t). \quad (67)$$

Applying Gronwall's inequality, we obtain

$$\begin{aligned}\tilde{E}_\varepsilon(t) &\leq \tilde{E}_\varepsilon(0)\exp\left(-\frac{1}{1+2\varepsilon(L^2+L^4)}t\right) \\ &= \left|\varepsilon\int_0^L (u_t - u_{xxt})\,u\,dx\right| \\ &\quad + \varepsilon\left(L_0 + L_1 + L^4\|q\|^2\right)\left(1 + 2\varepsilon(L^2+L^4)\right) \\ &\quad \cdot\left(1 - \exp\left(-\frac{1}{1+2\varepsilon(L^2+L^4)}t\right)\right). \end{aligned} \quad (68)$$

Since the given invariant set B is bounded, $\tilde{E}_\varepsilon(0)$ is also bounded. Then there exists $tB > 0$ large enough such that

$$\begin{aligned}&\left[1 - 2\varepsilon(L^2+L^4)\right]\tilde{E}(t) \\ &\leq \tilde{E}_\varepsilon(t) \leq \varepsilon\left(L_0 + L_1 + L^4\|q\|^2\right)\left(1 + 2\varepsilon(L^2+L^4)\right), \\ &\qquad \forall t > t_B. \end{aligned} \quad (69)$$

Then from (64) we have

$$\|(u(t), u_t(t))\|^2_{V\times U} \leq \frac{4\varepsilon\left(L_0 + L_1 + L^4\|q\|^2\right)\left(1 + 2\varepsilon(L^2+L^4)\right)}{1 - 2\varepsilon(L^2+L^4)}, \quad \forall t > t_B. \quad (70)$$

This shows that

$$B = \left\{(u, u_t) \in H_0 : \|(u, u_t)\|^2_{H_0} \leq \frac{4\varepsilon\left(L_0 + L_1 + L^4\|q\|^2\right)\left(1 + 2\varepsilon(L^2+L^4)\right)}{1 - 2\varepsilon(L^2+L^4)}\right\} \quad (71)$$

is an absorbing set for $S(t)$ in $H0$. The proof of Theorem 4 is ended.

A semigroup $S(t)$ is asymptotically smooth in H if for any bounded positive invariant set $B\subset H$, there exists a compact set $K \subset \overline{B}$ such that

$$\operatorname{dist}(S(t)B, K) \longrightarrow 0 \quad \text{as } t \longrightarrow \infty. \quad (72)$$

Then the following lemma is well known.

Lemma 5 (see [16], Theorem 2.3). Let (t) be a dissipative C_0- semigroup defined on a metric space H; then (t) has a compact global attractor in H if and only if it is asymptotically smooth in H.

The asymptotic smoothness can be verified from a result by Khanmamedov [17] and Chueshov and Lasiecka [16]. Assume that H is a Banach space.

Lemma 6 (see [16], Proposition 2.10). Assume that for any bounded positive invariant set $B\subset H$ and for any $\varepsilon>0$, there exists $T = (\varepsilon, B)$ such that

$$\|S(T)x - S(T)y\| \leq \varepsilon + \phi_T(x, y), \quad \forall x, y \in B, \quad (73)$$

where $\phi_T : H\times H \to R$ satisfies

$$\liminf_{k\to\infty}\liminf_{l\to\infty}\phi_T(z_k, z_l) = 0 \quad (74)$$

for any sequence (z_n) of B. Then (t) is asymptotically smooth in H.

Theorem 7. Assume the hypotheses of Theorem 2 and $\varrho = r =$; then the corresponding semigroup (t) of problem (1)–(4) is asymptotic compactness.

Proof. We are going to apply Lemma 6 to prove the asymptotic compactness. Given initial data (u^0, u^1) and (V^0, V^1) in a bounded invariant set $B\subset H_0$, let u, V be the corresponding weak solutions of problem (1)–(4). Then the difference $w = u - V$ is a weak solution of

$$\begin{aligned}&w_{tt} - w_{xxtt} + w_{xxxx} - \sigma\left(\|u_x\|^2\right)w_{xx} - \Delta\sigma v_{xx} \\ &\quad - \phi\left(\|u_x\|^2\right)w_{xxt} - \Delta\phi v_{xxt} = 0, \\ &w(0,t) = w_x(L\cdot t) = w_{xx}(0,t) = 0, \\ &\qquad w_{xxx}(L,t) = \Delta f + \Delta g, \end{aligned} \quad (75)$$

where

$$\begin{aligned}\Delta\sigma &= \sigma\left(\|u_x\|^2\right) - \sigma\left(\|v_x\|^2\right), \\ \Delta\phi &= \phi\left(\|u_x\|^2\right) - \phi\left(\|v_x\|^2\right), \\ \Delta f &= f(u(L,t)) - f(v(L,t)), \\ \Delta g &= g(u_t(L,t)) - g(v_t(L,t)). \end{aligned} \quad (76)$$

Let us assume

$$F(t) = \frac{1}{2}\|w_t\|^2 + \frac{1}{2}\|w_{xt}\|^2 + \frac{1}{2}\|w_{xx}\|^2 + \frac{1}{2}\sigma\left(\|u_x\|^2\right)\|w_x\|^2 \quad (77)$$

and define

$$\psi(t) = \int_0^L (w_t - w_{xxt})\,w\,dx. \quad (78)$$

As before, by density, we can assume formally that w is sufficiently regular. Then, multiplying the first equation in (75) by w_t and integrating over (0, L), we get

$$\frac{1}{2}\frac{d}{dt}\left(\|w_t\|^2+\|w_{xt}\|^2+\|w_{xx}\|^2+\sigma\left(\|u_x\|^2\right)\|w_x\|^2\right)$$
$$+\phi\left(\|u_x\|^2\right)\|w_{xt}\|^2+w_{xxx}(L,t)\,w_t(L,t)$$
$$=\sigma'\left(\|u_x\|^2\right)\int_0^L u_xu_{xt}dx\,\|w_x\|^2+\Delta\sigma\int_0^L v_{xx}w_tdx$$
$$+\Delta\phi\int_0^L v_{xxt}w_tdx. \tag{79}$$

Taking into account the third equation in (75) we see that

$$\frac{d}{dt}F(t)+\phi\left(\|u_x\|^2\right)\|w_{xt}\|^2+\Delta gw_t(L,t)$$
$$=\sigma'\left(\|u_x\|^2\right)\int_0^L u_xu_{xt}dx\,\|w_x\|^2+\Delta\sigma\int_0^L v_{xx}w_tdx$$
$$+\Delta\phi\int_0^L v_{xxt}w_tdx-\Delta fw_t(L,t). \tag{80}$$

By multiplying first equation in (75) by w and integrating over (0, L), we obtain that

$$\frac{d}{dt}\phi(t)-\|w_t\|^2-\|w_{xt}\|^2+\|w_{xx}\|^2$$
$$+\sigma\left(\|u_x\|^2\right)\|w_x\|^2$$
$$+\phi\left(\|u_x\|^2\right)\int_0^L w_xw_{xt}dx+w_{xxx}w(L,t)$$
$$-\Delta\sigma\int_0^L v_{xx}w\,dx-\Delta\phi\int_0^L v_{xxt}w\,dx=0. \tag{81}$$

Also taking into account the third equation in (75) we see that

$$\frac{d}{dt}\phi(t)+2F(t)$$
$$=-\phi\left(\|u_x\|^2\right)\int_0^L w_xw_{xt}dx-(\Delta f+\Delta g)\,w(L,t)$$
$$+\Delta\sigma\int_0^L v_{xx}w\,dx$$
$$-\Delta\phi\int_0^L v_{xt}w_xdx+2\|w_t\|^2+2\|w_{xt}\|^2. \tag{82}$$

Then summing (80) with η times (82) we obtain that

$$\frac{d}{dt}F(t)+\eta\frac{d}{dt}\phi(t)+\eta F(t)+\eta F(t)$$
$$+\phi\left(\|u_x\|^2\right)\|w_{xt}\|^2+\Delta gw_t(L,t)$$
$$=\sigma'\left(\|u_x\|^2\right)\int_0^L u_xu_{xt}dx\,\|w_x\|^2+\Delta\sigma\int_0^L v_{xx}w_tdx$$
$$+\Delta\phi\int_0^L v_{xxt}w_tdx-\Delta fw_t(L,t)$$
$$-\eta\phi\left(\|u_x\|^2\right)\int_0^L w_xw_{xt}dx-\eta(\Delta f+\Delta g)\,w(L,t)$$
$$+\eta\Delta\sigma\int_0^L v_{xx}w\,dx$$
$$-\eta\Delta\phi\int_0^L v_{xt}w_xdx+2\eta\|w_t\|^2+2\eta\|w_{xt}\|^2. \tag{83}$$

In view of assumption (17) of the function g, we obtain that

$$\Delta gw_t(L,t)\geq k_1\left|w_t(L,t)\right|^2. \tag{84}$$

Also since $(s)\geq\alpha+\beta s^\gamma$ $(\alpha,\beta>0,\gamma\geq 1)$, we see that

$$\phi\left(\|u_x\|^2\right)\|w_{xt}\|^2\geq\left(\alpha+\beta\|u_x\|^{2\gamma}\right)\|w_{xt}\|^2. \tag{85}$$

In the following, let us estimate the right hand side of (83). We recall that u, V, and w satisfy the estimate $\|\ \|u_t\|^2+\|u_{xt}\|^2+\|u_{xx}\|^2+\hat{\sigma}(\|u_x\|^2)\leq M_1$, then denoting by C_0 a generic positive constant which depends only on B we can simplify several notations.

Firstly, since $\sigma(\cdot),\phi(\cdot)\in C^1(R),\phi(0)=0,$ and $\|u_x\|^2\leq C_0$, we get $\sigma'(\|u_x\|^2)\leq C_0,\phi(\|u_x\|^2)\leq C_0$; then we have

$$\sigma'\left(\|u_x\|^2\right)\int_0^L u_xu_{xt}dx\,\|w_x\|^2\leq C_0\|w_x\|^2;$$
$$-\eta\phi\left(\|u_x\|^2\right)\int_0^L w_xw_{xt}dx\leq C_0\|w_x\|\,\|w_{xt}\|. \tag{86}$$

From the mean value theorem and noting that $\sigma(a^2)-\sigma(b^2)\leq\sigma'(\sup\{a^2,b^2\})|a+b||a-b|$ and $\phi(a^2)-\phi(b^2)\leq\phi'(\sup\{a^2,b^2\})|a+b||a-b|$, we have

$$\Delta\sigma\int_0^L v_{xx}w_t\,dx\leq C_0\|w_x\|\,\|w_t\|,$$
$$\eta\Delta\sigma\int_0^L v_{xx}w\,dx\leq C_0\|w_x\|^2,$$
$$\Delta\phi\int_0^L v_{xxt}w_tdx\leq C_0\|w_x\|\,\|w_{xt}\|,$$
$$-\eta\Delta\phi\int_0^L v_{xt}w_xdx\leq C_0\|w_x\|^2. \tag{87}$$

From assumption (19) of the function f and $\varrho=0$ and inequalities (44), we get

$$|-\Delta f w_t(L)| \le 3k\|w(t)\|_\infty |w_t(l,t)|$$
$$\le C_0\|w_x\| |w_t(L)|$$
$$\le C_0\|w_x\|^2 + \frac{k}{2}|w_t(L)|^2,$$
$$|-\eta w(L)\Delta f| \le \eta 3k\|w\|_\infty |w(L)|$$
$$\le \eta 3k\|w\|_\infty \sqrt{L}\|w_x\|$$
$$\le C_0\|w_x\|^2. \quad (88)$$

From assumption (18) of the function g and r=0 and inequalities (44), we get

$$|-\eta w(L,t)\Delta g| \le \eta 3k_2|w(L,t)||w_t(L,t)|$$
$$\le \eta\sqrt{L}\|w_x\| 3k_2|w_t(L,t)|$$
$$\le \frac{9\eta^2 Lk_2^2}{2k_1}\|w_x\|^2 + \frac{k_1}{2}|w_t(L,t)|^2. \quad (89)$$

Substituting (84)–(89) into (83) and using Schwarz inequality, we obtain that

$$\frac{d}{dt}F(t) + \eta\frac{d}{dt}\phi(t) + 2\eta F(t) + (\alpha + \beta\|u_x\|^\gamma)\|w_{xt}\|^2$$
$$\le \left(5C_0 + \frac{9\eta^2 Lk_2^2}{2k_1} + \frac{2C_0^2}{\eta}\right)\|w_x\|^2 + (3\eta L + 3\eta)\|w_{xt}\|^2. \quad (90)$$

With $\eta \le \alpha/(3L+3)$ sufficiently small enough, we have

$$\frac{d}{dt}F(t) + \eta\frac{d}{dt}\phi(t) + 2\eta F(t)$$
$$\le \left(5C_0 + \frac{9\eta^2 Lk_2^2}{2k_1} + \frac{2C_0^2}{\eta}\right)\|w_x\|^2. \quad (91)$$

Defining $(t) = (t) + \eta\phi(t)$ and considering (44)-(45), we have

$$|F_\eta(t) - F(t)| = |\eta\phi(t)|$$
$$= \left|\eta\int_0^L (w_t - w_{xxt})\,w\,dx\right|$$
$$\le \eta\|w_t\|\|w\| + \eta\|w_{xt}\|\|w_x\|$$
$$\le \eta\|w_t\| L^2\|w_{xx}\| + \eta\|w_{xt}\|\sqrt{L}\|w_{xx}\|$$
$$\le \eta L^2\left(\|w_t\|^2 + \|w_{xt}\|^2 + \|w_{xx}\|^2\right)$$
$$\le \eta L^2 F(t), \quad (92)$$

Which implies for $\eta < \min\{1/L^2, \alpha/(3L+3)\}$ sufficiently small enough that

$$(1-\eta L^2)F(t) \le F_\eta(t) \le (1+\eta L^2)F(t). \quad (93)$$

Let $C_1 = 5C_0 + 9\eta^2 Lk_2^2/2k_1 + 2C_0^2/\eta$; then inserting (93) into (91), we obtain that

$$\frac{d}{dt}F_\eta(t) + \frac{2\eta}{1+\eta L^2}F_\eta(t) \le C_1\|w_x\|^2, \quad t \ge 0. \quad (94)$$

From Gronwall's lemma, we get

$$F_\eta(t) \le F_\eta(0)e^{-(2\eta/(1+\eta L^2))t}$$
$$+ C_1\int_0^t e^{-(\eta/(1+\eta L^2))(t-s)}\|w_x(s)\|^2 ds. \quad (95)$$

On the other hand, we have

$$F_\eta(t) \ge (1-\eta L^2)F(t)$$
$$= \frac{(1-\eta L^2)}{2}\left(\|w_t\|^2 + \|w_{xt}\|^2 + \|w_{xx}\|^2\right.$$
$$\left. + \sigma(\|u_x\|^2)\|w_x\|^2\right)$$
$$\ge \frac{1-\eta L^2}{2}\left(\|w_t\|^2 + \|w_{xt}\|^2 + \|w_{xx}\|^2\right). \quad (96)$$

Therefore, combining (95) and (96), we can fix a constant $C_B > 0$, depending on the size of B but not on t>0, such that $E_1(0) \le C_B$; so

$$\|(w(t), w_t(t))\|_{V\times U} \le \frac{\sqrt{2}}{\sqrt{1-\eta L^2}}C_B\exp\left(-\frac{\eta}{1+\eta L^2}t\right)$$
$$+ \sqrt{\frac{2C_1}{1-\eta L^2}}\left[\int_0^t \|w_x(s)\|^2 ds\right]^{1/2}. \quad (97)$$

Given ε>0, we choose T large such that

$$\sqrt{\frac{2}{1-\eta L^2}}C_B\exp\left(-\frac{\eta}{1+\eta L^2}t\right) \le \varepsilon \quad (98)$$

and define $\phi_T : H0 \times H0 \to R$ as

$$\phi_T((u^0,u^1),(v^0,v^1))$$
$$= \sqrt{\frac{2C_1}{1-\eta L^2}}\left[\int_0^T \|u_x(t) - v_x(t)\|^2 dt\right]^{1/2}. \quad (99)$$

Then from (97)–(99), we get

$$\|S(T)(u^0,u^1) - S(T)(v^0,v^1)\|_{H_0}$$
$$\le \varepsilon + \phi_T((u^0,u^1),(v^0,v^1)) \quad (100)$$

for all (u^0, u^1), (V^0, V^1)$\in B$.

Let (u_n^0, u_n^1) be a given sequence of initial data in B. Then the corresponding sequence ((t), u_{tn}) of solutions of problem (1)–(4) is uniformly bounded in H_0, because B is bounded and positively invariant. So $\{u_n\}$ is bounded in $C([0,\infty),V) \cap C^1([0,\infty),U)$. Since $V \hookrightarrow H_0^1(\Omega)$ compactly, there exists a subsequence u_{nk} which converges strongly in $C([0,T],H_0^1(\Omega))$. Therefore

$$\lim_{k\to\infty}\lim_{l\to\infty}\int_0^T \|u_{xnk}(s) - u_{xnl}(s)\|^2 ds = 0,$$
$$\lim_{k\to\infty}\lim_{l\to\infty}\phi_T((u_{nk}^0,u_{nk}^1),(u_{nl}^0,u_{nl}^1)) = 0. \quad (101)$$

So (t) is asymptotically smooth in $H0$. That is, Lemma 6 holds. Thus Theorem 7 is proved.

Remark 8. The novelty and difficulty of Theorems 4 lie in the appropriate definition on the functions $\varphi(t) = \int_0^L (u_t - u_{xxt})u\,dx$, $\psi(t) = \int_0^L (w_t - w_{xxt})w\,dx$ and the relationship between $\tilde{E}_\varepsilon(t)$ and the energy function $\tilde{E}_\varepsilon(t)$ and the relationship between (t) and the function (t).

Lemma 9 ([18], Theorem 1.1 in Chapter I). One assumes that H is a metric space and that the operators $\{S(t)\}_{t\geq 0}$ are given and satisfy the usual group properties and $S(t)$ is continuous operator from H into itself and the semigroup $\{S(t)\}_{t\geq 0}$ is asymptotically compact. One also assumes that there exist an open set U and a bounded set $\mathbb{B}$ of U such that U is absorbing in $\mathbb{B}$. Then $(\mathbb{B})$ is a compact attractor

By view of Lemma 9, with Theorems 4 and 7, the main result of a global attractor reads as follows.

Theorem 10. The corresponding semigroup (t) of problem (1)–(4) has a compact global attractor in the phase space H_0.

CONFLICT OF INTERESTS

The authors declare that there is no conflict of interests regarding the publication of this paper.

ACKNOWLEDGMENTS

The project is supported by the National Natural Science Foundation of China (Grant no. 11172194 and no. 11401420), the Natural Science Foundation of Shanxi Province, China (Grant no. 2010011008), and the Natural Science Foundation for Young Scientists of Shanxi Province, China (Grant no. 2011021002-2).

REFERENCES

1. S.Woinowsky-Krieger, "The effect of axial force on the vibration of hinged bars," Journal of Applied Mechanics, vol. 17, pp. 35–36, 1950.
2. J. M. Ball, "Initial-boundary value problems for an extensible beam," Journal of Mathematical Analysis and Applications, vol. 42, pp. 61–90, 1973.
3. L. A. Medeiros, "On a new class of nonlinear wave equations," Journal of Mathematical Analysis and Applications, vol. 69, no. 1, pp. 252–262, 1979.
4. S. K. Patcheu, "On a global solution and asymptotic behaviour for the generalized damped extensible beam equation," Journal of Differential Equations, vol. 135, no. 2, pp. 299–314, 1997.
5. T. F. Ma and V. Narciso, "Global attractor for a model of extensible beam with nonlinear damping and source terms," Nonlinear Analysis. Theory, Methods & Applications, vol. 73, no. 10, pp. 3402–3412, 2010.
6. C. Giorgi, M. G. Naso, V. Pata, and M. Potomkin, "Global attractors for the extensible thermoelastic beam system," Journal of Differential Equations, vol. 246, no. 9, pp. 3496–3517, 2009.
7. Z. J. Yang, "On an extensible beam equation with nonlinear damping and source terms," Journal of Differential Equations, vol. 254, no. 9, pp. 3903–3927, 2013.
8. A. F. Pazoto and G. P. Menzala, "Uniform stabilization of a nonlinear beam model with thermal effects and nonlinear boundary dissipation," Funkcialaj Ekvacioj, vol. 43, no. 2, pp. 339–360, 2000.
9. T. F. Ma, "Boundary stabilization for a non-linear beam on elastic bearings," Mathematical Methods in the Applied Sciences, vol. 24, no. 8, pp. 583–594, 2001.
10. T. F. Ma, V. Narciso, and M. L. Pelicer, "Long-time behavior of a model of extensible beams with nonlinear boundary dissipations," Journal of Mathematical Analysis and Applications, vol. 396, no. 2, pp. 694–703, 2012.
11. P. G. Geredeli and I. Lasiecka, "Asymptotic analysis and upper semicontinuity with respect to rotational inertia of attractors to von Karman plates with geometrically localized dissipation and critical nonlinearity," Nonlinear Analysis :Theory, Methods & Applications, vol. 91, pp. 72–92, 2013.
12. G. Ji and I. Lasiecka, "Nonlinear boundary feedback stabilization for a semilinear Kirchhoff plate with dissipation acting only via moments-limiting behavior," Journal of Mathematical Analysis and Applications, vol. 229, no. 2, pp. 452–479, 1999.
13. I. Chueshov, "Global attractors for a class of Kirchhoff wave models with a structural nonlinear damping," Journal of Abstract Differential Equations and Applications, vol. 1, no. 1, pp. 86–106, 2010.
14. I. Chueshov, "Long-time dynamics of Kirchhoff wave models with strong nonlinear damping," Journal of Differential Equations, vol. 252, no. 2, pp. 1229–1262, 2012.
15. Z. Yang, P. Ding, and Z. Liu, "Global attractor for the Kirchhoff type equations with strong nonlinear damping and supercritical nonlinearity," Applied Mathematics Letters, vol. 33, pp. 12–17, 2014.
16. I. Chueshov and I. Lasiecka, "Long-time behavior of second order evolution equations with nonlinear damping," Memoirs of the American Mathematical Society, vol. 195, no. 912, 2008.
17. A. K. Khanmamedov, "Global attractors for von Karman equations with nonlinear interior dissipation," Journal of Mathematical Analysis and Applications, vol. 318, no. 1, pp. 92–101, 2006.
18. R. Teman, Infinite Dimensional Dynamic System in Mechanics and Physics, Springer, New York, NY, USA, 1998.

CHAPTER 5

SEISMIC BEHAVIOR OF SUBSTANDARD RC COLUMNS RETROFITTED WITH EMBEDDED ARAMID FIBER REINFORCED POLYMER (AFRP) REINFORCEMENT

Engin C. Seyhan, Caglar Goksu, Ahmet Uzunhasanoglu and Alper Ilki

Keywords: Aramid, flexure, FRP reinforcement, reinforced concrete, retrofitting, seismic

INTRODUCTION

Many substandard reinforced concrete (RC) structures should be seismically retrofitted to reduce their vulnerabilities against seismic actions. Financial constraints, disturbance to the occupants and disruption of functions of the structures are the main obstacles for proper seismic retrofitting of these substandard existing structures. In recent years, use of fiber reinforced polymers (FRP) in construction industry has become quite common. They offer feasible and innovative solutions for seismic retrofitting due to their lightweight, high tensile strength and noncorrosive character [1,2,3,4,5,6]. FRPs are also utilized as external tendons for strengthening concrete structures [7,8,9]. While there are many studies on external confinement of columns or joints [10,11,12,13,14,15,16,17,18,19,20,21,22,23,24,25,26], and on flexural retrofit with near surface mounted (NSM) FRP rods or pultruded strips under monotonic loading conditions [27,28,29,30,31,32,33,34,35,36,37,38,39,40,41,42], only few studies are available on flexural retrofitting using FRPs under repeated loading conditions [43,44,45]. According to the best knowledge of the authors', reversed cyclic flexural behavior of RC members strengthened with longitudinal FRP reinforcement was studied for the first time by Ilki and Kumbasar [46]. More recently, Bournas and Triantafillou [47], Goksu et al. [48], Fahmy and Wu [49], Vrettos et al. [50], Li et al. [51], Goksu et al. [52] and Faustino and Chastre [53] also studied this type of behavior. However, in all cases, except the studies of Goksu et al. [48,52], it is difficult to perceive the performance of the NSM FRP system and anchoring of NSM reinforcement into the footing in case of low strength concrete. In the study of Goksu et al. [48], which was the precursor of this study, the possibility of using carbon FRP (CFRP) longitudinal (rod, laminate, sheet) and transverse (sheet) reinforcement for the flexural seismic retrofit of low strength RC members under reversed cyclic loading conditions was investigated. The aim of flexural retrofitting by using FRP reinforcement, mainly in the longitudinal direction, was to obtain flexural strength enhancement under reversed cyclic loading, since many existing structures, among other deficiencies, suffer from lack of sufficient flexural strength in case of seismic loading. Furthermore, it was targeted not to sacrifice from drift capacity, since drift capacity is as important as strength in case of seismic actions. In the study of Goksu et al. [48], an enhancement in flexural capacity was obtained until large drift ratios (approximately 6% drift ratio) through flexural FRP retrofit. Moreover, after testing of performance of several anchorage types for connection of FRP reinforcement to the footing, the most effective anchorage detail was obtained, and was also utilized in the current testing program. The major differences between the current testing program and the work reported by Goksu et al. [48] are (i) the presence of axial load (20% of axial load capacity of the columns, whereas no axial load was applied on the specimens in the study of Goksu et al. [48]), (ii) the different type of FRP reinforcement utilized in the current study, aramid FRP pultruded laminates (AFRP), and (iii) intentional partial debonding of FRP reinforcement in the anchorage hole to allow more distributed FRP deformations at the critical interface between the column

Citation: Engin C. Seyhan, Caglar Goksu, Ahmet Uzunhasanoglu and Alper Ilki, Seismic Behavior of Substandard RC Columns Retrofitted with Embedded Aramid Fiber Reinforced Polymer (AFRP) Reinforcement, doi:10.3390/polym7121527.

and footing in the current study. The reason of utilizing AFRP reinforcement rather than CFRP or GFRP in the current study is the expectation of potentially better performance of AFRP due to its better toughness characteristics [54].

In this study, four cantilever RC columns were constructed using low strength concrete and plain reinforcing bars for representing relatively old substandard structures. According to the compressive tests of core specimens extracted from many different existing buildings constructed before 1990s in Turkey, the concrete compressive strengths are found to be between 5 and 13 MPa. The microstructural analyses of these core specimens revealed that the concretes were very porous and the water/cement ratio used for making them was excessively high [55]. The columns were then tested under reversed cyclic lateral and constant axial loads before and after retrofitting. Three different anchorage details were designed for embedding AFRP laminates to the existing footing. These different details constitute the main test variable. While the applied retrofitting technique is similar to NSM technique in terms of mechanical contribution of FRP reinforcement to the flexural strength, application technique of retrofitting is different than NSM technique, as has been introduced by Goksu et al. [48] before. As is well known, concrete cover is generally deteriorated due to corrosion of internal steel reinforcing bars of substandard structures constructed with low strength concrete. Typical appearances of columns of existing substandard RC frame buildings with corroded reinforcement are shown in Figure 1. In the applied retrofitting technique, firstly, deteriorated concrete cover is removed. Secondly, a layer of high strength structural repair mortar (SRM) is applied to obtain a flat surface for bonding AFRP reinforcement. After hardening of SRM, AFRP longitudinal reinforcement is bonded to the substrate by using epoxy adhesive. Then, a second layer of SRM is applied to form the new concrete cover. As the final stage of seismic retrofit, the strengthened column is externally jacketed with CFRP sheets in transverse direction. Consequently, by this technique, all retrofitting application was conducted within the thickness of the original concrete cover, which may be deemed as an advantage with respect to the conventional retrofitting techniques. Clearly, AFRP reinforcement, which is embedded into SRM, is expected to perform better than NSM application in case of poor quality concrete in terms of structural integrity.

Figure 1: The appearances of existing substandard columns of RC frame structures.

Test results, supported by the analytical calculations, revealed that the cyclic flexural capacity of the substandard RC members can be increased significantly by the use of AFRP pultruded laminates through the applied retrofitting method. Current technical guidelines/documents do not allow use of longitudinal FRP reinforcement for flexural retrofit under cyclic loading in potential plastic hinge regions of RC members [5,56,57]. This is basically due to concerns related with ductility. In this study, however, it was observed that the deformation capacity of the retrofitted columns was also acceptable for a satisfactory seismic performance.

It was observed from the seismic tests that the measured rupture strain of AFRP reinforcement, which was subjected to cyclic loading, was approximately 43% of design rupture strain of FRP reinforcement (design rupture strain of FRP reinforcement = environmental reduction factor × manufacturer declared ultimate rupture strain). For NSM applications under monotonic loading conditions, ACI 440-2R-08 [5] recommends to consider 70% of ultimate strain of FRP. In this study, it is demonstrated clearly that allowable strain limit for longitudinal FRP reinforcement should be lower in case of reversed cyclic loading conditions (only 43% of manufacturer declared ultimate strain of FRP reinforcement). It is also shown that the increased flexural capacity as well as nonlinear inelastic displacement response of columns after strengthening could be predicted satisfactorily using conventional RC theory.

EXPERIMENTAL PROGRAM

The aim of the applied retrofitting technique is the enhancement of column flexural capacity using AFRP reinforcement under cyclic lateral loading in the presence of constant axial load without a significant sacrifice from deformation capacity. On the other hand, like many existing substandard RC columns, the spacing of the transverse reinforcement of the columns was insufficient. This might have caused deficiencies in terms of ductility and shear strength after flexural strength enhancement through FRP retrofitting. Therefore, after the intervention made for flexural strength enhancement through longitudinal AFRP reinforcement, the columns were also jacketed externally with CFRP sheets in transverse direction. As known, while external FRP confinement has a remarkable contribution to deformability and shear strength, its contribution to flexural strength (which is the main issue investigated in this study) is marginal with respect to the contribution of longitudinal AFRP reinforcement. Nevertheless, both contributions were taken into account during the analytical calculations presented in this paper.

Description of Test Specimens

Four substandard symmetrically reinforced cantilever RC columns were tested under reversed cyclic lateral and constant axial load. The main features of the specimens are shown in Table 1. While lateral loads were applied through displacement cycles in pushing and pulling directions with increasing amplitudes, axial load was kept constant. All columns were identical (before retrofitting) and flexure-critical.

Table 1: Main Features of the Specimens

Specimen	Main longitudinal AFRP [a] reinforcement	Additional anchorage reinforcement [a]	Anchorage type [a]	Transverse CFRP [b] reinforcement
REF	-	-	-	-
LAM	2 × 2 (each 42 mm x 1.4 mm)	-	Fully bonded	2 plies × 0.166 mm
LAM-PB	2 × 2 (each 42 mm x 1.4 mm)	-	Partially bonded	2 plies × 0.166 mm
LAM-LAM	2 × 2 (each 42 mm × 1.4 mm)	2 × 2 (each 42 mm × 1.4 mm)	Fully bonded	2 plies × 0.166 mm

[a] Aramide based, pultruded laminates; [b] Carbon based, unidirectional sheets fully wrapped around the retrofitted specimens. REF: reference specimen; LAM, LAM-PB, LAM-LAM: retrofitted specimens.

The geometry and the reinforcement details of the columns are presented in Figure 2. In this figure, it can be observed that the longitudinal bars were (4ϕ14) continuous from the bottom of the footing to the top of the column. The spacing of transverse bars was 200 mm in the testing zone (ϕ10/200). All internal reinforcing bars were plain round bars, which had been used commonly until 2000s in Turkey.

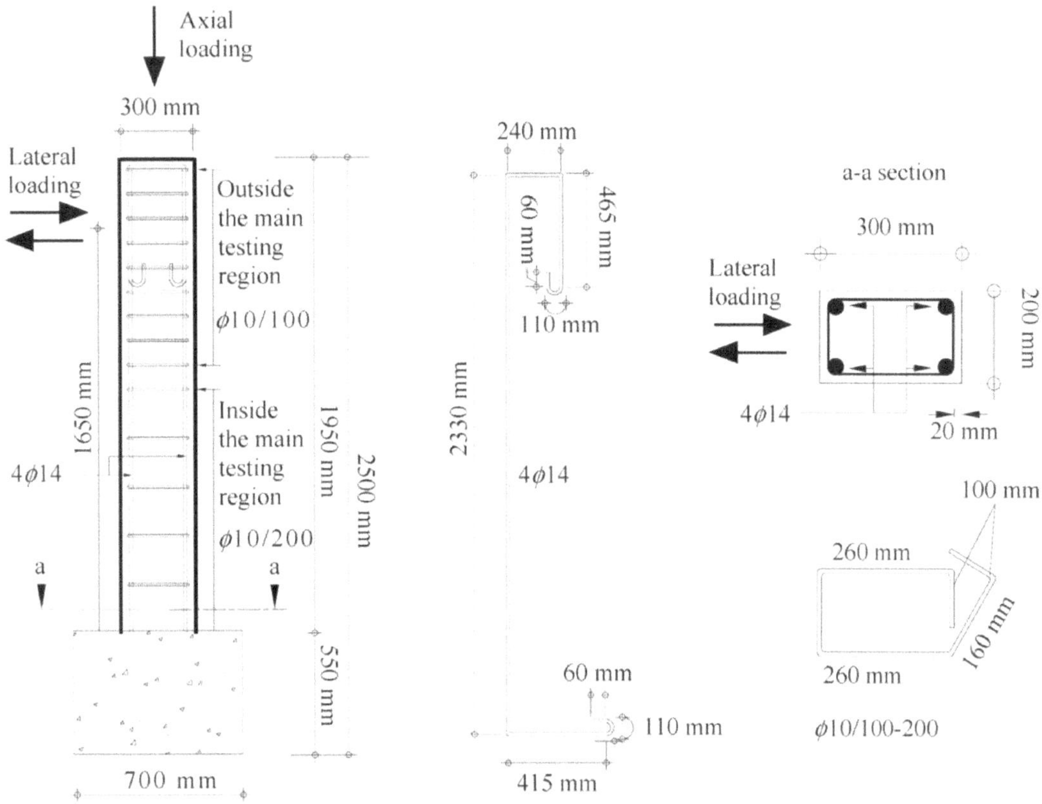

Figure 2: The geometry and reinforcement details of the columns.

The reference specimen, denoted as REF, was tested without any retrofit. The other columns (LAM, LAM-LAM and LAM-PB) were tested after they were retrofitted with AFRP reinforcement in longitudinal and CFRP reinforcement in transverse directions. The longitudinal AFRP reinforcement embedded in the columns was anchored directly to the footing in all retrofitted specimens, while additional AFRP anchorage reinforcement was used for the specimen LAM-LAM to achieve a better performance by avoiding potential failures of FRP reinforcement at the interface of the column and the footing. Contrary to the fully bonded anchorages of longitudinal FRP reinforcement in case of the specimens LAM and LAM-LAM, the anchorage of the FRP reinforcement in case of the specimen LAM-PB was intentionally partly debonded from concrete using insulating tape. Partial debonding of AFRP reinforcement in the anchorage zone was for preventing the localized damage of the AFRP reinforcement at the interface of column and footing by avoiding localization of stresses at this critical zone. Details on each retrofitting scheme are explained below and presented in Figure 3.

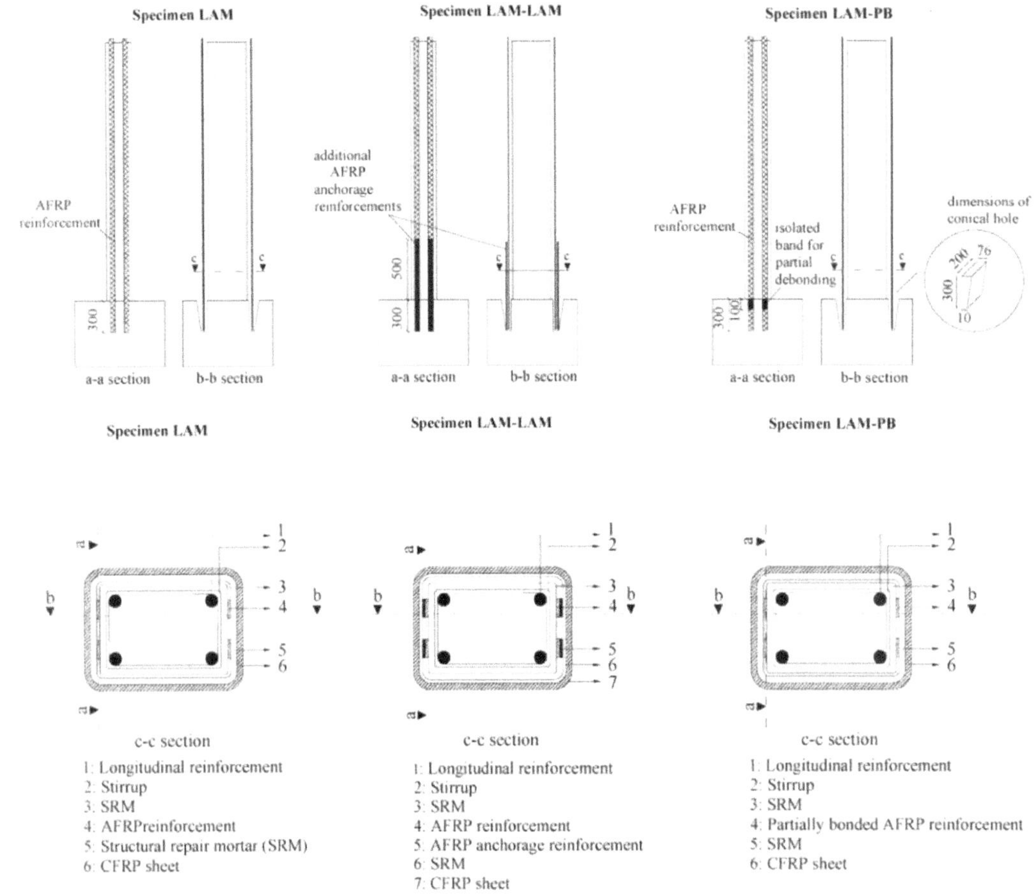

Figure 3: The retrofit schemes of the specimens (Note: Dimensions in mm).

Characteristics of Materials

Low strength concrete and plain reinforcing bars were used for the construction of the columns for representing relatively old, existing substandard structures. All RC columns were cast at the same day, from the same concrete production batch. The 28th-day average compression strength was measured as 10.3 MPa. An additional compressive strength test, performed on the 750th day, yielded an average compressive strength of 15.2 MPa. The compressive strength, at the time of testing of columns (between 57th and 64th days), was calculated as 12.0 MPa, according to the procedure described in Eurocode 2-Part 1 (2004) [58]. For the numerical analysis, this value was multiplied with 0.85 to account for the size effect in concrete columns. The concrete mix-proportions of cement: aggregate (0–4 mm): aggregate (5–12 mm): sand: water were 1:3.8:2.2:4.7:1 (by weight). The yield stress (f_y) and tensile strength (f_{max}) of longitudinal bars were 296 and 400 MPa, while those of transverse bars were 315, and 400 MPa, respectively. The compression tests of concrete specimens and tension tests of steel bars were carried out in accordance with TS-EN Turkish Standards 12390-3 [59] and TS 708 [60], respectively. Two different types of FRP reinforcement were used in retrofitting; AFRP pultruded laminates in longitudinal and CFRP sheets in transverse direction. The appearance and geometrical and mechanical properties of FRP reinforcement are presented inFigure 4 and Table 2, respectively. In this table, t_f, w_f, E_f and ε_{fu}* are the effective thickness, the effective width, the tensile elastic modulus and the ultimate rupture strain of FRP reinforcement declared by the manufacturer (BASF Construction Chemicals, Augsburg, Germany). The compressive strengths of SRM (cement based structural repair mortar, EN-1504-3 [61]) used for forming the new concrete cover, the epoxy adhesive used for bonding the AFRP reinforcement to the column surface (EN-1504-4 [62]), the epoxy adhesive used for bonding CFRP sheets in transverse direction to the column surface, and the epoxy grout used for anchoring AFRP reinforcement in the footing were 50, 75, 60 and 80 MPa (obtained from 40 mm× 40 mm × 160 mm prism test results at 7 days of age), respectively. The compressive characteristics of all materials were determined according to TS-EN 196-1 [63], and the mechanical characteristics of CFRP sheets were determined according to ASTM D3039 [64].

Table 2: Properties of FRP reinforcement

FRP reinforcement	E_f (N/mm^2)	t_f (mm)	w_f (mm)	ε_{fu}*
AFRP (laminate)	60,000	1.4	42	0.023
CFRP (sheet)	230,000	0.166	500	0.015

FRP, fiber reinforced polymers; AFRP, aramide reinforced polymers; CFRP, carbon reinforced polymers; E_f, tensile elastic modulus of FRP reinforcement; t_f, effective thickness of FRP reinforcement; ; w_f, effective width of FRP reinforcement; ε_{fu}*, ultimate rupture strain of FRP reinforcement declared by manufacturer.

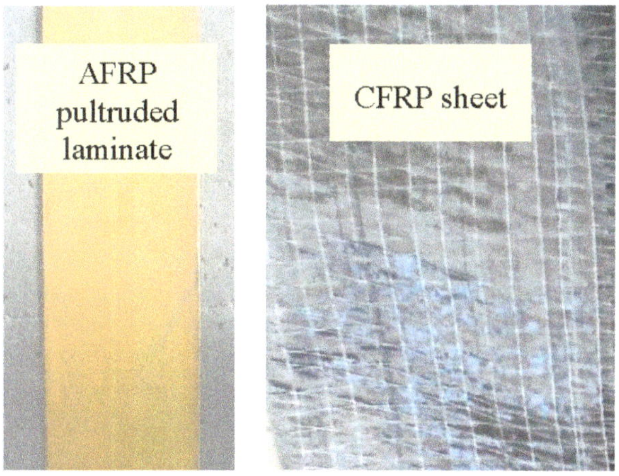

Figure 4: The appearances of longitudinal AFRP and transverse CFRP reinforcement.

Retrofitting Procedure

For enhancing the column flexural capacity, three specimens were retrofitted with longitudinal AFRP reinforcement, which was embedded within the concrete cover (after removal of poor concrete cover), and anchored to the existing footing. The retrofitting details and the application stages are shown in Figure 3 and Figure 5, respectively. As a first step, the concrete cover was removed for more effective utilization of AFRP reinforcement until the longitudinal steel bars were exposed (Figure 5a). This allowed AFRP reinforcement to be embedded within the cover thickness. This application may also cover a corrosion repair procedure, which is generally required in case of rehabilitation of old buildings, which were constructed with poor quality concrete. Then, a thin layer of SRM was applied to obtain a sound and smooth substrate over the internal steel reinforcing bars (Figure 5b). Afterwards, an epoxy-based primer was applied over the SRM to increase the adhesion between the AFRP reinforcement and SRM. Then, two pultruded AFRP laminates of 1.4 mm thickness and 42 mm width were placed symmetrically on each side of the columns in the strong direction, with an anchorage length of 300 mm (Figure 3 and Figure 5c), and the laminates were bonded to the substrate by using an epoxy adhesive. For the connection of the AFRP longitudinal reinforcement to the footing, conical holes were dug in the footing (Figure 3). The longitudinal AFRP reinforcement embedded in the columns was anchored directly to the footing in all retrofitted specimens. Different from the specimen LAM, for the specimen LAM-PB, the top 100 mm long part of the AFRP reinforcement in the conical hole was wrapped with a plastic insulating band in order to create a partially debonded anchor (Figure 3 and Figure 5d). Therefore, total bonded lengths of the AFRP reinforcement in the footing were 200 mm (rather than 300 mm) for the specimen LAM-PB. Additional AFRP anchorage reinforcement of 800 mm length was used at the column-footing interface in case of the specimen LAM-LAM (Figure 3 and Figure 5e). The 300 mm long part of the additional AFRP anchorage reinforcement was embedded into the conical hole. AFRP anchorage reinforcement and the main AFRP reinforcement were bonded to each other by using the same epoxy adhesive, and they were anchored to the footing (in the pre-dug conical hole) together by using epoxy grout. Same epoxy grout was used to fill the conical holes in the footings of specimens LAM and LAM-PB. After all, a layer of SRM was applied to bring the column cross-section to its original dimensions (Figure 5f). It should be noted that, buckling of NSM FRP bars was observed during the reversed cyclic tests carried out by Bournas and Triantafillou [47]. Therefore, after installation of longitudinal AFRP reinforcement and formation of concrete cover with SRM, as a final step, CFRP sheets were wrapped around the columns in transverse direction to enhance the deformability through confinement action and to avoid potential shear damages due to increased flexural strength as well as to prevent buckling of AFRP reinforcement in compression (Figure 5g). Additionally, CFRP confinement is believed to enhance the bond between the core concrete, SRM and AFRP reinforcement. Consequently, the external dimensions of the reference and retrofitted specimens were identical (200 mm × 300 mm). In other words, all retrofitting application was conducted within the thickness of the original concrete cover. Naturally, during practical applications, the safety of the footing and soil under the footing should be checked to confirm that they have also sufficient safety margin against increased internal forces.

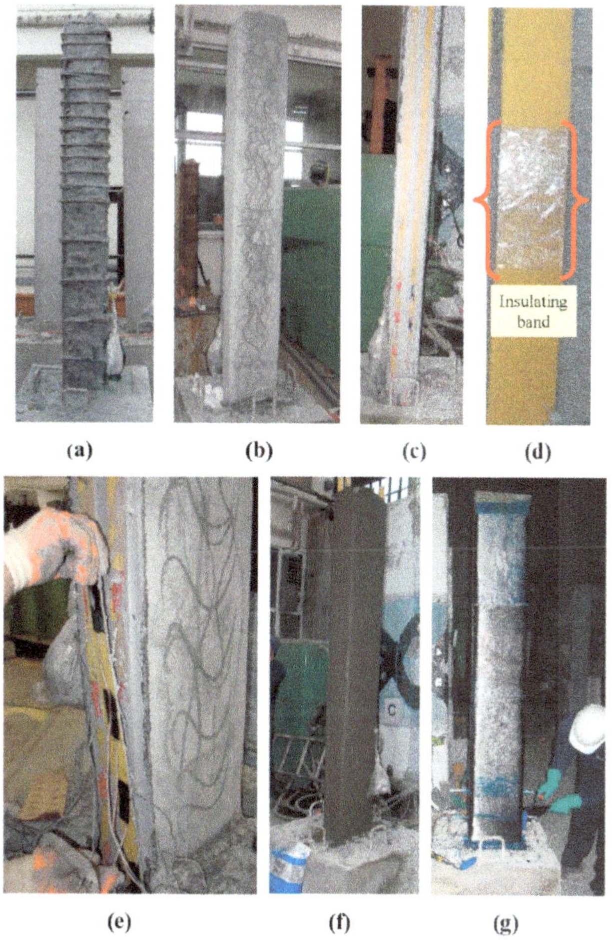

Figure 5: (**a**) Removal of cover concrete, (**b**) application of a thin layer of cement based SRM to obtain a flat surface, (**c**) installation of the AFRP pultruded laminates, (**d**) laminate with isolated band (only for the specimen LAM-PB), (**e**) application of additional AFRP laminate anchorages (only for the specimen LAM-LAM), (**f**) application of last layer of structural repair mortar, and (**g**) wrapping with CFRP sheets in transverse direction.

Test Setup

The specimens were tested under constant axial load and reversed cyclic lateral loads in a quasi-static displacement-controlled manner. A displacement-based loading protocol was applied during testing of all specimens. Similar loading patterns were used by Matamoros and Sozen [65], Iacobucci et al. [13], Ilki et al. [20], Bournas and Triantafillou [66], and Goksu et al. [48] during testing of RC columns. Target lateral drift ratios were calculated as the ratio of the lateral displacement to the specimen height (at which actuator is connected) in both pushing and pulling directions (Figure 6). The axial load of 120 kN, which corresponded to 20% and 18% of the axial load capacity of the reference column without and with consideration of the capacity of the internal steel reinforcement, respectively, was applied to all specimens (Figure 7a). Other than a large number of displacement transducers, a number of strain gauges were also used on steel bars and AFRP reinforcement in longitudinal direction and on stirrups in transverse direction (Figure 7b).

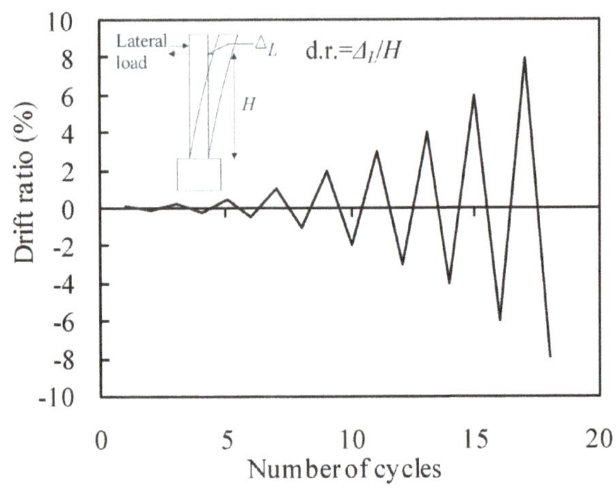

Figure 6: Loading protocol of the columns.

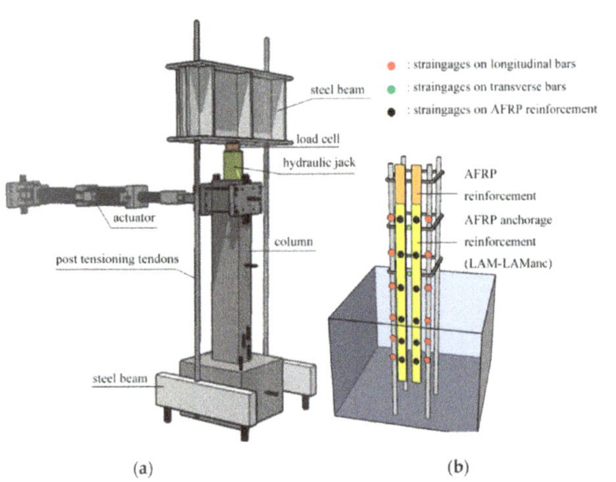

Figure 7: (**a**) The testing setup; and (**b**) Strain gauges on steel bars and AFRP reinforcement.

ANALYTICAL CONSIDERATIONS

Theoretical Flexural Strength and Displacement Capacity before and after Retrofit

The theoretical strength and local deformation characteristics are determined through moment–curvature relationships obtained by fiber analysis approach using the stress–strain relationships of materials representing monotonic increasing stresses. The monotonic moment–curvature relationships are obtained using XTRACT (2007) [67] software at the critical sections of each column for predicting the load–displacement relationships of the columns. Therefore, the theoretical load–displacement relationships approximately represent the envelopes of hysteretic curves. The effect of axial load is considered while obtaining the moment–curvature relationships. In the moment–curvature analysis, steel reinforcing bars are assumed to behave in an elasto–plastic manner with strain hardening. The stress–strain relationship of the steel reinforcing bar is defined according to actual uniaxial tension test results. Longitudinal AFRP reinforcement is taken into account as linear elastic material in tension since the buckling and debonding of AFRP reinforcement during cyclic loading is prevented by SRM and CFRP confinement. Nevertheless, the contribution of the AFRP reinforcement in compression is neglected. Since the retrofitted specimens, thereby the AFRP applications, were exposed to laboratory conditions for a very short time prior to testing, the design rupture strain of AFRP reinforcement (ε_{fu}) is assumed to be equal to ε_{fu}*, which is the manufacturer declared ultimate strain of FRP reinforcement (Equation (1)). In other words, the environmental reduction factor, C_E, is taken as 1.0. While ACI 440-2R-08 [5] recommends to reduce ε_{fu} by a factor of 0.70 for NSM FRP applications under monotonic flexural loading, ε_{fu} is reduced by a factor of 0.43 during the analysis presented here (the reason for this assumption is explained in the section titled "Strains of Steel and AFRP Reinforcement"). For obtaining the stress–strain relationship of CFRP confined concrete, the model proposed by Ilki et al. [68] is used. The stress–strain relationship of CFRP confined concrete is obtained for two different parts of the confined cross–section (for core concrete confined by transverse steel bars and external CFRP, and for SRM as concrete cover confined by only external CFRP). The stress–strain relationships of unconfined concrete, core concrete, SRM, internal steel and AFRP reinforcement used in calculation of moment capacities are presented in Figure 8. The marks in the figure correspond to the analytical values of the specimens during analysis. It should be noted that SRM causes an increase of 7% in theoretical moment capacity, while both SRM and AFRP cause an increase of 51% and 97% for the specimens LAM and LAM-LAM, respectively, in theoretical moment capacity in comparison with the reference column. The moment–curvature analysis is terminated when longitudinal FRP reinforcement ruptures (Figure 9).

$$\varepsilon_{fu} = C_E \cdot \varepsilon_{fu}^{*} \quad (1)$$

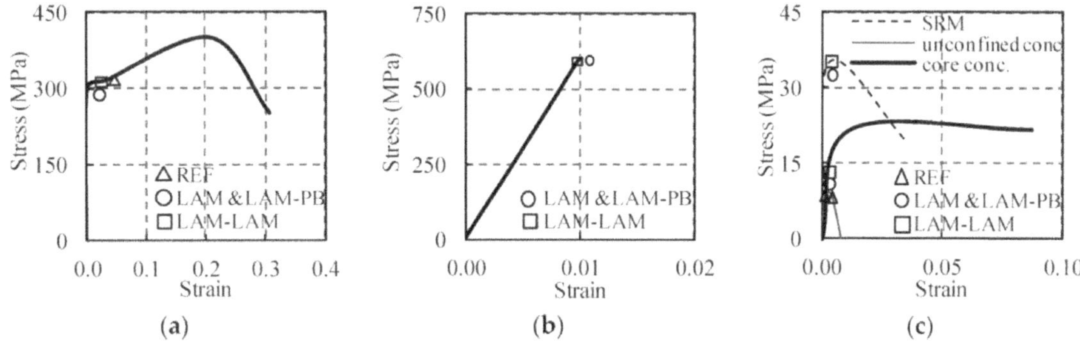

Figure 8: The theoretical stress–strain relationships of the (**a**) Internal steel reinforcement; (**b**) AFRP reinforcement; and (**c**) Concrete (conc., concrete).

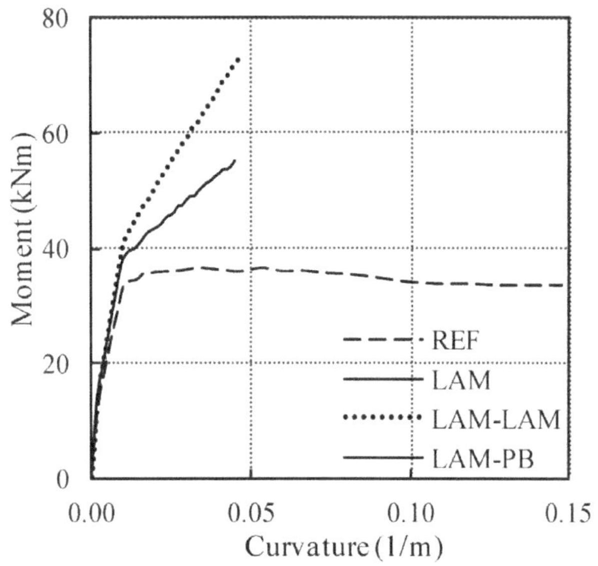

Figure 9: Theoretical moment–curvature relationships. (Note: The moment–curvature relationships for the specimens LAM and LAM–PB are the same).

After obtaining the moment–curvature relationships through fiber analysis approach with summarized assumptions, the total top displacements (δ_t) of the columns are determined considering the elastic and inelastic deformations. It should be noted that the shear deformations are neglected since they are very small with respect to contribution of flexure. As a first step, for calculating the yield displacements (δ_y) of the columns using Equation (2), the columns are discretized into 5 elements for considering effective flexural stiffness, rather than the gross cross-sectional stiffness, at different heights of the column as a function of bending moment when the longitudinal reinforcing bars yield at the critical maximum moment region (Figure 10). In Equation (2), χ_i is the elastic curvature (corresponding to respective bending moment on that discretized part), l_i is the length of the respective part, and h_i is the distance between the center of the respective part and the tip of the column, where the lateral load is applied. Sequentially, for obtaining the plastic component of the total top displacements (δ_p) of the columns, conventional plastic hinge approach is used (Equation (3)). In Equation (3), χ_p is the plastic curvature at the plastic hinge of the column, l_p is the plastic hinge length, H is the distance between the interface of the column and the footing and the point of application of the lateral load. The length of plastic hinge (l_p) of the column is assumed as h/2 (150 mm) according to TSDC [57], where h refers to the depth of the cross-section of the column. Finally, the total top displacements (δ_t) of the columns are determined using Equation (4). The theoretical lateral load–displacement relationships are then obtained and compared with the experimental load–displacement relationships in the Section "Results and Discussions".

$$\delta_y = \sum_{i=1}^{5} \chi_i \cdot l_i \cdot h_i \quad (2)$$

$$\delta_p = \chi_p \cdot l_p \cdot \left(H - \frac{l_p}{2}\right) \quad (3)$$

$$\delta_t = \delta_y + \delta_p \quad (4)$$

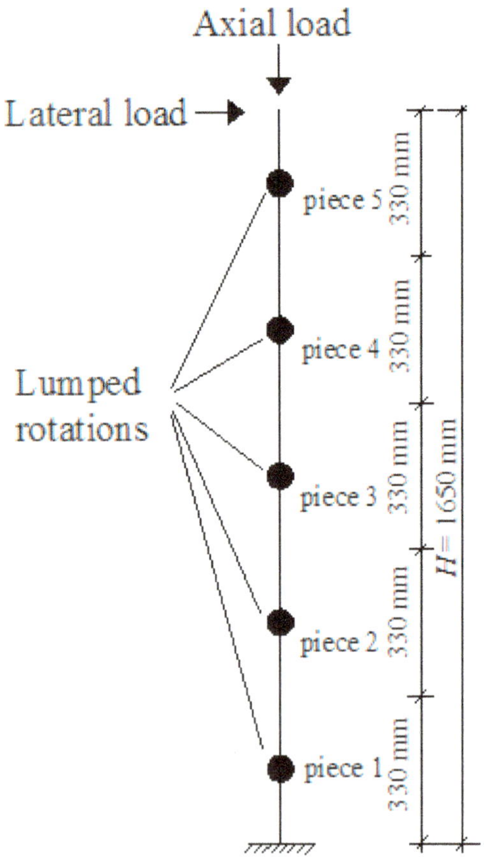

Figure 10: The locations of lumped rotations.

Shear Strength before and after Retrofit

According to ACI 318M-14 [69], before retrofit, the shear strength, which includes contributions of concrete and transverse reinforcement, can be obtained using Equations (5) and (6). In these equations, V_s is the nominal shear strength provided by transverse reinforcement, A_v is the area of the transverse reinforcement, f_{yt} is the design yield strength of transverse reinforcement, d is the effective depth, s is the spacing of shear reinforcement, V_c is the nominal shear strength provided by concrete, N_u is the factored axial load normal to cross-section (to be taken as positive for compression), A_g is the gross area of the column section, f'_c is the design compressive strength of concrete, b_w is the width of the column,

$$V_s = \frac{A_v f_{yt} d}{s} \qquad (5)$$

$$V_c = 0.166\left(1 + \frac{N_u}{14A_g}\right)\cdot\sqrt{f'_c}\,b_w d \qquad (6)$$

According to ACI-440-2R-08 [5], the contribution of FRP reinforcement to shear strength can be obtained using Equations (7) and (8).

$$V_f = \frac{A_{fv}\cdot\varepsilon_{fe}\cdot E_f\cdot d_{fv}}{s_f} \qquad (7)$$

$$A_{fv} = 2n\cdot t_f\cdot w_f \qquad (8)$$

where V_f is the FRP contribution to shear strength, A_{fv} is the area of FRP shear reinforcement, ε_{fe} is the effective strain of FRP sheet attained at failure, which was assumed to be 0.004 as recommended by ACI-440-2R-08 [5] and TSDC [57], E_f is the tensile modulus of elasticity of FRP shear reinforcement, d_{fv} is the effective depth of FRP shear reinforcement, s_f is the spacing of FRP reinforcement plies, n is the number of plies of FRP reinforcement, t_f is nominal thickness of one ply of FRP reinforcement, and w_f is the width of FRP reinforcement plies.

According to ACI-440-2R-08 [5], the total shear strength of the cross-section (V_n) resisted with FRP reinforcement, concrete and transverse reinforcement can be obtained using Equation (9).

$$V_n = V_c + V_s + \psi_f V_f \qquad (9)$$

where ψ_f is the FRP strength reduction factor and was assumed to be 0.95 as recommended by ACI-440-2R-08 [5] for shear fully wrapped sections.

Since the shear strengths of the columns (before and after retrofit) were remarkably higher than their flexural strengths, as expected, flexural effects dominated the behavior of the specimens (Table 3). This table includes the observed and predicted failure modes, which are consistent.

Table 3: Theoretical and experimental lateral load capacities and consecutive failure modes of the columns.

Specimens	REF	LAM and LAM-PB	LAM-LAM
Theoretical lateral load capacity (kN)	21.2	32.1	41.8
Experimental lateral load capacity (kN)	24.5	33.8 (for the specimen LAM) 34.6 (for the specimen LAM-PB)	45.1
Theoretical moment capacity (M_0) (kNm)	36.5	55.0	72.4
Theoretical failure mode	Y before C	Y, T	Y, then T
Observed failure mode	Y before C	Y, F	Y, FA
Shear strength (kN)	108.0 ($V_{c+}V_s$)	260.6 ($V_{c+}V_{s+}\psi_f\cdot V_f$)	

Y: Tensile reinforcement yields; C: Concrete crushes; T: AFRP reinforcement reaches its ultimate tensile capacity; F: AFRP reinforcement fractured; FA: Both AFRP reinforcement and anchorage fractured.

Anchorage Length for FRP Reinforcement in Footing

In order to utilize the full capacity of the FRP reinforcement, the failure of FRP reinforcement was desired to be due to rupture of the reinforcement rather than pullout, concrete splitting, concrete cone and spacing and edge cone failures, which have been listed as five primary failure modes in ACI-355-1R-91 [70]. Among these failures, the possibility of concrete cone, concrete splitting and edge cone failures was significantly reduced due to the constraining effect of steel reinforcing bars in the footing, where the conical anchor holes were dug in. Therefore, only possibilities of pullout failure and fracture of FRP reinforcement were taken into account by comparing the tensile strength of FRP reinforcement and the adhesion between concrete and epoxy grout (Equations (10) and (11)). In these equations, F_{bond} is the bond capacity, τ_b is the uniform bond strength along the anchorage length, l_{db} is the embedment length, while u is the perimeter, F_{frp} is the tensile capacity of the FRP reinforcement, ε_{fu} is the design rupture strain of FRP reinforcement, and A_{frp} is the cross-sectional area of the FRP reinforcement. For the selected embedment length (300 mm), the calculated value of F_{bond} (1012 kN) was far over F_{frp} (277 kN), which clearly eliminated the possibility of pullout failure. It should be noted that while calculating F_{bond}, τ_b was calculated to be 6.3 N/mm² based on ACI-355-1R-91 [70]. Moreover, test results of the retrofitted columns also confirmed that the embedment length of FRP reinforcement was sufficient. No debonding problem or damages resulting from concrete cone, concrete splitting, edge cone and pullout failures were observed during the tests.

$$F_{bond} = \tau_b \cdot l_{db} \cdot u \quad (10)$$

$$F_{frp} = E_f \cdot \varepsilon_{fu} \cdot A_{frp} \quad (11)$$

RESULTS AND DISCUSSIONS

The test results were outlined through hysteretic load–displacement curves, envelopes of these curves, strain distributions along longitudinal, transverse steel and longitudinal AFRP reinforcement, energy dissipation capacities, average section rotations in different gauge lengths and damage progression and distribution. Furthermore, experimental data was compared with the theoretical findings.

Lateral Load–Displacement Curves and Failure Modes

The hysteretic lateral load–displacement curves and their envelopes are presented in Figure 11 and Figure 12, respectively. The damage progression of the columns with increasing drifts is summarized in Table 4.

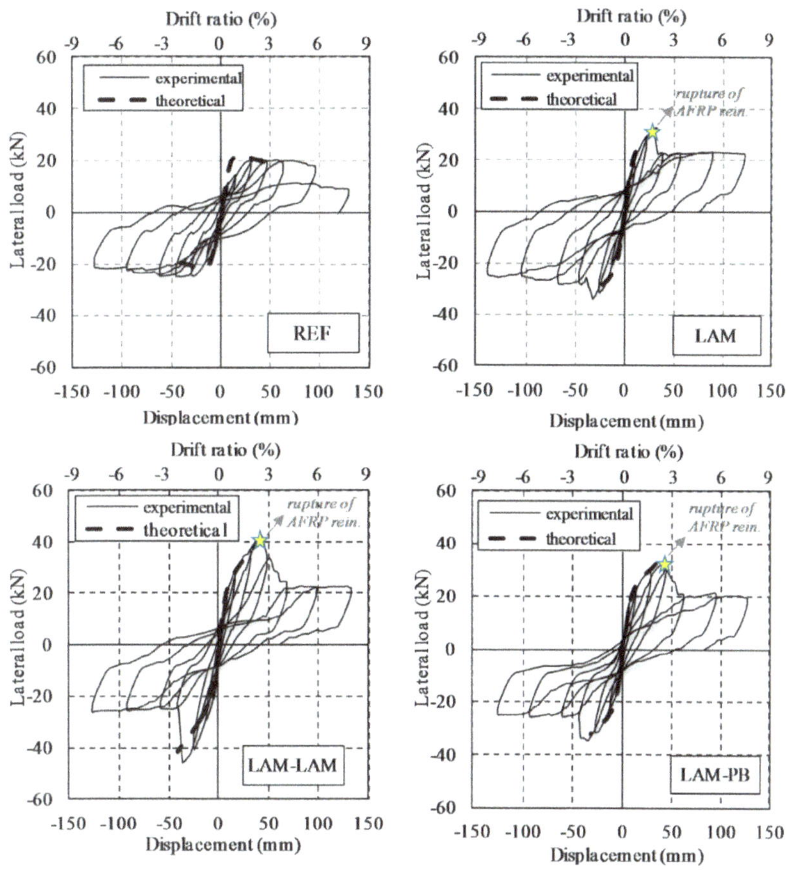

Figure 11: Experimental and theoretical lateral load–displacement curves for all specimens (rein., reinforcement).

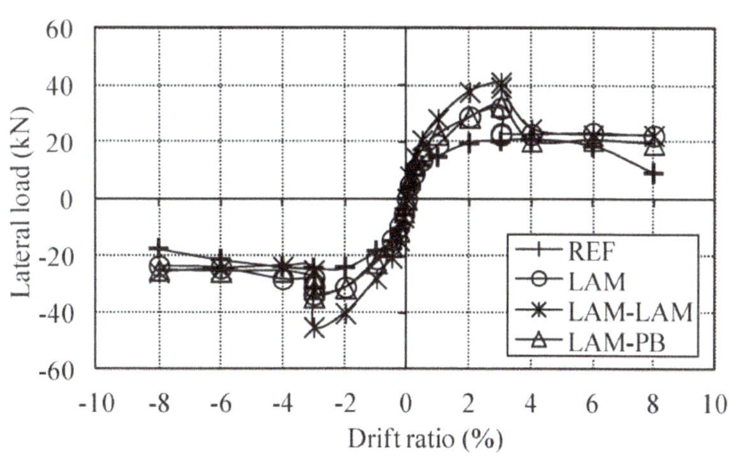

Figure 12: The envelopes of load–displacement relationships.

Table 4: Damage progression of the columns (The numbers are drift ratios in %).

Damage mechanisms	Specimens			
	REF	LAM	LAM-LAM	LAM-PB
Yielding of longitudinal steel bars	2	3	3	3
Crushing of concrete cover	3	–	–	–
Spalling of concrete cover	~6	–	–	–
Fracture of AFRP reinforcement	–	3	3	3

As seen in Figure 11 and Figure 12 and Table 3, the reference column (REF) reached its theoretical flexural capacity and exhibited a ductile behavior. The column LAM, retrofitted with AFRP reinforcement, experienced an enhancement in strength up to the drift ratio around 3%. At this drift, the enhancement in strength was around 38% with respect to the column REF. The sudden remarkable loss of strength upon exceeding the drift ratio around 3% was due to the fracture of AFRP reinforcement at the interface of the column and the footing (Figure 13a). As seen in Figure 11 and Figure 12, the column LAM-LAM exhibited a remarkably superior performance with respect to the reference and the other retrofitted columns by resisting lateral loads, approximately 1.9 and 1.3 times the loads resisted by the reference and other retrofitted columns, respectively. As seen in Figure 11, the column LAM-LAM also sustained its lateral load capacity until the drift ratio of 3%, around where the AFRP reinforcement and AFRP anchorage reinforcement fractured at the same section (Figure 13b). The higher strength of the column LAM-LAM is due to the contribution of additional AFRP anchorage reinforcement to the flexural capacity. The column LAM-PB behaved similar to the column LAM in terms of maximum lateral load and fracture of AFRP reinforcement around 3% drift ratio. At this drift ratio, the enhancement in strength was around 41% with respect to the column REF. It was observed that, while pulling, both AFRP reinforcements under tension were fractured upon exceeding the drift ratio of 3%. On the other hand, while pushing, only one AFRP reinforcement under tension fractured, whereas the other one did not fracture, instead, decomposed locally just above the isolated height (Figure 13c). Decomposition is the separation of the individual fibers after fracture of the epoxy matrix. Contribution of decomposed AFRP reinforcement to the behavior at larger drifts limited plastic residual deformations of the column LAM-PB. As seen inFigure 11, marginal residual displacements remained after unloading branches in pulling direction up to 8% drift ratio. Therefore, this type of anchorage system can be used to limit the residual displacement (and consequently residual damage). The retrofitted columns behaved similar in terms of strength degradation after the rupture of FRP reinforcement as the specimens tested by Ilki and Kumbasar [46], Bournas and Triantafillou [47], and Vrettos et al. [50], where the drift ratios at failure were approximately 2%, 4% and 2.5%, respectively. It is important to note that all significant damage was accumulated at the base of the columns, since the columns were wrapped with CFRP sheets in transverse direction along the full height. Consequently, the crack width reached several centimeters at the intersection of the column and the footing. This type of damage may be quite disadvantageous in case of earthquakes since the distribution of plastic deformations through the potential plastic hinge length is prevented due to presence of a rigid transverse CFRP jacket. The accumulation of a remarkable portion of plastic deformations only at the interface of the column and the footing may significantly reduce the overall drift capacity of the column. Since drift capacity is essentially important for a satisfactory seismic performance, this kind of reduction in drift capacity should be avoided. This problem can be overcome by application of transverse sheets in strips allowing distribution of damage (and plastic deformations) in the unconfined zones between transverse FRP strips.

Figure 13: Fracture of AFRP reinforcement during the test (**a**) the column LAM at 3% drift ratio, (**b**) the column LAM-LAM at 3% drift ratio, and (**c**) the column LAM-PB at 3% drift ratio.

Average experimental moment–curvature relationships obtained for critical sections of columns are presented in Figure 14. In this figure M (Second order effect is considered) is the experimental moment, M_0 is the theoretical moment capacity calculated per specimen, which is obtained by fiber analysis approach. The theoretical moment capacities (M_0) are presented in Table 3. For the calculation of experimental moment–curvature relationships, the average curvature values obtained along 0–20 mm, 20–150 mm, and 150–300 mm heights above the footing were taken into account. As seen in Figure 14, for the retrofitted columns, the average curvature values of the column sections measured along 0–20 mm height were in the order of 4.5×10^{-3} (1/mm), while the curvatures measured in 20–150 mm height above the footing were in the order of 4.5×10^{-4}(1/mm). Since the curvature values of the columns measured along 150–300 mm height were even smaller, these results are not presented in Figure 14. In agreement with these measurements, the damage was also accumulated particularly in 20 mm height of the column from top of the footing. During the autopsy after the tests, no concrete crushing and no cracking (other than the main crack at the interface of the column and the footing) were observed on the retrofitted columns. On the other hand, the deformations (as well as cracks) were distributed through the column height in case of the reference column.

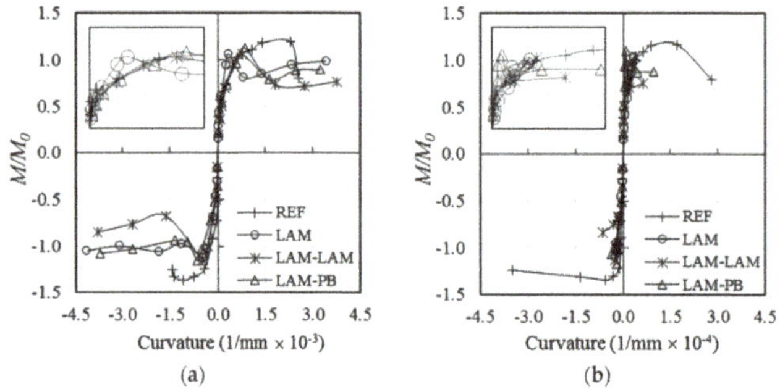

Figure 14: Moment–curvature relationships obtained in columns (**a**) 0–20 mm, and (**b**) 20–150 mm above the footing. (The 0.0–1.0 region of the figure is blown up at the left hand side of the figure).

Strains of Steel and AFRP Reinforcement

The strain values were obtained by making use of the strain gauges on longitudinal and transverse steel, and longitudinal AFRP reinforcement (Figure 15). The efficiency of AFRP reinforcement on flexural capacity can also be seen through the measured strains of longitudinal AFRP reinforcement. The maximum strains of longitudinal AFRP reinforcements in tension were measured to be in the range of 0.01 right before the rupture of the AFRP reinforcement. This value corresponded to approximately 43% of design rupture strain of FRP reinforcement. ACI 440-2R-08 [5] recommends to reduce ε_{fu} by using a factor of 0.70 for NSM application technique under monotonic flexural loading. This value is quite unconservative for the applied retrofitting technique in this study. It shall be much lower for cyclic loading based on the test results. In the study of Vrettos et al. [50], where the longitudinal CFRP sheets together with CFRP anchors were used for flexural retrofitting of RC columns, the measured strains were also far behind (approximately 38% of design rupture strain of FRP reinforcement) the strain reduction factor recommended by ACI 440-2R-08 [5]. It should be noted that, in the study of Vrettos et al. [50], the ultimate rupture strain of FRP reinforcement was not measured directly through pre-installed strain gauges but calculated through analysis of the cross-section at the column base by using conventional RC force equilibrium, strain compatibility, and material constitutive relationships based on the test results. As shown in Figure 15c, f, right after the rupture of the AFRP reinforcement, the strains decrease within the AFRP reinforcement, and corresponding stresses are transferred to the longitudinal reinforcing bars (the strains of steel bars increased remarkably along the measured height). The longitudinal steel bars of the column REF yielded at drift ratio of 2%, while that of retrofitted columns yielded at drift ratio around 3% right after the columns experienced strength degradation due to rupture of AFRP reinforcement. Since steel reinforcing bars in tension already reached their yield strain at around 3% drift ratio before rupture of AFRP reinforcement, the brittle failure of the column was avoided. It should be noted that drift ratios exceeding 3% may only be expected if substandard RC structures are subjected to severe earthquakes. Nevertheless, after AFRP reinforcements were fractured at around 3% drift ratio, the columns still resisted considerable lateral forces due to contribution of steel reinforcing bars until extremely large drift ratios (~6% to 8%) and kept their vertical load resisting capacities. According to strain gauge measurements, strains of longitudinal steel bars did not reach tensile strain corresponding to strain hardening. This observation is consistent with the nearly horizontal branch of the load–displacement relationships after fracture of AFRP reinforcement. On the other hand, none of the stirrups yielded (strains are in the range of 0.00004 and 0.0006) and no damage was observed on CFRP sheets in transverse direction due to relatively low level of applied axial load (20% axial load capacity of the column) and shear stresses.

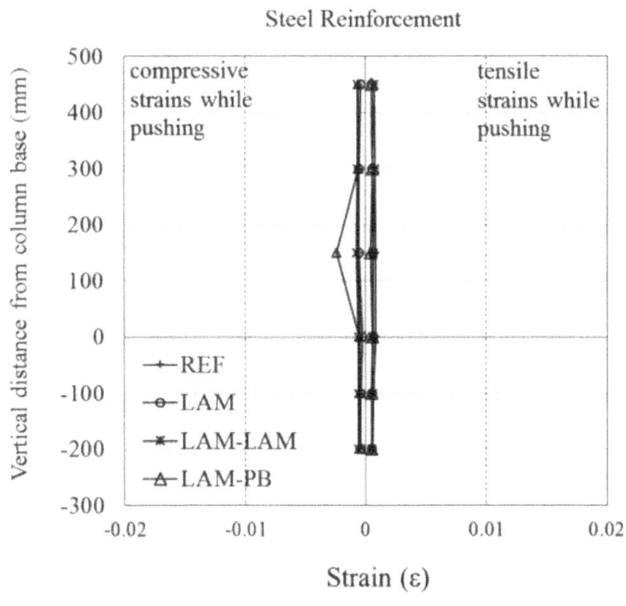

(a)

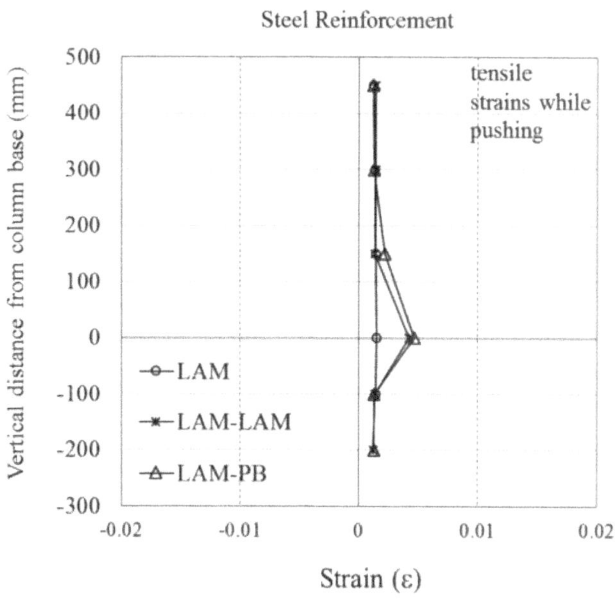

(b)

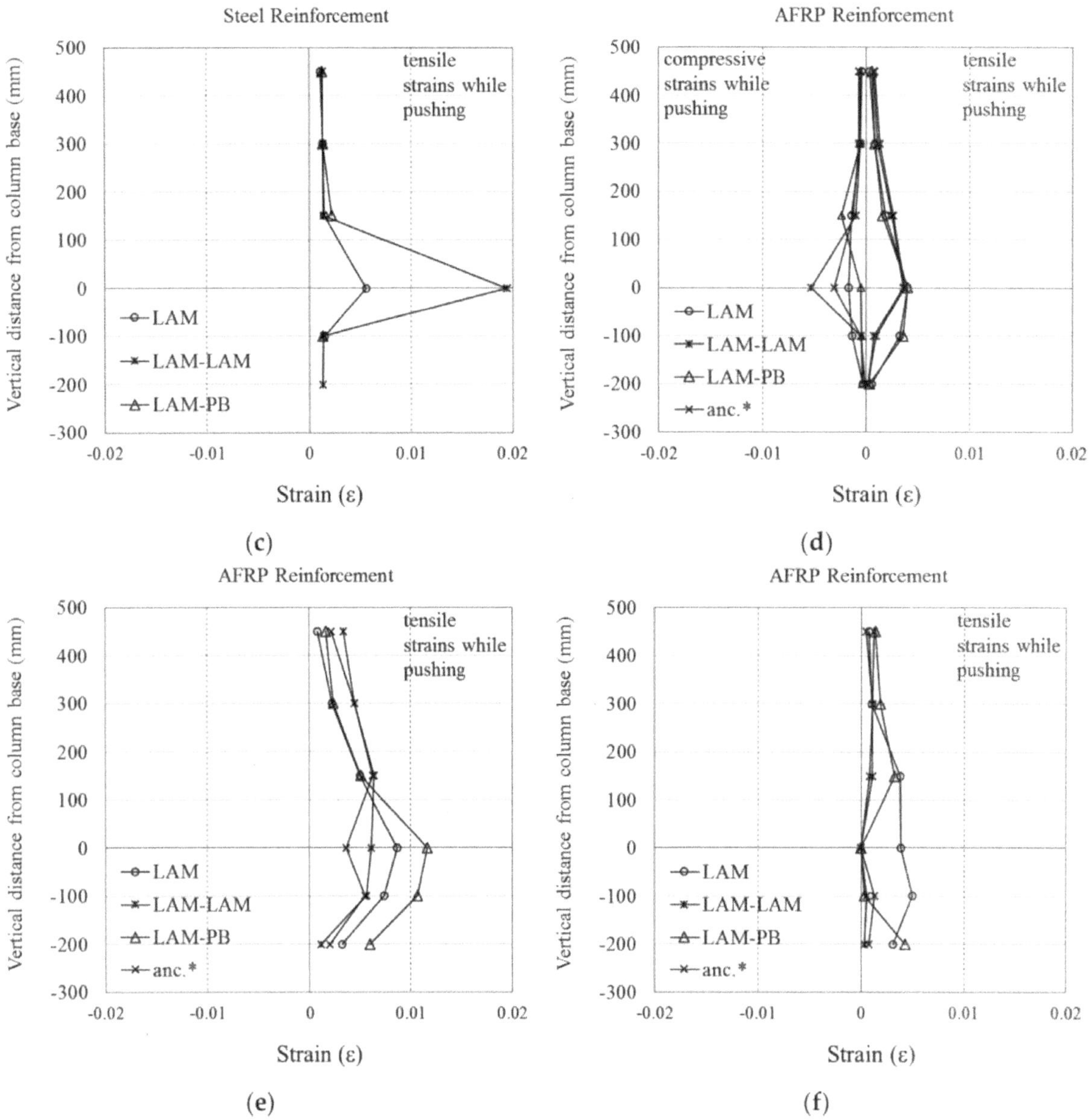

Figure 15: Strain distribution of longitudinal steel reinforcement (**a**) at ±0.01 drift ratio, (**b**) right before rupture of AFRP reinforcement (at around 3% drift ratio), (**c**) right after rupture of AFRP reinforcement (at around 3% drift ratio), and AFRP reinforcement (**d**) at ±0.01 drift ratio, (**e**) right before rupture of AFRP reinforcement (at around 3% drift ratio), (**f**) right after rupture of AFRP reinforcement (at around 3% drift ratio). * strain measurements obtained from additional AFRP anchorage reinforcements for the column LAM-LAM

Energy Dissipation Capacities

Energy dissipation capacities of the tested columns are calculated as the areas enclosed by the load–displacement hysteresis loops and presented in Figure 16. As seen in this figure, until the rupture of the AFRP reinforcement, the column LAM-LAM shows the highest energy dissipation capacity due to its higher lateral load capacity with respect to other columns. Moreover, the energy dissipation capacity of the column LAM is higher than the column LAM-PB. This is due to larger elastic deformations along the embedded unbonded length of AFRP reinforcement in case of column LAM-PB. After the rupture of the AFRP reinforcement, energy dissipation capacities of the columns LAM and LAM-LAM are approximately same, since the columns behave like the reference column in terms of sustained lateral load. As seen inFigure 16, the dissipated energy is significantly less for column LAM-PB due to elastic behavior of unbonded AFRP reinforcement. Nevertheless, with the gradual damaging of AFRP reinforcement, the contribution of AFRP reinforcement to the elastic behavior is diminished and plastic deformations increased causing more significant residual displacements with respect to previous loading cycles.

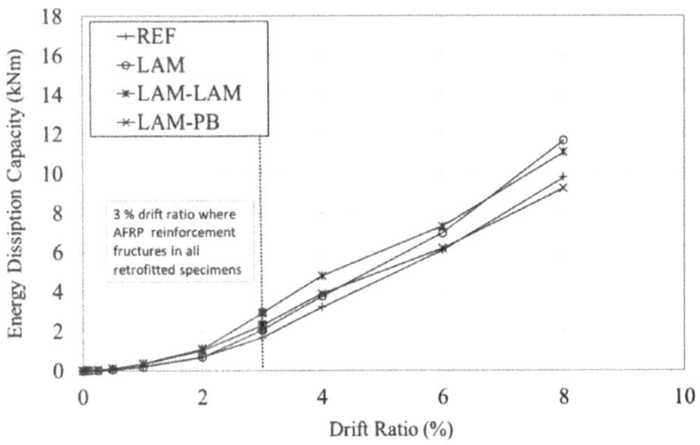

Figure 16: Energy dissipation capacities of the columns.

CONCLUSIONS

In this study, the effects of using anchored and embedded AFRP laminate longitudinal reinforcement for the flexural retrofit of substandard RC columns were investigated. The tests were conducted under constant axial load and reversed cyclic lateral loads. Based on the experimental and theoretical studies, the following conclusions/observations can be listed:

The reference column reached its theoretical flexural capacity, and exhibited a ductile behavior as foreseen during design.

The retrofitted columns (LAM, LAM-LAM, LAM-PB) failed due to the rupture of AFRP reinforcement at around 3% drift ratio. At this drift ratio, the enhancement in strength was around 38%, 90% and 41% for the columns LAM, LAM-LAM, and LAM-PB, respectively, with respect to the reference column. It should be noted that drift ratios exceeding 3% may only be expected if substandard RC structures are subjected to severe earthquakes. Nevertheless, after AFRP reinforcements were fractured at around 3% drift ratio, the columns still resisted considerable lateral forces due to contribution of steel reinforcing bars until extremely large drift ratios (~6% to 8%). The column LAM-LAM exhibited a remarkably superior performance with respect to the reference and the other retrofitted columns due to additional AFRP anchorage reinforcement. When compared with the column LAM, the proposed anchorage type used for the column LAM-PB, limited the residual plastic deformations, while did not affect the lateral load capacity adversely. Therefore, this type of anchorage system can be utilized to limit the residual displacement and corresponding damage. It was observed that the predicted results through theoretical calculations satisfactorily captured the tested results by means of lateral load and deformation capacities, particularly, when 43% of the design rupture strain of FRP reinforcement was considered as the rupture strain of the FRP reinforcement during theoretical calculations.

The AFRP reinforcement of the retrofitted columns ruptured at the interface of the column and the footing. The experimentally measured rupture strain of AFRP reinforcement was observed to be approximately 43% of design rupture strain of FRP reinforcement. Therefore, the strain reduction factor (70%) recommended by ACI 440-2R-08 [5] for NSM applications under monotonic loading conditions may lead to unconservative predictions of lateral load-deformation capacity for the proposed retrofitting technique in this study under cyclic loading conditions. Clearly, further investigations are also needed to accurately identify the theoretical seismic behavior of the columns retrofitted using NSM techniques under cyclic loading conditions.

It can be inferred from the results of this study that the proposed retrofitting method is promising for improving the seismic performance of substandard columns built with low quality concrete under reversed cyclic flexure and axial force. It should be noted that these conclusions are mainly based on the test results of this study, therefore, they are limited to the ranges of test parameters and details considered in this study.

ACKNOWLEDGMENTS

The financial support of the ITU Scientific Research Department (Scientific Research Project No:326024), BASF and ART-YOL Companies and the assistance of staff of Structural and Earthquake Engineering Laboratory and Building Materials Laboratory of Istanbul Technical University are acknowledged gratefully.

CONFLICTS OF INTEREST

The authors declare no conflict of interest.

REFERENCES

1. CEB-FIB. Externally Bonded FRP Reinforcement for RC Structures; Technical report for International Federation for Structural Concrete: Lausanne, Switzerland, 2001.
2. Pessiki, S.; Harries, K.A.; Kestner, J.T.; Sause, R.; Ricles, J.M. Axial behavior of reinforced concrete columns confined with FRP jackets. J. Compos. Constr. 2001, 5, 237–245.
3. Bakis, C.E.; Bank, L.C.; Brown, V.L.; Cosenza, E.; Davalos, J.F.; Lesko, J.J.; Machida, A.; Rizkalla, S.H.; Triantafillou, T.C. Fiber-reinforced polymer composites for construction-State-of-the-Art Review. J. Compos. Constr. 2002, 6, 73–87.
4. Lam, L.; Teng, J.G. Design-oriented stress-strain model for FRP-confined concrete. Constr. Build. Mater. 2003, 17, 471–489.
5. ACI 440.2R-08. Guide for design and construction of externally bonded FRP systems for strengthening concrete structures. American Concrete Institute 440: Farmington Hills, MI, USA, 2008.
6. Bank, L.C. Progressive failure and ductility of FRP composites for construction: Review. J. Compos. Constr. 2013, 17, 406–419.
7. Bennitz, A.; Schmidth, J.W.; Nilimaa, J.; Taljsten, B.; Goltermann, P.; Ravn, D.L. Reinforced concrete T-beams externally prestressed with unbonded carbon fiber-reinforced polymer tendons. Struct. J. 2012, 109, 521–530.
8. Lou, T.; Lopes, S.M.R.; Lopes, A. Numerical analysis of behaviour of concrete beams with external FRP tendons. Constr. Build. Mater. 2012, 35, 970–978.
9. Lou, T.; Lopes, S.M.R.; Lopes, A. External CFRP tendon members: Secondary reactions and moment

redistribution.Compos. Eng. 2014, 57, 250–261.

10. Seible, F.; Priestley, M.J.; Hegemier, G.A.; Innamorato, D. Seismic retrofit of RC columns with continuous carbon fiber jackets. J. Compos. Constr. 1997, 1, 52–62.
11. Sheikh, S.; Yau, G. Seismic behavior of concrete columns confined with steel and fiber-reinforced polymers. Struct. J.2002, 99, 72–80.
12. Antonopoulos, C.P.; Triantafillou, T.C. Experimental investigation of FRP-strengthened RC beam-column joints. J. Compos. Constr. 2003, 7, 39–49.
13. Iacobucci, R.D.; Sheikh, A.S.; Bayrak, O. Retrofit of square concrete columns with carbon fiber-reinforced polymer for seismic resistance. Struct. J. 2003, 100, 785–794.
14. Xiao, Y. Applications of FRP composites in concrete columns. Adv. Struct. Eng. 2004, 7, 335–343.
15. Prota, A.; Nanni, A.; Manfredi, G.; Cosenza, E. Selective upgrade of underdesigned reinforced concrete beam-column joints using carbon fiber polymers. Struct. J. 2004, 101, 699–707.
16. Bousias, S.N.; Triantafillou, T.C.; Fardis, M.N.; Spathis, L.; O'Regan, B.A. Fiber-reinforced polymer retrofitting of rectangular reinforced concrete columns with or without corrosion. Struct. J. 2004, 101, 512–520.
17. Chang, S.; Li, Y.; Loh, C. Experimental study of seismic behaviors of as-built and carbon fiber reinforced plastics repaired reinforced concrete bridge columns. J. Bridge Eng. 2004, 9, 391–402.
18. Memon, M.S.; Sheikh, S.A. Seismic resistance of square concrete columns retrofitted with glass fiber-reinforced polymer. Struct. J. 2005, 102, 774–783.
19. Tsonos, A.G. Effectiveness of CFRP-jackets in post-earthquake and pre-earthquake retrofitting of beam-column subassemblages. Struct. Eng. Mech. 2007, 27, 393–408.
20. Ilki, A.; Demir, C.; Bedirhanoglu, I.; Kumbasar, N. Seismic retrofit of brittle and low-strength RC columns using fiber reinforced polymer and cementitious composites. Adv. Struct. Eng. 2009, 12, 325–347.
21. Garcia, R.; Hajirasouliha, I.; Pilakoutas, K. Seismic behaviour of deficient RC frames strengthened with CFRP composites. Eng. Struct. 2010, 32, 3075–3085.
22. Ilki, A.; Bedirhanoglu, I.; Kumbasar, N. Behavior of FRP retrofitted joints built with plain bars and low-strength concrete. J. Compos. Constr. 2011, 15, 312–326.
23. Wang, Z.; Wang, D.; Smith, S.T.; Lu, D. CFRP-confined square RC columns. I: Experimental investigation. J. Compos. Constr. 2011, 16, 150–160.
24. Jirawattanasomkul, T.; Zhang, D.W.; Ueda, T. Prediction of the post-peak behavior of reinforced concrete columns with and without FRP-jacketing. Eng. Struct. 2013, 56, 1511–1526.
25. Parvin, A.; Brighton, D. FRP composites strengthening of concrete columns under various loading conditions. Polymers 2014, 6, 1040–1056.
26. Demir, C.; Darılmaz, K.; Ilki, A. Cyclic stress–strain relationships of FRP confined concrete members. Arabian J. Sci. Eng. 2015, 40, 363–379.
27. Nanni, A.; Alkhrdaji, T.; Chen, G.; Barker, M.; Yang, X.; Mayo, R. Overview of testing to failure program of a highway bridge strengthened with FRP composites. In Proceedings of Fourth International Symposium on FRP for Reinforcement of Concrete Structures, Baltimore, MD, USA, November 1999; pp. 69–80.
28. De Lorenzis, L.; Nanni, A. Bond between near surface mounted fiber reinforced polymer rods and concrete in structural strengthening. Struct. J. 2002, 99, 123–132.
29. De Lorenzis, L.; Lundgren, K.; Rizzo, A. Anchorage length of near-surface mounted fiber reinforced polymer bars for concrete strengthening-experimental investigation and numerical modelling. Struct. J. 2004, 101, 269–278.
30. Hassan, T.K.; Rizkalla, S.H. Bond mechanism of near-surface-mounted fiber-reinforced polymer bars for flexural strengthening of concrete structures. Struct. J. 2004, 101, 830–839.
31. El-Hacha, R.; Rizkalla, S.H. Near-surface-mounted fiber-reinforced polymer reinforcements for flexural strengthening of concrete structures. Struct. J. 2004, 101, 717–726.
32. Barros, J.A.O.; Dias, S.J.E. Near surface mounted CFRP laminates for shear strengthening of concrete beams. Cem. Concr. Compos. 2006, 28, 276–292.
33. Nordin, H.; Täljsten, B. Concrete beams strengthened with prestressed near surface mounted CFRP. J. Compos. Constr.2006, 10, 60–68.
34. Liu, I.S.T.; Oehlers, D.J.; Seracino, R. Tests on the ductility of reinforced concrete beams retrofitted with FRP and steel near-surface mounted plates. J. Compos. Constr. 2006, 10, 106–114.
35. Teng, J.G.; de Lorenzis, L.; Wang, B.; Li, R.; Wong, T.N.; Lam, L. Debonding failures of RC beams strengthened with near surface mounted CFRP strips. J. Compos. Constr. 2006, 10, 92–105.
36. De Lorenzis, L.; Teng, J.G. Near-surface mounted FRP reinforcement: An emerging technique for strengthening structures. Compos. Eng. 2007, 38, 119–143.
37. Seracino, R.; Raizal Saifulnaz, M.R.; Oehlers, D.J. Generic debonding resistance of EB and NSM plate-to-concrete joints. J. Compos. Constr. 2007, 11, 62–70.
38. Castro, E.K.; Melo, G.S.; Nagato, Y. Flexural strengthening of RC "t" beams with near surface mounted (NSM) FRP reinforcements. In Proceedings of International Symposium on FRP Reinforcement of Concrete Structures (FRPRCS-8), Patras, Greece, 16–18 July 2007.
39. Chikh, N.; Foret, G.; Bousalem, B.; Firas, S. Analysis of reinforced concrete members strengthened with near-surface-mounted CFRP. In Proceedings of 11DBMC Conference on Durability of Building Materials and Components, Istanbul, Turkey, 11–14 May 2008.
40. El-Maaddawy, T.A.; El-Dieb, A.S. Near-surface-mounted composite system for repair and strengthening of reinforced concrete columns subjected to axial load and biaxial bending. J. Compos. Constr. 2011, 15, 602–

614.

41. Sharaky, I.A.; Torres, L.; Comas, J.; Barris, C. Flexural response of reinforced concrete (RC) beams strengthened with near surface mounted (NSM) fibre reinforced polymer (FRP) bars. Compos. Struct. 2013, 109, 8–22.
42. Bilotta, A.; Ceroni, F.; Barros, J.; Costa, I.; Palmieri, A.; Szabó, Z.; Nigro, E.; Matthys, S.; Balazs, G.; Pecce, M. Bond of NSM FRP-strengthened concrete: Round robin test initiative. J. Compos. Constr. 2015.
43. Sena Cruz, J.M.; Barros, J.A.O.; Gettu, R.; Azevedo, A.F.M. Bond behavior of near surface mounted CFRP laminate strips under monotonic and cyclic loading. J. Compos. Constr. 2006, 10, 295–303.
44. Badawi, M.; Soudki, K. Fatigue behavior of RC beams strengthened with NSM CFRP rods. J. Compos. Constr. 2009, 13, 415–421.
45. Ceroni, F. Experimental performances of RC beams strengthened with FRP materials. Constr. Build. Mater. 2010, 24, 1547–1559.
46. Ilki, A.; Kumbasar, N. The behavior of damaged reinforced concrete members strengthened by carbon fiber reinforced polymer composites. Tech. J. Turk. Chamb. Civil Eng. 2002, 13, 2597–2616, (In Turkish with English abstract).
47. Bournas, D.A.; Triantafillou, T.C. Flexural strengthening of reinforced concrete columns with near surface mounted FRP or stainless steel. Struct. J. 2009, 106, 495–505.
48. Goksu, C.; Polat, A.; Ilki, A. Attempt for seismic retrofit of existing sub-standard RC members under reversed cyclic flexural effects. J. Compos. Constr. 2012, 16, 286–299.
49. Fahmy, M.F.M.; Wu, Z. Retrofitting of existing RC square bridge columns using basalt FRP rebars. Available online: http://www.iifc-hq.org/proceedings/CICE_2012/02_FRP%20in%20Seismic%20Retrofitting/02_495_Fahmy,Wu_RETROFITTING%20OF%20EXISTING%20RC%20SQUARE.pdf (accessed on 27 November 2015).
50. Vrettos, I.; Kefela, E.; Triantafillou, T.C. Innovative flexural strengthening of reinforced concrete columns using carbon-fiber anchors. Struct. J. 2013, 110, 63–70.
51. Li, X.; Lv, H.L.; Zhang, G.C.; Sha, S.Y.; Zhou, S.C. Seismic retrofitting of rectangular reinforced concrete columns using fiber composites for enhanced flexural strength. J. Reinf. Plast. Compos. 2013, 32, 619–630.
52. Goksu, C.; Uzunhasanoglu, A.; Seyhan, E.C.; Ilki, A. Seismic retrofit of sub-standard RC columns with embedded aramid FRP reinforcement. In Proceedings of SMAR 2013 Second Conference on Smart Monitoring, Assessment and Rehabilitation of Civil Structures, Istanbul, Turkey, 9–11 September 2013.
53. Faustino, P.; Chastre, C. Flexural strengthening of columns with CFRP composites and stainless steel: Cyclic behavior.J. Struct. Eng. 2015.
54. CNR-DT 200/2004. Guide for the design and construction of externally bonded FRP systems for strengthening existing structures. Available online: http://www.cnr.it/documenti/norme/IstruzioniCNR_DT200_2004_eng.pdf (accessed on 27 November 2015).
55. Akkaya, Y.; Demir, N. Durability issues in existing buildings and new infrastructure projects. In Proceedings of Joint Symposium on Concrete Engineering (JSCE), Istanbul, Turkey, 9 June 2010.
56. Eurocode 8-Part 3. Assessment and retrofitting of buildings. European Committee for Standardization: Brussels, Belgium, 2005.
57. TSDC. Regulations for buildings to be constructed in earthquake prone areas. Turkish Seismic Design Code: Ankara, Turkey, 2007.
58. Eurocode 2: Design of concrete structures. Part 1. General rules and rules for buildings. European Committee for Standardization: Brussels, Belgium, 2004.
59. TS EN 12390–3. Testing hardened concrete-Part 3: Compressive strength of test specimens. Turkish Standards Institute: Ankara, Turkey, 2003.
60. TS 708. Steel bars for concrete. Turkish Standards Institute: Ankara, Turkey, 1996.
61. Part 3: Structural and non-structural repair. In Products and Systems for the Protection and Repair of Concrete Structures. Definitions, Requirements, Quality Control and Evaluation of Conformity; BS EN 1504–3:2005; British Standard Institute: London, UK, 2006.
62. Part 4: Structural bonding. In Products and Systems for the Protection and Repair of Concrete Structures. Definitions, Requirements, Quality Control and Evaluation of Conformity; EN 1504–4:2004; British Standard Institute: London, UK, 2004.
63. TS-EN 196–1. Methods of testing cement-Part 1: Determination of strength. Turkish Standards Institute: Ankara, Turkey, 2009.
64. ASTM D3039. Standard test method tensile properties of polymer matrix composite materials. ASTM International: West Conshohocken, PA, USA, 2014.
65. Matamoros, A.B.; Sozen, M.A. Drift limits of high-strength concrete columns subjected to load reversals. J. Struct. Eng.2003, 129, 297–313.
66. Bournas, D.A.; Triantafillou, T.C. Bond strength of lap spliced bars in concrete confined with composite jackets. J. Compos. Constr. 2011, 15, 156–167.
67. XTRACT 3.0.8. Cross-sectional structural analysis of components. TRC: Rancho Cordova, CA, USA, 2007.
68. Ilki, A.; Peker, O.; Karamuk, E.; Demir, C.; Kumbasar, N. FRP retrofit of low and medium strength circular and rectangular reinforced concrete columns. J. Mater. Civil Eng. 2008, 20, 169–188.
69. ACI-318M-14. Metric building code requirements for structural concrete and commentary. American Concrete Institute 318: Farmington Hills, MI, USA, 2014.
70. ACI-355–1R-91. State-of-the-art report on anchorage to concrete. American Concrete Institute 355: Farmington Hills, MI, USA, 1991.

CHAPTER 6

DETERMINATION OF THE IMPACT BEHAVIOR OF CONCRETE AND REINFORCED CONCRETE BEAMS

Engin Gücüyen, R. Tuğrul Erdem, Erkan Kantar and Muhiddin Bağcı

Keywords: Impact Loads, Concrete and Reinforced Concrete Beam, ABAQUS

INTRODUCTION

Structures and structural members are under the effect of various loads in their economic lives. Structural members as column, beam, slab and wall are subjected to earthquake and wind forces as well as person, property, machine and their self-loads. The structure is generally under the effect of two main loads as static and dynamic loads. While static loads are permanent ones, dynamic loads occur suddenly and effect structures for a length of time. Since, destructive damages may happen due to dynamic loads, the response of the structure against these loads shall be estimated correctly. Impact effects are the changes on mechanical properties of members due to stresses occurring at the strike moment between objects. Car, helicopter, plane strikes, ship strikes to abutments, water structures and petrol platforms, explosions in military establishments, projectile and missile strikes, crane accidents while carrying members and stone and rock falls to structures located in roadsides are the examples of impact incidents.

Impact tests on structural members are based on the investigation of impact loads by using different test devices and determination of the damages due to weight falling testing apparatuses. While standardization studies about impact tests are proceeding, there is not a standard in respect of test methods and procedures yet. However, a big progress is made in studies about testing apparatuses and impact limits in ASTM (American Society for Testing Materials) E23 Standard [1]. Performed studies about impact effect are determining the mechanical properties of members under high stress values. On the other hand, testing apparatuses are designed to overcome the deficiencies occurred by dynamic effects. Impact reactions, crack failures, energy capacities and damage situations are determined by the help of these studies in the literature [2-11]. In this study, concrete and reinforced concrete beams having 150x150x710 mm sizes are produced in the laboratory. Afterwards, these two beams are tested under the effect of impact loads. Falling numbers, crack progresses, time varying acceleration, velocity, displacement and impact force values are obtained after tests. Furthermore, testing apparatus is modeled in ABAQUS [12] finite elements analysis program. Finally, test and analysis results are compared and suggestions are proposed.

TESTING APPARATUS

There is not an available standard about testing apparatus for impact tests yet. For this reason, testing apparatus and measuring devices are decided after literature review. The studies show that weight falling testing apparatus is the best way to investigate the shear, bending failures and deformations in structural members under impact effect. The testing apparatus for this study is designed in accordance with this purpose. This apparatus is based on free falling of a steel hammer and striking to different sized members from various heights. The devices as accelerometers, ring force sensor, data logger, connection wires, optic photocells and a computer are used in the tests as well as testing apparatus. Key features of weight falling devices are investigated after studying the performed articles and the apparatus is created. With this apparatus, different sized and

Citation: Engin Gücüyen, R. Tuğrul Erdem , Erkan Kantar and Muhiddin Bağcı, Determination of the Impact Behavior of Concrete and Reinforced Concrete Beams, doi:10.3390/mca18030502.

weighted members can be tested under impact loads. In this way, structural members can be compared according to various parameters and support conditions. It is stated that eccentricity is one of the most important factors effecting the results in tests. Secondary effects appear if the hammer does not hit the center of gravity of the member. To avoid this situation, eccentricity is set to zero while producing the testing apparatus. Working mechanism of the apparatus is given in Figure 1.

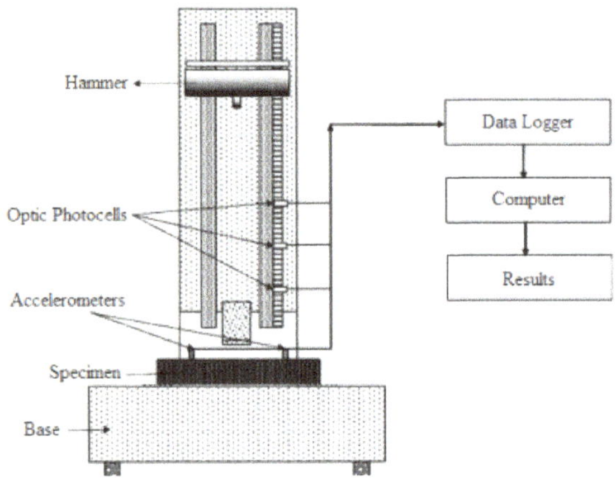

Figure 1: Testing apparatus

The basic approach in weight falling test apparatus is changing the potential energy to kinetic one at impact moment. This energy conversion causes rapid stress concentration on the members. Loss of energy during free falling is gained by the member. 1000x1000x70 mm sized apparatus base is made of steel plates and it performs as an absorber. Optic photocells are located on the apparatus and they measure time from the first movement of the hammer to the impact moment.

EXPERIMENTAL STUDY

150x150x710 mm sized concrete and reinforced concrete beams are produced in the experimental part of the study. Concrete class is determined as C30 after testing the strength of cubic concrete samples at the end of 28 days. While longitudinal reinforcements are 2ϕ6 on the top, they are 2ϕ8 at the bottom of the beam. 5 stirrups are used in the section and their diameters are 8 mm. Concrete covers are taken as 20 mm. Height of the hammer is 80 mm from the beam and weight of the hammer is 10 kg. Beam section and reinforcements are given in Figure 2.

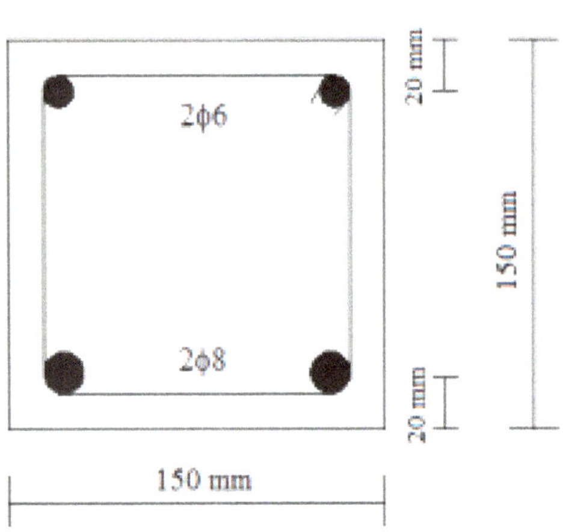

Figure 2: Beam section and reinforcements

Beam molds are manufactured from plywood forms. First of all, reinforcements and concrete covers are located and concrete is poured into the molds. Afterwards vibration is performed for a while. Finally, concrete surfaces are straightened with a trowel. Production of beam members is seen in Figure3.

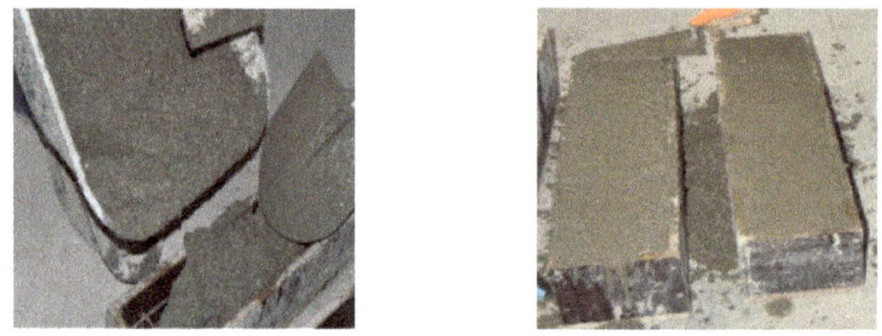

Figure 3: Beam members

Beams are located into curing pool for 28 days to complete the setting duration. Then, the beams are painted by using a ceiling paint to observe the crack and damage failures much better. After determining the accelerometers places on the beams, 4 holes have been made by a slow speed drill without causing any damage on the members. Brass devices are placed in the holes by steel dowels. These special devices transfer all kinds of effects to the members almost without any loss. In addition, brass devices restrain the movements of the accelerometers. In impact tests, point loading is applied by steel hammer to the member. However, the hammer does not fall to the member directly. Because, when it hits the member, point loading appears and it makes a hole on the member. In addition, collapsing occurs and it makes impossible to take any measurement from the member. So, a steel plate shall be used to obstruct this situation. The inner reaction is distributed along the member. Rubber layer delays the occurrence of inner peak load. By this way, the member gains time to completely take the load applied by steel hammer. A neoprene rubber layer takes part between the steel plate and the beam member in the tests. Geometrical placement of the steel plate, rubber and accelerometers on the beam are given in Figure 4.

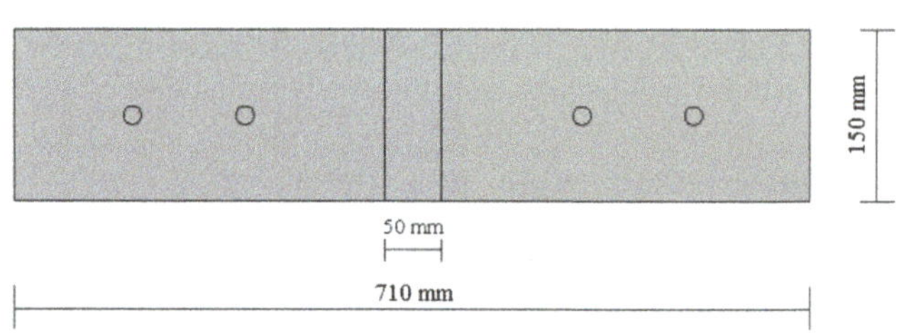

Figure 4: Testing equipment on the beam

After support conditions are provided for the beam, accelerometers and force sensor are placed. All measurement devices are connected to the data logger with particular cables. Strict attention is paid to prevent the cables from any damage. Finally, data logger is connected to computer to transfer the test information. Prepared concrete beam member is given as an example in Figure 5.

Figure 5: Concrete beam member on testing apparatus

Friction factor which effects the results during tests is almost set to zero. The connection between the hammer and shafts is provided by disk wheeled designed members. Optic photocells placed on the apparatus make available to see the time of free falling movement from beginning to the end. By this way, velocity of the hammer can be calculated. Moreover, total falling number is seen on the screen. During the tests, data obtained from accelerometers is saved by the data logger and transferred to the computer. The results are evaluated by using special software. Concrete and reinforced concrete beam members are tested until collapsing from 80 cm height and until the effect of 10 kg weighted hammer. Inner effects are determined by using accelerometers for each fall. Time varying acceleration, velocity, displacement and impact load values of the beams are investigated.

FINITE ELEMENTS ANALYSIS

In this part, obtained experimental results are compared with ABAQUS finite elements analysis program which is widely used in academic studies by researchers. For this purpose, testing apparatus and beam members are completely modeled in the program as seen in Figure 6.

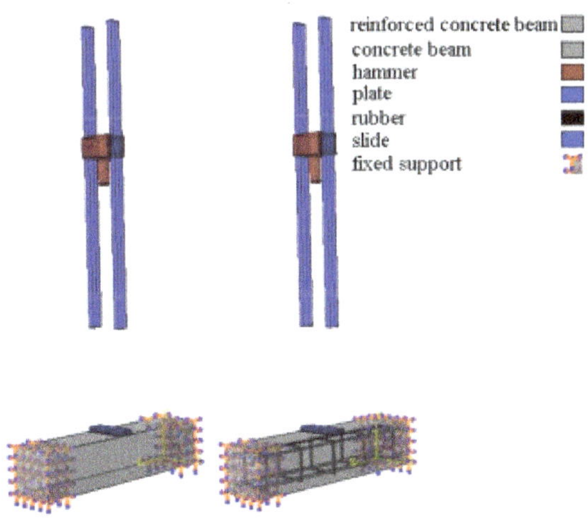

Figure 6: ABAQUS models of beams

Quadratic tetrahedral elements of type C3D10M are utilized in all parts of the models. 32944 number of nodes and 21856 number of elements for hammer, 117450 number of nodes and 81055 number of elements for concrete beam, 2x529 number of nodes and 2x240 number of elements for slides, 1038 number of nodes and 461 number of elements for rubber, 1623 number of nodes and 900 number of elements for plate, 185821 number of nodes and 130312 number of elements for reinforced concrete beam and 9176 number of nodes and 3374 number of elements for rubber are used in the numerical analysis. Finally, maximum experimental and analysis results for concrete and reinforced concrete beams after first falling are given in the following figure

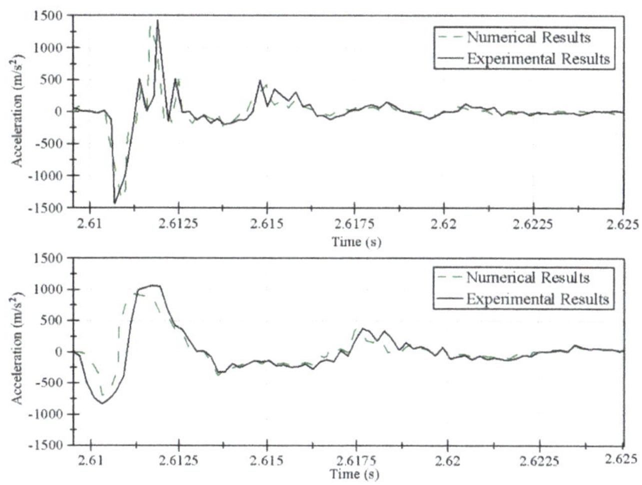

Figure 7: Maximum acceleration-time values for beams

Top diagram belongs to concrete beam in Figure 7. Acceleration values for concrete beam are bigger than reinforced concrete one as given in Figure 7.

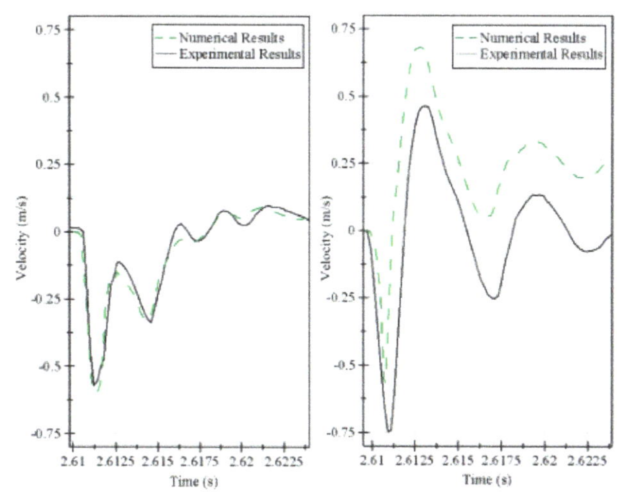

Figure 8: Velocity-time values for beams

Left diagram in Figure 8 pertains to concrete beam. Velocity values are obtained by integrating acceleration values and the difference between numerical and experimental results are bigger for reinforced concrete beam.

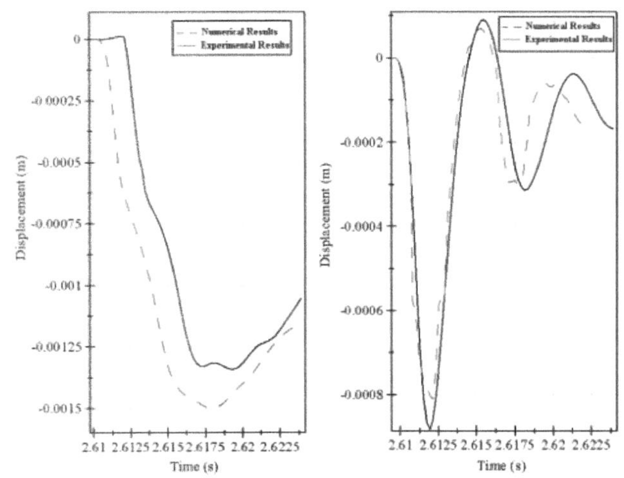

Figure 9: Displacement-time values for beams

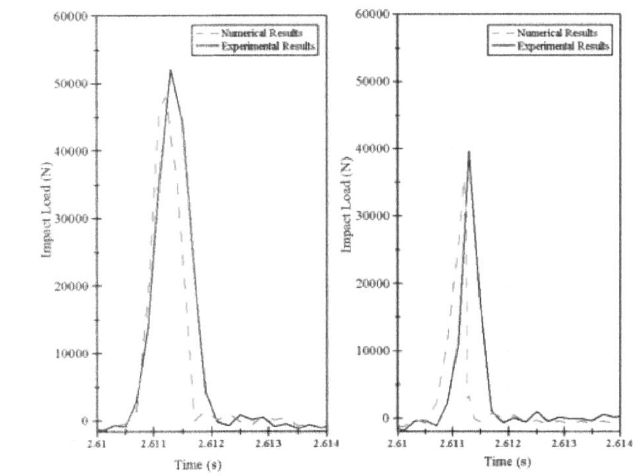

Figure 10: Impact load-time values for beams

Left diagrams in both Figure 9 and 10 belong to concrete beam. Displacement and impact load values are bigger for concrete beam.

RESULTS

In this study, concrete and reinforced concrete beams which have150x150x710 mm sizes are tested under the effect of impact loads. For this purpose, the beams are produced in the first place. Afterwards, these beams are prepared for tests. Testing apparatus which is designed to investigate impact effects is used for experimental part of the study. There are four accelerometers are placed on the members and the results are transferred to the computer by data logger. In the finite elements analysis part, ABAQUS is used for solutions. Quadratic tetrahedral elements of type C3D10M are used in all parts of the models. Beam members, testing apparatus, rubber and steel plate are modeled in the program. Since stress transfer and strength values are transferred in a short span of time, analysis and time steps are studiously determined. Maximum values that are after performing nonlinear dynamic analyses are compared with experimental results. After acceleration values are obtained velocity and displacements values are calculated by integration operations The beams whose compressive strengths are 30 MPa, are tested under the effect of 10 kg hammer from 80 mm height. Acceleration-time, velocity-time, displacement time and impact load-time diagrams are obtained for each beam. It is seen that bigger values are obtained for concrete beam in both analyses and a good relationship is obtained between test and finite elements analysis results. Due to the longitudinal reinforcements and stirrups, there are differences in falling numbers. More fallings are needed for reinforced concrete beam. Besides that, since reinforced concrete beam is more ductile, more rebounds are observed after each hit in experimental and finite elements analyses. Rebound numbers are bigger in the first hits. While damages and cracks on the beams increase, rebound numbers on the members decrease. This situation shows that, damages and inner cracks of the beam gradually increase after each falling. First cracks occur in the front face of the beams and start from top to bottom. The cracks on the surfaces occur due to falling numbers. Finally, this study can be improved by testing different sized concrete and reinforced concrete members for different parameters.

REFERENCES

1. T. A. Siewert, M.P. Manahan, M. P. McCowan, J. M. Holt, F. J. Marsh and E. A. Ruth, The history and importance of impact testing, Pendulum Impact Testing: A Century of Progress ASTM STP 1380, West Conshohocken PA, 3-15, 1999.
2. E. Kantar, R.T. Erdem and Ö. Anıl, Nonlinear finite element analysis of impact behaviour of concrete beam, The 1st International Symposium on Computing in Science & Engineering, 892-898, 2010.
3. N. P. Banthia, Impact resistance of concrete, PhD Thesis, The University of British Columbia Department of Civil Engineering, 1987.
4. M. A. Erki and U. Meier, Impact loading of concrete beams externally strengthened with cfrp laminates, Journal of Composites for Construction 3(3), 117-124, 1999.
5. S. Saatci and F. J. Vecchio, Effects of shear mechanisms on impact behavior of reinforced concrete beams, ACI Structural Journal 106(1), 78-86, 2009.
6. S. Saatci and F. J. Vecchio, Nonlinear finite element modeling of reinforced concrete structures under impact loads, ACI Structural Journal 106(5), 717-725, 2009.
7. S. Tachibana, H. Masuya and S. Nakamura, Experimental study on the impact behavior and performance of reinforced concrete beam with some absorbing materials, 31st Conference on OUR WORLD IN CONCRETE & STRUCTURES, 16-17 August 2006.
8. Y. Sawamoto, H. Tsubota, Y. Kasai, N. Koshika and H, Morikawa, Analytical studies on local damage to reinforced concrete structures under impact loading by discrete element method, Nuclear Engineering and Design 179, 157–177, 1998.
9. A. Q. Bhatti, N. Kishi and H. Mikami, An applicability of dynamic response analysis of shearfailure type RC beams with lightweight aggregate concrete under falling-weight impact loading Materials and Structures 44, 221–231, 2011.
10. V. Korotkov, D. Poprygin, K. Ilin and S. Ryzhov, Determination of dynamic reaction in concrete floors

of civil structures of nuclear power plant in accidental drops of heavy objects, Abaqus Users' Conference, 399–408, 2004.

11. E. Kantar and Ö. Anıl, Low velocity impact behavior of concrete beam strengthened with CFRP strip, Steel and Composite Structures 12(3), 207-230, 2012.
12. ABAQUS User Manual, Version 6.10, 2010.

CHAPTER 7

NEURAL NETWORK MODEL FOR MOMENT-CURVATURE RELATIONSHIP OF REINFORCED CONCRETE SECTIONS

Muhiddin Bağcı

Keywords: Moment-Curvature, Neutral Network

INTRODUCTION

The moment curvature for a cross-section envelope describes the changes in force capacity with deformation during a nonlinear analysis. The relationship between moment and curvature demonstrates the strength, ductility, energy dissipation capacity and rigidity of the section under question. To obtain the moment-curvature relationship of reinforced concrete section, various researchers have investigated using different models. Parviz [1] used firstly the filament method. Ersoy and Özcebe [2] presented a computer program to determine moment-curvature relationships of confined concrete sections. Artificial neural networks (ANN) are one of the artificial intelligence (AI) applications which have recently been used widely to model some of human interesting activities in many areas of science and engineering. The generalized delta rule algorithm of artificial neural networks is employed to predict the flexural behavior of Steel Fibre Reinforced Concrete (SFRC) T-beams using a computer program developed using C++ by Patodi and Purani [3]. For some other examples of ANN applications in structural analysis, the reader is referred to Jadid et al. [4] ; Berke et al. [5] ; Lee et al. [6]; Avdelas et al [7]; Abdalla and Stavroulakis [8]; Karlık et al. [9]. As far as structural analysis and design are concerned, Hajela et al. [10] used BPNN to represent the forcedisplacement relationship in static structural analysis. Jenkins considered the application of neural nets to approximate structural analysis and especially to a comparatively simple structure [11]. Mukherjee et al. [12] mapped the relationship between the slenderness ration, the modulus of elasticity and the buckling load for columns. As the input taken directly from the experimental results, factors affecting the buckling load of columns are automatically incorporated in the model to a great extent. Adeli defined the learning parameters as a function of iteration number of the training [13].

In this study, the behavior values of reinforced concrete sections subjected to flexure and axial load were obtained by using an analytical solution named the filament model, and then the required data for the network training were prepared. To obtain the behavior of confined concrete, several data points were used in training a multi-layer, feed-forward and back propagation artificial neural network (ANN) algorithm. The behavior values were calculated using the neural network and were compared with those obtained from the analytical results. Finally, the reliability of the ANN solution was validated by comparing experimental values with modeled values.

MATERIAL MODELS

Moment-curvature analysis for a reinforced concrete section, indicating the available flexural strength and ductility can be carried out provided that the stress-strain relationships for the concrete and steel reinforcements are known. Typical stress–strain curves for concrete are shown in Fig. 1.

Citation: Muhiddin Bağcı, Neural Network Model for Moment-Curvature Relationship of Reinforced Concrete Sections, doi:10.3390/mca15010066.

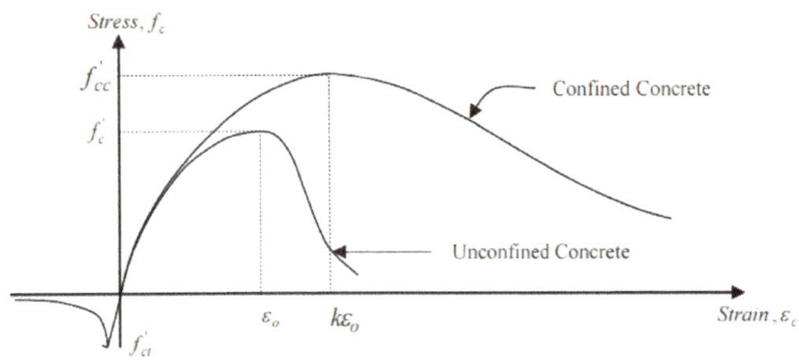

Figure 1: Typical concrete stress-strain curve

The material model for concrete used in this analysis is based on a model suggested by Modified Kent–Park [14]. This model takes into account the different stress-strain curves for unconfined and confined concrete as shown in Fig. 2. The general shape of curve is modeled by a second degree parabola for the ascending branch up to the maximum stress which corresponds to strain level of 0.002 and linear horizontal part leading to the ultimate strain.

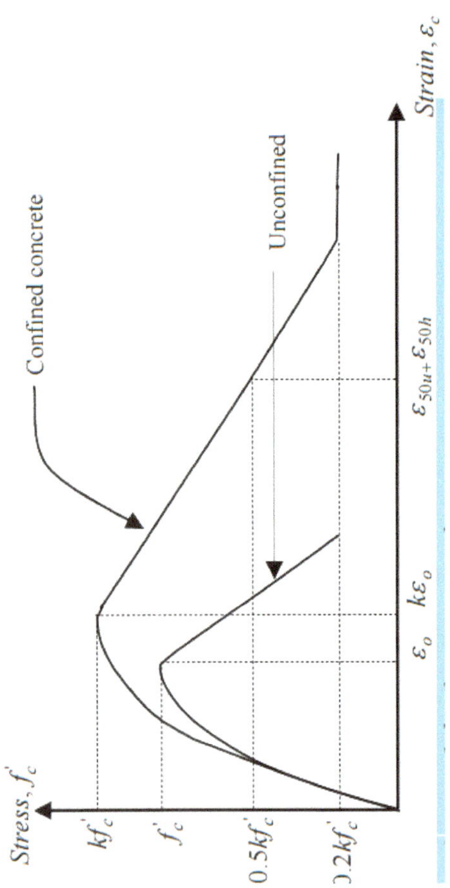

Figure 2: Modified Kent-Park concrete stress-strain curve.

In the ascending branch, the concrete compressive stress, c_f, at a given strain, c_ε is given by:

$$f_c = kf'_c\left[2\left(\frac{\varepsilon_c}{\varepsilon_o k}\right)-\left(\frac{\varepsilon_c}{\varepsilon_o k}\right)^2\right]; \qquad \varepsilon_c \le k\varepsilon_o \qquad (1)$$

where f'_c is the concrete compressive strength and ε_0 is concrete strain at the maximum stress assumed to be 0.002. It can be seen from Eq. (1) that concrete reaches a maximum stress of kf'_c at a strain of $k\,\varepsilon_0$. k is a factor which accounts for the strength increase due to the confinement. The value of the parameter k is obtained from;

$$k = 1 + \frac{\rho_s f_{yh}}{f'_c} \qquad (2)$$

where f_{yh} is the yield strength of stirrups, and ρ_s is the ratio of the volume of hoop reinforcement to the volume of concrete core measured to the outside of stirrups. The descending branch of the stress–strain curve is described as follows;

$$f_c = kf'_c\left[1 - Z_m(\varepsilon_c - \varepsilon_o k)\right] \ge 0.2f'_c; \qquad \varepsilon_c > k\varepsilon_o \qquad (3)$$

where

$$Z_m = \frac{0.5}{\varepsilon_{50u} + \varepsilon_{50h} - \varepsilon_o k} \qquad (4)$$

$$\varepsilon_{50u} = \frac{3 + 0.285 f'_c}{1.42 f'_c - 1000} \qquad (5)$$

$$\varepsilon_{50h} = 0.75\rho_s\sqrt{b'' / s} \qquad (6)$$

The term ε_{50u} defines the slope of the falling branch of the unconfined concrete, which identifies the strain at which the stress has fallen to $0.5kf'_c$. The ε_{50h} is the additional ductility in concrete which is provided by transverse reinforcement. The "b" is the width of the confined core measured to the outside of stirrups, and s is the center to center spacing of stirrups or hoop sets. At large strains, the value of compressive stress is kept constant at $0.2kf'_c$ to account for the ability of concrete to support load at large strains.

When a reinforced concrete member is subjected to tensile strains less than the cracking strain of concrete, the stress-strain relationship is approximately linear. A bilinear model is used for concrete in tension. Rüsch [15] recommends the following relationship:

$$f_{c\,tention} = -E_{ct}\varepsilon_{ct}; \qquad 0 \le \varepsilon_t \le \varepsilon_{ct} \qquad (7)$$

$$f_{ctension} = f_{ct} - 5000 f_{ct}(\varepsilon_{ct} - 1000); \qquad \varepsilon_t \ge \varepsilon_{ct} \qquad (8)$$

where E_{ct} is the modulus of elasticity in tension, ε_{ct} is the tensile strain. c_1 is taken as 0.5, ε_{cto} as 0.0001 and ε_{ctu} as 0.0002. The relationships given in Eq. (7) and Eq.(8) are shown in Fig. 3.

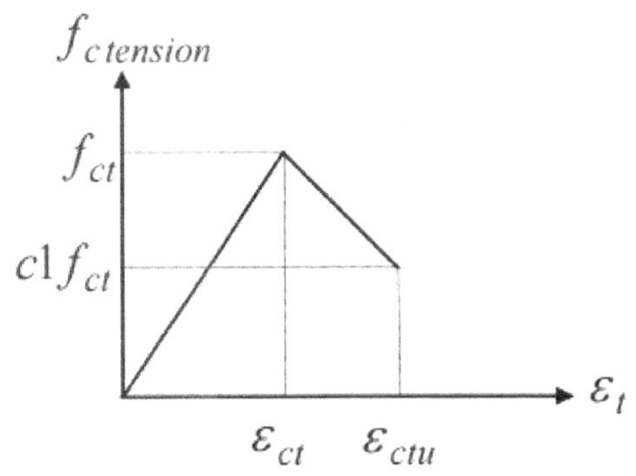

Figure 3: Tension model for RC

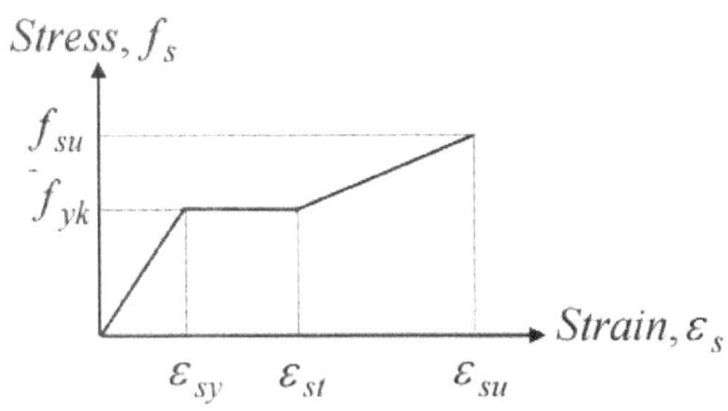

Figure 4: Idealized reinforcement stress-strain curve

A sample reinforcement stress–strain relationship is shown in Fig.4. The constitutive model used for steel reinforcement is a simple elastic-plastic three linear model. These are the linear segment, the yield plateau, and the strain hardening segment. There are some other parameters of the reinforcement stress–strain relationship, such as the reinforcement yield strength , f_{yk}, the ultimate reinforcement strength, f_{su} , the reinforcement yield strain, ε_{sy}, hardening strain , ε_{st}, the ultimate strain, ε_{su} , and modulus of elasticity , E_s .

METHOD OF ANALYSIS

The reinforced concrete section is modeled using filament method. As can be seen from Fig. 5, the cross-section is divided into 40 filaments to determine a momentcurvature relationship. For each filament, confined core and unconfined cover areas are defined. For a given strain at the extreme fiber in compression, the depth of neutral axis satisfying the force equilibrium is found by trial. For each filament, the average stresses are calculated at the centroids of unconfined and confined portions of the filament. To achieve this, first the strain at the centroids of the filament is calculated using the compatibility requirements. This centroidal strain is later used along with the appropriate concrete models to calculate stresses acting on the unconfined and confined portions of the filament. Finite concrete forces for the confined and unconfined portions of the filament (ΔF_{cc} and ΔF_{cu} respectively) are given multiplying the stress with the corresponding areas as follows:

$$\Delta F_{cu} = f_{cui} \cdot A_{cui} \tag{9}$$

and

$$\Delta F_{cc} = f_{cci} \cdot A_{cci} \tag{10}$$

where f_{cci} and f_{cui} are the concrete stresses for the confined and unconfined portions of the layer i.

Stress in the reinforcement at a given level is found by entering the f-ε diagram of steel with the strain value found from the compatibility requirements. Steel force at that level is given by multiplying the stress found with the area of the reinforcement at that level as:

$$F_{si} = \sigma_{si} A_{si} \tag{11}$$

This algorithm is demonstrated in Fig. 5 where only some typical finite forces are shown.

The moment-curvature relationship for a given axial load is determined by step by step incrementing concrete strain in the extreme compression fibre ε_{cm} . For each value of ε_{cm}, the strain gradient, i.e. the curvature φ, is obtained by satisfying the force equilibrium equation. The bending moment M corresponding to chosen value of ε_{cm} and axial load N are determined by taking moments of the internal forces about the geometric centroids of the section.

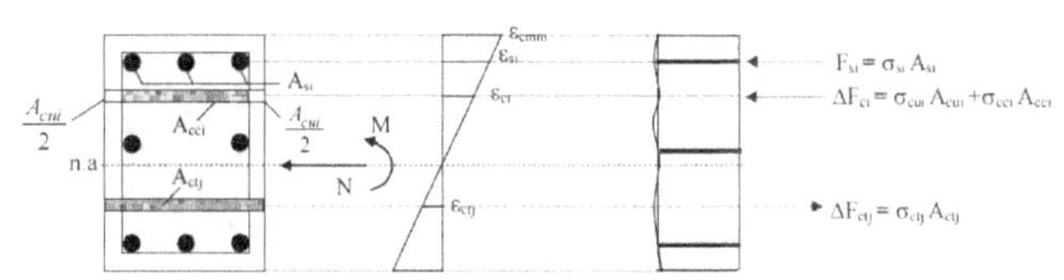

Figure 5: Strains and finite forces in the cross-section.

PARAMETRIC STUDY

In this section the effect of different variables on flexural behavior are investigated using analytical solutions developed to predict the moment-curvature relationship of reinforced concrete cross-sections shown in Fig. 6.

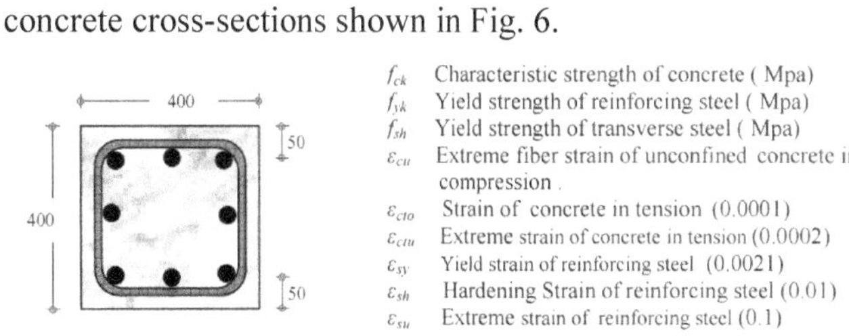

Figure 6: The cross-section considered in analyses.

Table 1: Results according to different variables

	Variable properties							Curvature (rad/m)					
No	f_{ck} (Mpa)	N/N_o (N)	f_{sh} (Mpa)	s (cm)	Ø (mm)	ρ	f_{yk} (Mpa)	TY	TH	CvC	CoC	ε_{cm}	M (kN-m)
1	30	0	420	15	8	0.02	420	0.0085	0.0365	0.0321	0.0492	0.0125	241.0
2	20	0	420	15	8	0.02	420	0.0105	-	0.0245	0.0350	0.0125	228.2
3	16	0	420	15	8	0.02	420	0.0093	0.0363	0.0212	0.0323	0.0125	222.4
4	30	0.25	420	15	8	0.02	420	0.0112	-	0.0120	0.0141	0.0028	350.1
5	20	0.25	420	15	8	0.02	420	0.0171	-	0.0115	0.0125	0.0032	296.2
6	16	0.25	420	15	8	0.02	420	0.0167	-	0.0112	0.0118	0.0032	236.9
7	30	0.5	420	15	8	0.02	420	0.0251	-	0.0079	0.0089	0.0032	342.4
8	20	0.5	420	15	8	0.02	420	0.0264	-	0.0077	0.0081	0.0040	279.2
9	16	0.5	420	15	8	0.02	420	0.0263	-	0.0076	0.0079	0.0040	253.1
10	30	0.75	420	15	8	0.02	420	-	-	0.0078	0.0071	0.0028	246.1
11	20	0.75	420	15	8	0.02	420	-	-	0.0054	0.0065	0.0036	217.2
12	16	0.75	420	15	8	0.02	420	-	-	0.0048	0.0059	0.0036	173.8
13	30	0	420	15	8	0.011	420	0.0082	0.0338	0.0391	0.0648	0.0100	135.6
14	30	0	360	15	8	0.011	420	0.0082	0.0338	0.0391	0.0647	0.0100	135.5
15	30	0	300	15	8	0.011	420	0.0082	0.0337	0.0390	0.0646	0.0100	135.3
15	30	0	220	15	8	0.011	420	0.0082	0.0337	0.0390	0.0645	0.0100	135.0
17	30	0.25	420	15	8	0.011	420	0.0115	0.0520	0.0133	0.0151	0.0034	214.3
18	30	0.25	360	15	8	0.011	420	0.0115	0.0518	0.0133	0.0151	0.0032	213.7
19	30	0.25	300	15	8	0.011	420	0.0115	0.0515	0.0133	0.0150	0.0032	213.2
20	30	0.25	220	15	8	0.011	420	0.0114	0.0534	0.0132	0.0150	0.0032	212.6
21	30	0.50	420	15	8	0.011	420	0.0194	-	0.0082	-	0.0038	229.1
22	30	0.50	360	15	8	0.011	420	0.0193	-	0.0082	-	0.0038	227.2
23	30	0.50	300	15	8	0.011	420	0.0210	-	0.0081	-	0.0038	225.3
24	30	0.50	220	15	8	0.011	420	0.0208	-	0.0081	-	0.0038	222.7
25	30	0.75	420	15	8	0.011	420	-	-	0.0057	-	0.0030	172.7
...	.....	.....	.....	....	.....	.....	....	.....	.....	.....	.....	....	.....
50	30	0.5	420	15	8	0.011	220	0.0104	-	0.0084	0.0094	0.0028	196.7
51	30	0.75	420	15	8	0.011	320	0.0151	-	0.0058	0.0065	0.0030	163.7
52	30	0.75	420	15	8	0.011	220	0.0124	-	0.0059	0.0067	0.0030	157.8

allow the following conclusions to be drawn.

1- As the compressive concrete strength increases, the tendency toward a brittle, sudden failure also increases. One of the disadvantages of a high-strength concrete is that it is more brittle than a concrete of a lower strength. The increasing compressive strength causes a decrease in ductility. The compressive strength, f_{ck}, does not have any effects on the behavior in the case of pure bending. The compressive strength becomes effective with increasing axial load. The maximum moment capacity changes ±25% due to ±25% compressive strength variation

2- The ductility decreases as level of the axial load increases. The variation of ductility with the level of axial load is quite significant. It is interesting to note that, although the sections considered are well confined, the behavior becomes very brittle under high levels of axial load. The upper limits imposed on axial loads in seismic codes roots from such considerations.

3- It is found that yield strength of transverse reinforcement, f_{sh}, has no effect on the behavior at all levels of the axial load.

4- The most important parameters for obtaining a ductile behavior are spacing of the confinement and the reinforcement configuration. Generally, closer confinement spacing and a denser reinforcement configuration does not contribute to a higher load capacity. The results presented in this study show that for a well-confined cross-section it is an advantage to use a higher grade of steel, while for a lightly confined section it is not. Table 1 shows that closer confinement spacing has little effect on maximum load. However, by decreasing the reinforcement confinement spacing a less brittle behavior can be achieved. From these tables it can be seen that the greatest effect of confinement is gained in pure compression.

5- To achieve ductility, the transverse reinforcement volume ratio needs to be increased and the reinforcement configuration should be designed to provide high confinement. As can be seen from Table 1, the increase in ductility with transverse reinforcement diameter has no significant effect on moment capacity. The crushing of core concrete delays with an increase in the diameter of transverse reinforcement. The diameter of transverse reinforcement becomes effective with the increasing axial load.

6- The reinforcement volumetric ratio, ρ has an important effect on the behavior of the confined section. The reinforcement volumetric ratio has significant effect on the behavior at low level axial load. The ultimate moment capacity increases from 10 % up to 30 % with the reinforcement volumetric ratio. The moment capacity decreases with the higher axial load. The reinforcement volumetric ratio is not effective on ductility.

7- The ductility increases remarkably when the reinforcement yield strength is increased with reinforcement configuration. The reinforcement yield strength, f_{yk}, , is an effective parameter in case of pure bending. The ultimate moment capacity changes from ±10 % up to ±30 % with the reinforcement yield strength.

ANN MODELING

Artificial Neural Networks (ANNs) approach is used to determine the behavior of confined concrete sections in this study. ANNs do not require an explicit understanding of the mechanism underlying the process, which is the main advantage. It has the capacity to learn the relationship between input and output provided that sufficient data are available for its training. The analytical results available for the confined sections were used to prepare the training and testing data sets for the network. The present study is concerned with the prediction of a confined section using ANN. In this study a neural network program model developed by Karlık [9] in PASCAL was used. The data for training and testing were formed using parametric results. For generating the data analytically, filament method is used. The database consists of 52 sets of results, of which 45 sets were used for training the network, and the remaining 7 were used for testing in Table 1

The training patterns should be normalized before they are applied to the neural network so as to limit the input and output values within a specified range. This is due to the large difference in the values of the data provided to the neural network. Besides, the activation function used in the back propagation neural network is a sigmoid function. The lower and upper limits of this function are 0 and 1, respectively. The following formula is used to pre-process the input data sets whose values are between 0 and 1.

$$v = \frac{x - x_{min}}{x_{max} - x_{min}} \quad (12)$$

Since the output value of the sigmoid function is between 0 and 1, the following function might be used. The combinations momentum rates {0, 0.3, 0.5, 0.7, and 0.9} are used to investigate their effects on the behaviour of the neural network convergence. The results are shown in Fig.7. The effects of all the given learning parameters and momentum rates on the convergence epoch and generalization of the neural network are shown in Table.2.

$$o = \frac{t - t_{min}}{t_{max} - t_{min}} \quad (13)$$

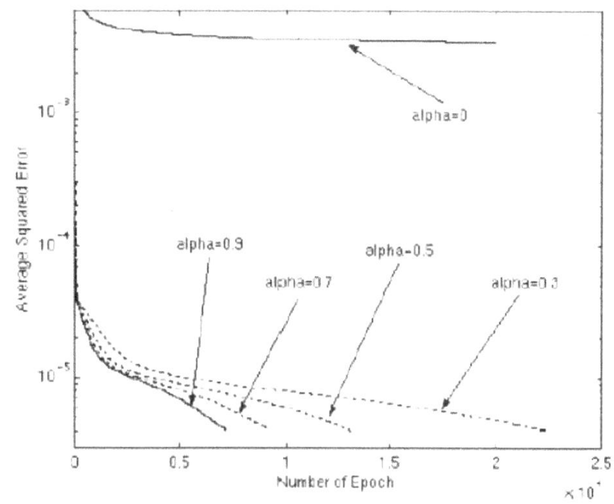

Figure 7: Effect of Momentum Rate on the Training of Neural Network

Table.2: Effect of Learning and Momentum Rate on the Behaviour of Neural Network

Momentum factor	Behaviour of Neural Network	Learning rate %
0.9	Training of the neural network	7.82
	Generalizations for test patterns	3.7

The parametric study was conducted to find out the optimum number of hidden layers as well as the number of nodes for the present problem. The results of the parametric study conducted were shown in Fig.8. Training for all these network configurations was carried out initially for one thousand cycles with error tolerance value of 0.025. When the number of hidden layers was made two, only the architecture 12-13 reached the smallest error tolerance in 1000 cycles. With one hidden layer, the architecture was not able to attain the required error tolerance of 0.0065 within 1000 cycles. Hence, for the problem under consideration, the network with 2 hidden layers having the 12-13 architecture was chosen since it reaches the required error tolerance with the least number of cycles, which in turn would reduce the CPU time requirement.

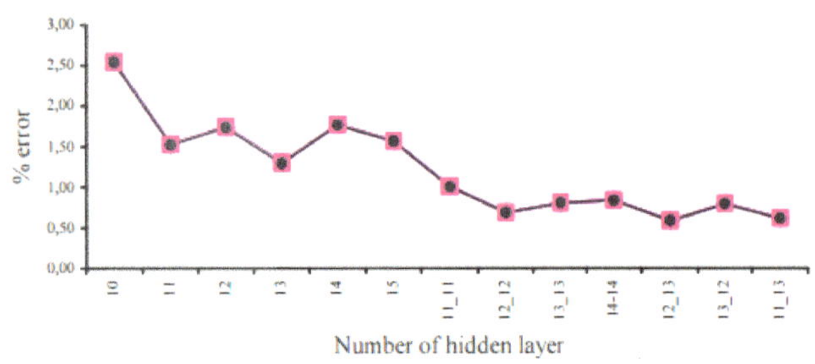

Figure 8: The error changes due to the number of nodes in the hidden layer at 1000 iterations.

Using the 7-12-13-6 architecture in Fig.9, the network was trained and then tested. For training the network, totally 45 data set were used which were listed under Table 3. These input data sets were analytically generated using the filament model. The network, after being trained, was tested with 7 data sets. These 7 input data sets were formerly generated using the filament model. The remaining data sets used for testing the network are shown in Table 4.

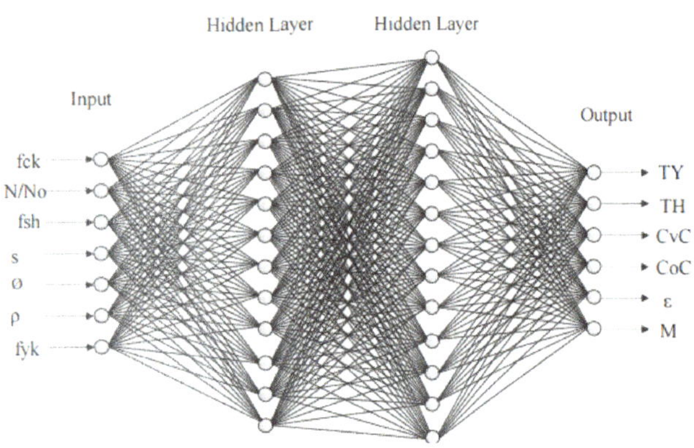

Figure 9: ANN architecture

Finally, the least required error convergence for 7-12-13-6 architecture was reached within 5000 cycles. A numerical study of training and testing of the network was conducted keeping the error tolerance values as 0.1, and 0.001. For an error tolerance of 0.1, the number of cycles required is less; but the results are less accurate. In the case of 0.001, even though the accuracy is high, the numbers of cycles required are very high. Hence, keeping in mind the number of cycles required for convergence together with the accuracy needed for training and testing, the minimum error tolerance was chosen as 0.7% in Fig. 10

The training results predicted using ANN is compared with the parametric values in Table 2. In these cases, results represent a one to one correspondence, that is, the predicted and the parametric values are identical. The average error $\left(\frac{|analytical - ANN| / ANN}{number\ of\ solution}\right)$ between the analytical and the ANN values produced is less than 0.2 %. The maximum differences between the analytical and ANN for TY, TH, CvC, CoC,ε and M are the outputs 0.965, 0.978, 1.039, 0.961, 0.962, and 0.976, respectively. Technically speaking, these errors are regarded to be sufficiently low.

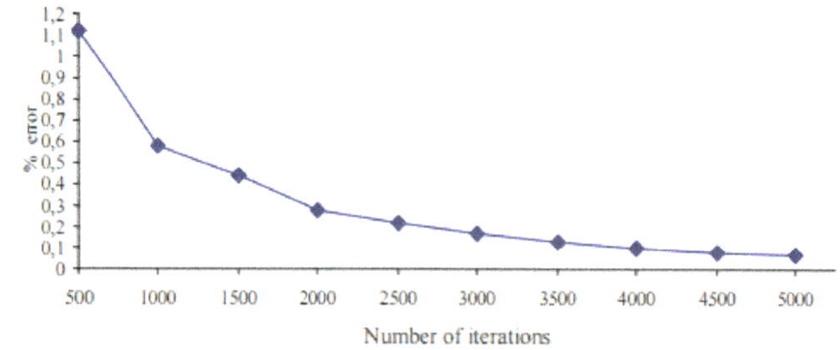

Figure 10. The error change at optimum ANN architecture (7:12:13:6)

Table 3: Training process and results

No	Meth.	TY	Anal / ANN	TH	Anal / ANN	CvC	Anal / ANN	CoC	Anal / ANN	ε	Anal / ANN	M	Anal / ANN
1	Analy. ANN	0.0085 0.0083	1,016	0.0365 0.0369	0,987	0.0321 0.0313	1,023	0.0492 0.0490	1,004	0.0125 0.0124	1,006	241.0 241.99	0,996
2	Analy. ANN	0.0105 0.0104	1,001	- -	-	0.0245 0.0243	1,006	0.0350 0.0351	0,995	0.0125 0.0126	0,988	228.2 228.86	0,997
5	Analy. ANN	0.0171 0.0170	1,004	- -	-	0.0115 0.0111	1,027	0.0125 0.0128	0,977	0.0032 0.0031	1,005	296.2 295.29	1,003
6	Analy. ANN	0.0167 0.0167	0,999	- -	-	0.0112 0.0115	0,966	0.0118 0.0118	0,993	0.0032 0.0032	0,993	236.9 238.05	0,995
8	Analy. ANN	0.0264 0.0264	0,999	- -	-	0.0077 0.0074	1,029	0.0081 0.0082	0,983	0.0040 0.0039	1,002	279.2 280.52	0,995
9	Analy. ANN	0.0263 0.0263	1,000	- -	-	0.0076 0.0076	0,996	0.0079 0.0077	1,023	0.0040 0.0040	1,000	253.1 252.7	1,002
10	Analy. ANN	- -	-	- -	-	0.0078 0.0078	0,992	0.0071 0.0073	0,969	0.0028 0.0028	0,994	246.1 246.14	1,000
11	Analy. ANN	- -	-	- -	-	0.0054 0.0054	0,996	0.0065 0.0064	1,009	0.0036 0.0035	1,004	217.2 217.03	1,001
13	Analy. ANN	0.0082 0.0082	0,998	0.0338 0.0334	1,011	0.0391 0.0388	1,005	0.0648 0.0659	0,983	0.0100 0.0099	1,005	135.6 135.74	0,999
14	Analy. ANN	0.0082 0.0082	0,991	0.0338 0.0335	1,008	0.0391 0.0387	1,010	0.0647 0.0655	0,987	0.0100 0.0100	1,000	135.5 135.04	1,003
16	Analy. ANN	0.0082 0.0083	0,985	0.0337 0.0334	1,006	0.0390 0.0386	1,010	0.0645 0.0653	0,987	0.0100 0.0100	0,999	135.0 133.11	1,014
18	Analy. ANN	0.0115 0.0114	1,008	0.0518 0.0513	1,008	0.0133 0.0135	0,984	0.0151 0.0153	0,981	0.0032 0.0032	0,988	213.7 212.29	1,007
19	Analy. ANN	0.0115 0.0113	1,010	0.0515 0.0517	0,994	0.0133 0.0133	0,995	0.0150 0.0153	0,977	0.0032 0.0031	1,026	213.2 211.2	1,009
20	Analy. ANN	0.0114 0.0115	0,985	0.0534 0.0518	1,011	0.0132 0.0130	1,013	0.0150 0.0151	0,988	0.0032 0.0031	1,028	212.6 212.25	1,002
21	Analy. ANN	0.0194 0.0196	0,985	- -	-	0.0082 0.0081	1,011	- -	-	0.0038 0.0038	0,994	229.1 230.16	0,995
22	Analy. ANN	0.0193 0.0199	0,968	- -	-	0.0082 0.0081	1,012	- -	-	0.0038 0.0038	0,991	227.2 228.16	0,996
...	.....	......	......	......	.....	....	.....	......	....	......	......	......	......
50	Analy. ANN	0.0104 0.0104	0,995	- -	-	0.0084 0.0082	1,022	0.0094 0.0092	1,018	0.0028 0.0028	0,984	196.7 195.85	1,004
51	Analy. ANN	0.0151 0.0150	1,005	- -	-	0.0058 0.0060	0,963	0.0065 0.0067	0,961	0.0030 0.0030	0,997	163.7 164.71	0,994
52	Analy. ANN	0.0124 0.0125	0,992	- -	-	0.0059 0.0059	0,995	0.0067 0.0068	0,972	0.0030 0.0029	1,006	157.8 157.09	1,005

The trained model was tested for a different set of input parameters and the output values were compared with the values based on analytical results. Seven different input values were applied to the model for testing the training network and the results were obtained in milliseconds. A comparison of the test and analytical values is given in Table 4. The average error $\left(\frac{|analytical - ANN| / ANN}{number\ of\ solution}\right)$ obtained is about 0.33 %. The maximum differences (analytical / ANN) for TY, TH, CvC, CoC, ε and M are about 0.967, 0.966, 0.972, 0.968, 0.991, and 0.992, respectively. Therefore, the results can be said to indicate that the trained NN models have achieved good performance.

Table 4: Testing process and results

No	Method	TY	Anal / ANN	TH	Anal / ANN	CvC	Anal / ANN	CoC	Anal / ANN	ε	Anal / ANN	M	Anal / ANN
3	Analy. ANN	0.0093 0.0094	0.987	0.0363 0.0375	0.966	0.0212 0.0206	1.025	0.0323 0.0313	1.031	0.0125 0.0125	0.999	222.4 224.1	0.992
4	Analy. ANN	0.0112 0.0110	1.014	- -	-	0.0120 0.0122	0.976	0.0141 0.0145	0.968	0.0028 0.0027	1.004	350.1 348.1	1.005
7	Analy. ANN	0.0251 0..025	1.001	- -	-	0.0079 0.0081	0.972	0.0089 0.0086	1.032	0.0032 0.0032	0.997	342.4 342.3	1.000
12	Analy. ANN	- -	-	- -	-	0.0048 0.0046	1.024	0.0059 0.0057	1.021	0.0036 0.0035	1.002	173.8 174.0	0.994
15	Analy. ANN	0.0082 0.0083	0.986	0.0337 0.0335	1.004	0.0390 0.0386	1.009	0.0646 0.0654	0.988	0.0100 0.0100	0.999	135.3 134.1	1.008
17	Analy. ANN	0.0115 0.0115	0.992	0.0520 0.0513	1.012	0.0133 0.0135	0.979	0.0151 0.0154	0.975	0.0034 0.0034	0.991	214.3 214.7	0.998
24	Analy. ANN	0.0208 0.0215	0.967	- -	-	0.0081 0.0082	0.978	- -	-	0.0038 0.0038	0.991	222.7 224.4	0.992

Compared to conventional digital computing techniques, neural networks are advantageous because of their special features, such as the massively parallel processing, distributed storing of information, low sensitivity to error, their very robust operation after training, generalisation and adaptability to new information.

CONCLUDING REMARKS

In this study, a back-propagation neural network model was employed to predict the influence of various parameters on the behavior of reinforced concrete sections. A neural network model was applied to the data derived from the analytical solutions. The analytical model is based on a filament modeling technique and capable of taking into account the crushing of cover and core concrete, the strain hardening of steel and the effect of confinement on core concrete.

To reduce the computing time of microprocessor of the system, a new computer model, which replies in milliseconds, was developed based on ANN method. A multilayer, back propagation and feed-forward ANN algorithm was used to train the data. The ANN algorithms are not able to replace the conventional analytical techniques completely since they need some key values for training. However, in the determination of reinforced cross-section behavior, they can be implemented as an efficient supplementary tool to reduce the computational cost drastically. Modeling process in neural network is more direct since there is no necessity to specify a mathematical relationship between the input and the output variables. The trained ANN was able to produce quick results for the reinforced cross-section behavior with the same degree of accuracy as the filament model analysis achieved under flexure and axial load. Therefore, the trained ANN may be used in practice for determining the reinforced cross-sections behavior as an alternative to the time consuming filament model analysis.

The ANN applications presented in this study have demonstrated the viability and feasibility of using analytical results for the reinforced confined sections' behavior. The obtained results have shown that the neural network model is successful in modeling the non-linear relationship between different input and output parameters even when it involves a relatively smaller number of training patterns. It is envisaged that the model developed may be used in practical structural engineering applications.

REFERENCES

1. S. Parviz, S. Jongsung and W. H . Jer, Axial / Flexural Behavior of Reinforced Concrete Sections: Effects of Design Variables, ACI, 88, 17-21, 1991.
2. U. Ersoy and G. Özcebe, Moment-Curvature Relationship of Confined Concrete Sections, First Japan-Turkey Workshop On Earthquake Engineering, Ankara, Turkey, 10-21,1997.
3. S. C. Patodi and V. S. Purani, Modeling flexural behavior of steel fibre reinforced concrete beams using neural networks, Journal of New Building Materials and Construction World, 4, 28-35, 1998.
4. M. N. Jadid and D.R. Fairbairn, Neural-network applications in predicting momentcurvature parameters from experimental data, Engineering Applications of Artificial Intelligence, 9, 309-319,1996.
5. L. Berke and P. Hajela, Applications of artificial neutral nets in structural mechanics, Structural Optimizations, 4, 90-9, 1992.
6. Y. Lee, S. H. Oh, H. K. Hong and M. W. Kim, Design rules of multi-layer perception, Proc. SPIE, Science of Artificial Neutral Nets in Structural Mechanics. Structural Optimization, 1710, 329-339, 1992
7. A. V. Avdelas, P. D. Panagiotopoulos and S. Kortesis, Neutral networks for computing in the elastoplastic analysis of structures, Meccanica, 30,1-15, 1995.
8. K. M. Abdalla and G. E. Stavroulakis, A Back propagation neutral network model for semi-rigid steel connections, Microcomputers in Civil Engineering, 10, 77-87, 1995.
9. B. Karlık, E. Ozkaya, S. Aydın, S. and M. Pakdemirli, Vibration of beam -mass systems using artificial neutral networks, Computers & Structures, 69, 339-347, 1998.
10. P. Hajela, and L. Berke, Neurobiological computational models in structural analysis and Design, Computers & Structures, 41, 657-667, 1991.
11. W. M. Jenkins, An Introduction to Neural Computing for the Structural Engineer. The Structural Engineer, 75, 38-4,. 1997.
12. Mukherjee, J. M. Deshpande and J. Anmala, Prediction of Buckling Load of Columns Using Artificial Neural Networks. Journal of Structural Engineering, 122, 1385-1387, 1996.
13. H. Adeli and H.S. Park, Counter propagation Neural Networks in Structural Engineering. Journal of Structural Engineering, .121, 1205-1212, 1995.
14. R. Park, M. J. N. Priestly and W. D. Gill, Ductility of square confined concrete columns, Journal of Structural Division, ASCE, 108, 929–950, 1992.
15. H. Rüsch and H. Hilsdrof, Verfomungei-genschaften von beton unter zentrishcen zugspannangen, Materialprüfungsemt für das bauwesen der technischen hochshule München, 1963.

CHAPTER 8

AN APPROXIMATE SOLUTION FOR BOUNDARY VALUE PROBLEMS IN STRUCTURAL ENGINEERING AND FLUID MECHANICS

A. Barari, M. Omidvar, D. D. Ganji, and Abbas Tahmasebi Poor

INTRODUCTION

This paper discusses the analytical approximate solution for fourth-order equations with nonlinear boundary conditions involving third-order derivatives. The general form of the equation for a fixed positive integer n, $n \geq 2$, is a differential equation of order 2n:

$$y^{(2n)} + f(x, y) = 0 \tag{1.1}$$

subject to the boundary conditions

$$y^{(2j)}(a) = A_{2j}, \quad y^{(2j)}(b) = B_{2j}, \quad j = 0(1)n-1, \tag{1.2}$$

where $-\infty < a \leq x \leq b < \infty$, A_{2j}, B_{2j}, $j = 0(1)n - 1$ are finite constants.

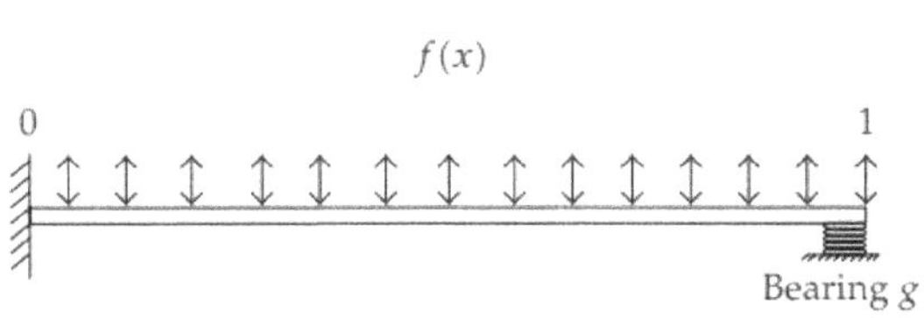

Figure 1: Beam on elastic bearing.

Citation: A. Barari, M. Omidvar, D. D. Ganji, and Abbas Tahmasebi Poor, "An Approximate Solution for Boundary Value Problems in Structural Engineering and Fluid Mechanics," Mathematical Problems in Engineering, vol. 2008, Article ID 394103, 13 pages, 2008. doi:10.1155/2008/394103.

It is assumed that y is sufficiently differentiable and that a unique solution of 1.1 exists. Problems of this kind are commonly encountered in plate-deflection theory and in fluid mechanics for modeling viscoelastic and inelastic flows [1–3]. Usmani [1, 2] discussed sixth order methods for the linear differential equation $y^{(4)} + P(x)y = q(x)$ subject to the boundary conditions $y(a) = A_0, y''(A) = A_2, y(b) = B_0, y''(b) = B_2$. The method described in [1] leads to five diagonal linear systems and involves p', p'', q', q'' at a and b, while the method described in [2] leads to nine diagonal linear systems.

Ma and Silva [4] adopted iterative solutions for (1.1) representing beams on elastic foundations. Referring to the classical beam theory, they stated that if u = u(x) denotes the configuration of the deformed beam, then the bending moment satisfies the relation M = −EIu//, where E is the Young modulus of elasticity and I is the inertial moment. Considering the deformation caused by a load f = f(x), they deduced, from a free-body diagram, that $f = -v'$ and $v = M' = -EIu'''$ where v denotes the shear force. For u representing an elastic beam of length L = 1, which is clamped at its left side x = 0, and resting on an elastic bearing at its right side x =1, and adding a load f along its length to cause deformations Figure 1, Ma and Silva [4] arrived at the following boundary value problem assuming an EI = 1:

$$u^{(iv)}(x) = f(x, u(x)), \quad 0 < x < 1, \tag{1.3}$$

the boundary conditions were taken as

$$u(0) = u'(0) = 0, \tag{1.4}$$

$$u''(1) = 0, \qquad u'''(1) = g(u(1)), \tag{1.5}$$

where $f \in C([0,1] \times \mathbb{R})$ and $g \in C(\mathbb{R})$ are real functions. The physical interpretation of the boundary conditions is that u///(1) is the shear force at x = 1, and the second condition in 1.5 means that the vertical force is equal to g(u(1)), which denotes a relation, possibly nonlinear, between the vertical force and the displacement u(1). Furthermore, since u//(1) = 0 indicates that there is no bending moment at x = 1, the beam is resting on the bearing g.

Solving (1.3) by means of iterative procedures, Ma and Silva [4] obtained solutions and argued that the accuracy of results depends highly upon the integration method used in the iterative process.

With the rapid development of nonlinear science, many different methods were proposed to solve differential equations, including boundary value problems (BVPS). These two methods are the homotopy perturbation method HPM [5–7] and the variational iteration method VIM [8–17]. In this paper, it is aimed to apply the variational iteration method proposed by He [14] to different forms of 1.1 subject to boundary conditions of physical significance.

BASIC IDEA OF HE'S VARIATIONAL ITERATION METHOD

To clarify the basic ideas of He's VIM, the following differential equation is considered:

$$L[u(t)] + N[u(t)] = g(t), \tag{2.1}$$

where L is a linear operator, N is a nonlinear operator, and gt is an inhomogeneous term. According to VIM, a correction functional could be written as follows:

$$u_{n+1}(t) = u_n(t) + \int_0^t \lambda(\tau)(Lu_n(\tau) + N\tilde{u}_n(\tau) - g(\tau))d\tau, \tag{2.2}$$

where λ is a general Lagrange multiplier which can be identified optimally via the variational theory. The subscript n indicates the nth approximation and $\tilde{u}_n$ is considered as a restricted variation, that is, $\delta\tilde{u}_n = 0$.

For fourth-order boundary value problem with suitable boundary conditions, Lagrangian multiplier can be identified by substituting the problem into (2.2), upon making it stationary leads to the following:

$$\frac{d^4}{d\tau^4}\lambda = 0,$$
$$-\lambda''' + 1|_{\tau=x} = 0,$$
$$\lambda''|_{\tau=x} = 0. \tag{2.3}$$

Solving the system of (2.3) yields

$$\lambda = \frac{1}{6}(\tau - x)^3 \tag{2.4}$$

and the variational iteration formula is obtained in the form

$$u_{n+1}(x) = u_n(x) + \int_0^x \frac{1}{6}(\tau - x)^3(u_n^{(4)}(\tau) + f(\tau, u_n, u_n', u_n'', u_n'''))d\tau. \tag{2.5}$$

THE APPLICATIONS OF VIM METHOD

In this section, the VIM is applied to different forms of the fourth-order boundary value problem introduced in through (1.1).

Example 3.1. Consider the following linear boundary value problem:

$$u^{(4)}(x) = 4e^x + u(x), \quad 0 < x < 1, \tag{3.1}$$

subject to the boundary conditions

$$u(0) = 1, \quad u'(0) = 2, \quad u(1) = 2e, \quad u'(1) = 3e. \tag{3.2}$$

The exact solution for this problem is

$$u(x) = (1 + x)e^x. \tag{3.3}$$

According to (2.5), the following iteration formulation is achieved:

$$u_{n+1}(x) = u_n(x) + \int_0^x \frac{1}{6}(\tau - x)^3(u_n^{(4)}(\tau) - u_n(\tau) - 4e^\tau)\,d\tau. \tag{3.4}$$

Now it is assumed that an initial approximation has the form

$$u_0(x) = ax^3 + bx^2 + cx + d, \tag{3.5}$$

where a, b, c, and d are unknown constants to be further determined.

By the iteration formula 3.4, the following first-order approximation may be written:

$$\begin{aligned} u_1(x) &= u_0(x) + \int_0^x \frac{1}{6}(\tau - x)^3(u_0^{(4)}(\tau) - u_0(\tau) - 4e^\tau)d\tau \\ &= ax^3 + bx^2 + cx + d + \int_0^x \frac{1}{6}(\tau - x)^3(-a\tau^3 - b\tau^2 - c_\tau - d - 4e^\tau)d\tau \\ &= \frac{1}{840}ax^7 + \frac{1}{360}bx^6 + \frac{1}{120}cx^5 + \frac{1}{24}dx^4 + \left(-\frac{2}{3} + a\right)x^3 + (b-2)x^2 + (c-4)x + 4e^x + d - 4. \end{aligned} \tag{3.6}$$

Incorporating the boundary conditions 3.2, into $u_1(x)$, the following coefficients can be obtained:

$$a = -\frac{2289756}{301681} + \frac{916440}{301681}e, \quad b = \frac{4575063}{301681} - \frac{1516680}{301681}e, \quad c = 2, \quad d = 1. \tag{3.7}$$

Therefore, the following first-order approximate solution is derived:

$$u_1(x) = \left(-\frac{27259}{3016810} + \frac{1091}{301681}e\right)x^7 + \left(\frac{1525021}{36201720} - \frac{4213}{301681}e\right)x^6$$
$$+\frac{1}{60}x^5 + \frac{1}{24}x^4 + \left(-\frac{7472630}{905043} + \frac{916440}{301681}e\right)x^3 + \left(\frac{3971701}{301681} - \frac{1516680}{301681}e\right)x^2 - 2x - 3 + 4e^x. \tag{3.8}$$

Comparison of the first-order approximate solution with exact solution is tabulated in Table 1, showing a remarkable agreement.

Similarly, the following second-order approximation is obtained:

$$u_2(x) = u_1(x) + \int_0^x \frac{1}{6}(\tau - x)^3 (u_1^{(4)}(\tau) - u_1(\tau) - 4e^\tau) d\tau$$
$$= \frac{1}{6652800}ax^{11} + \frac{1}{1814400}bx^{10} + \frac{1}{362880}cx^9 + \frac{1}{40320}dx^8 + \left(\frac{1}{840}a - \frac{1}{1260}\right)x^7$$
$$+\left(\frac{1}{360}b - \frac{1}{180}\right)x^6 + \left(-\frac{1}{30} + \frac{1}{120}c\right)x^5 + \left(-\frac{1}{6} + \frac{1}{24}d\right)x^4$$
$$+\left(-\frac{4}{3} + a\right)x^3 + (b-4)x^2 + (c-8)x - 8 + 8e^x + d,$$
$$a = -\frac{12706529114180}{681628862391} + \frac{85535681616000}{12042109902241}e, \quad c = 2,$$
$$b = \frac{8416302814865}{227209620797} - \frac{157452726614400}{12042109902241}e, \quad d = 1. \tag{3.9}$$

Table 1: Comparison of the first-order approximate solution with exact solution.

x	U_E	U_1	Error
0	1.000000000	1.000000000	$0.0000E+000$
0.1	1.215688010	1.215681524	$6.4860E-006$
0.2	1.465683310	1.465660890	$2.2420E-005$
0.3	1.754816450	1.754773923	$4.2527E-005$
0.4	2.088554577	2.088492979	$6.1598E-005$
0.5	2.473081906	2.473007265	$7.4641E-005$
0.6	2.915390080	2.915312734	$7.7346E-005$
0.7	3.423379602	3.423312592	$6.7010E-005$
0.8	4.005973670	4.005929404	$4.4266E-005$
0.9	4.673245911	4.673229891	$1.6020E-005$
1.0	$2e$	$2e$	$0.0000E+000$

Table 2: Comparison of the second-order approximate solution with exact solution

x	U_E	U_2	Error
0	1.000000000	1.000000000	$0.0E+000$
0.1	1.215688010	1.215688008	$2.0E-009$
0.2	1.465683310	1.465683305	$5.0E-009$
0.3	1.754816450	1.754816444	$6.0E-009$
0.4	2.088554577	2.088554566	$1.1E-008$
0.5	2.473081906	2.473081902	$4.0E-009$
0.6	2.915390080	2.915390064	$1.6E-008$
0.7	3.423379602	3.423379600	$2.0E-009$
0.8	4.005973670	4.005973650	$2.0E-008$
0.9	4.673245911	4.673245930	$1.9E-008$
1.0	$2e$	$2e$	$0.0E+000$

Therefore, the second-order approximate solution may be written as

$$u_2(x) = \left(-\frac{57756950519}{20612456798703840} + \frac{12857095}{12042109902241}e\right)x^{11}$$
$$+\left(\frac{1683260562973}{82449827194815360} - \frac{86779501}{12042109902241}e\right)x^{10} + \frac{1}{181440}x^9$$
$$+\frac{1}{40320}x^8 + \left(-\frac{731163797543}{31809346911580} + \frac{101828192400}{12042109902241}e\right)x^7$$
$$+\left(\frac{7961883573271}{81795463486920} - \frac{437368685040}{12042109902241}e\right)x^6 - \frac{1}{60}x^5$$
$$-\frac{1}{8}x^4 + \left(-\frac{13615367597368}{681628862391} + \frac{85535681616000}{12042109902241}e\right)x^3$$
$$+\left(\frac{7507464331677}{227209620797} - \frac{157452726614400}{12042109902241}e\right)x^2 - 6x - 7 + 8e^x. \tag{3.10}$$

Again, the obtained solution is of distinguishing accuracy, as indicated in Table 2 and Figure 2.

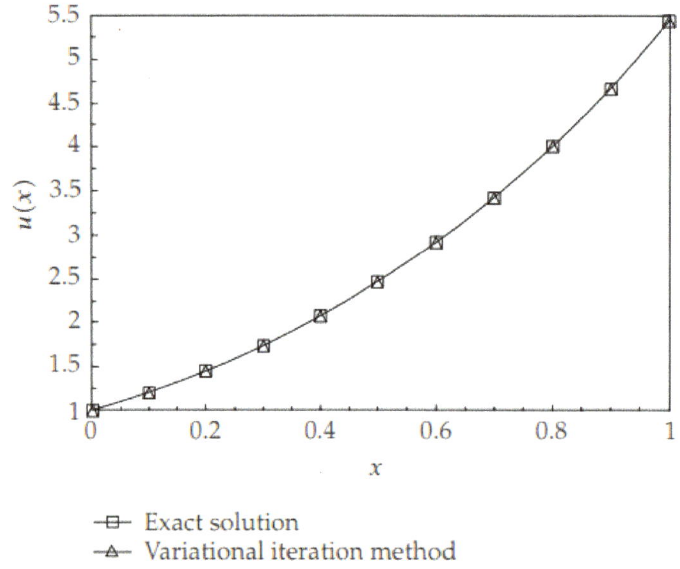

Figure 2: Comparison between different solutions.

Example 3.2. Consider the following linear boundary value problem:

$$u^{(4)}(x) = u(x) + u''(x) + e^{x}(x-3), \quad 0 < x < 1, \tag{3.11}$$

subject to the boundary conditions

$$u(0) = 1, \quad u'(0) = 0, \quad u(1) = 0, \quad u'(1) = -e. \tag{3.12}$$

The exact solution for this problem is

$$u(x) = (1-x)e^{x}. \tag{3.13}$$

According to (2.5), the iteration formulation may be written as

$$u_{n+1}(x) = u_n(x) + \int_0^x \frac{1}{6}(\tau - x)^3\left(u_n^{(4)}(\tau) - u_n(\tau) - u_u''(\tau) - e^{\tau}(\tau-3)\right)d\tau. \tag{3.14}$$

Now it is assumed that an initial approximation has the form

$$u_0(x) = ax^3 + bx^2 + cx + d. \tag{3.15}$$

Where a, b, c, and d are unknown constants to be further determined.

By the iteration formula 3.14, the following first-order approximation is developed:

$$\begin{aligned} u_1(x) &= u_0(x) + \int_0^x \frac{1}{6}(\tau - x)^3\left(u_0^{(4)}(\tau) - u_0(\tau) - u_0''(\tau) - e^{\tau}(\tau-3)\right)d\tau \\ &= ax^3 + bx^2 + cx + d + \int_0^x \frac{1}{6}(\tau - x)^3(-a\tau^3 - b\tau^2 - (6a+c)\tau - 2b - d - e^{\tau}(\tau - 3))d\tau \\ &= \frac{1}{840}ax^7 + \frac{1}{360}bx^6 + \left(\frac{1}{20}a + \frac{1}{120}c\right)x^5 + \left(\frac{1}{12}b + \frac{1}{24}d\right)x^4 + \left(\frac{2}{3} + a\right)x^3 + \left(b + \frac{5}{2}\right)x^2 \\ &\quad + \left(e^{x} + 6 + c\right)x - 7e^{x} + 7 + d. \end{aligned} \tag{3.16}$$

Table 3: Comparison of the first-order approximate solution with exact solution.

x	u_E	u_1	Error
0	1.0000000000	1.0000000000	0.0000000E + 000
0.1	0.9946538262	0.9947931547	1.3932850E - 004
0.2	0.9771222064	0.9775949040	4.7269760E - 004
0.3	0.9449011656	0.9457776230	8.7645740E - 004
0.4	0.8950948188	0.8963297250	1.2349062E - 003
0.5	0.8243606355	0.8258087440	1.4481085E - 003
0.6	0.7288475200	0.7302919280	1.4444080E - 003
0.7	0.6041258121	0.6053240800	1.1982679E - 003
0.8	0.4451081856	0.4458625400	7.5435440E - 004
0.9	0.2459603111	0.2462193000	2.5898890E - 004
1.0	0.0000000000	0.0000000000	0.0000000E + 000

Incorporating the boundary conditions 3.12, into $u_1(x)$, it can be written as a

$$a = \frac{7904470}{323149} - \frac{2950080}{323149}e, \qquad b = -\frac{12770295}{323149} + \frac{4640400}{323149}e, \qquad c = 0, \qquad d = 1. \tag{3.17}$$

Therefore, the following first-order approximate solution is obtained:

$$\begin{aligned} u_1(x) &= \left(\frac{112921}{3877788} - \frac{3512}{323149}e\right)x^7 + \left(-\frac{851353}{7755576} + \frac{12890}{323149}e\right)x^6 \\ &+ \left(\frac{790447}{646298} - \frac{147504}{323149}e\right)x^5 + \left(-\frac{25217441}{7755576} + \frac{386700}{323149}e\right)x^4 \\ &+ \left(\frac{24359708}{969447} - \frac{2950080}{323149}e\right)x^3 + \left(-\frac{23924845}{646298} + \frac{4640400}{323149}e\right)x^2 \\ &+ (6 + e^x)x + 8 - 7e^x. \end{aligned} \tag{3.18}$$

Comparison of the first-order approximate solution with exact solution is tabulated in Table 3, again showing a clear agreement. Even higher accurate solutions could be obtained without any difficulty.

Similarly, the following second-order approximation can be written as

$$\begin{aligned} u_2(x) &= u_1(x) + \int_0^x \frac{1}{6}(\tau - x)^3\left(u_1^{(4)}(\tau) - u_1(\tau) - u_1''(\tau) - e^\tau(\tau - 3)\right)d\tau \\ &= \frac{1}{6652800}ax^{11} + \frac{1}{1814400}bx^{10} + \left(\frac{1}{362880}c + \frac{1}{30240}a\right)x^9 + \left(\frac{1}{10080}b + \frac{1}{40320}d\right)x^8 \\ &+ \left(\frac{1}{5040}c + \frac{1}{420}a + \frac{1}{1260}\right)x^7 + \left(\frac{1}{720}d + \frac{1}{144} + \frac{1}{180}b\right)x^6 + \left(\frac{1}{12} + \frac{1}{20}a + \frac{1}{120}c\right)x^5 \\ &+ \left(\frac{1}{2} + \frac{1}{24}d + \frac{1}{12}b\right)x^4 + (3 + a)x^3 + \left(b + \frac{21}{2}\right)x^2 + (24 + 3e^x + c)x + 27 - 27e^x + d. \end{aligned} \tag{3.19}$$

Incorporating the boundary conditions, 3.12, into $u_2(x)$, yields

$$\begin{aligned} a &= \frac{381804789300110}{4289712004667} - \frac{140985028800000}{4289712004667}e, \qquad c = 0, \\ b &= -\frac{629495301082065}{4289712004667} + \frac{230790037363200}{4289712004667}e, \qquad d = 1. \end{aligned} \tag{3.20}$$

The following second-order approximate solution is then achieved in the following form:

$$
\begin{aligned}
u_2(x) = {} & \left(\frac{3470952630001}{259441782042260160} - \frac{63575500}{12869136014001}e\right)x^{11} \\
& + \left(-\frac{41966353405471}{518883564084520320} + \frac{381597284}{12869136014001}e\right)x^{10} \\
& + \left(\frac{38180478930011}{12972089102113008} - \frac{13986610000}{12869136014001}e\right)x^{9} \\
& + \left(-\frac{2513691492323593}{172961188028173440} + \frac{22895837040}{4289712004667}e\right)x^{8} \\
& + \left(\frac{1149704079904997}{5405037125880420} - \frac{335678640000}{4289712004667}e\right)x^{7} \\
& + \left(-\frac{415373822050043}{514765440560040} + \frac{1282166874240}{4289712004667}e\right)x^{6} \\
& + \left(\frac{233372585584733}{51476544056004} - \frac{7049251440000}{4289712004667}e\right)x^{5} \\
& + \left(-\frac{1203224346103459}{102953088112008} + \frac{19232503113600}{4289712004667}e\right)x^{4} \\
& + \left(\frac{394673925314111}{4289712004667} - \frac{140985028800000}{4289712004667}e\right)x^{3} \\
& + \left(-\frac{1168906650066123}{8579424009334} + \frac{230790037363200}{4289712004667}e\right)x^{2} \\
& + (3e^{x} + 24)x + 28 - 27e^{x}.
\end{aligned} \tag{3.21}
$$

The obtained solution is of evident accuracy, as shown in Table 4 and Figure 3.

Example 3.3. Consider the following nonlinear boundary value problem:

$$u^{(4)}(x) = u^2(x) + g(x), \quad 0 < x < 1, \tag{3.22}$$

subject to the boundary conditions

$$u(0) = 0, \quad u'(0) = 0, \quad u(1) = 1, \quad u'(1) = 1, \tag{3.23}$$

where

$$g(x) = -x^{10} + 4x^9 - 4x^8 - 4x^7 + 8x^6 - 4x^4 + 120x - 48. \tag{3.24}$$

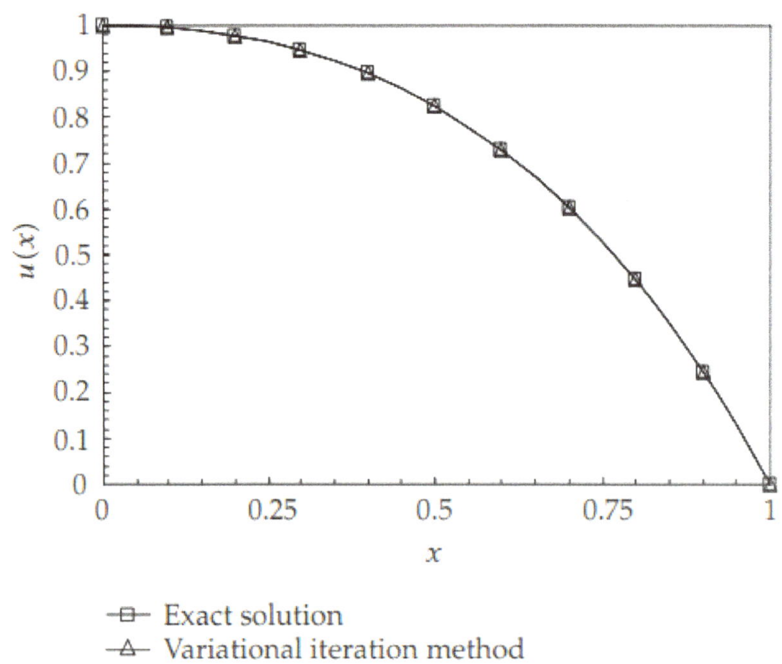

Figure 3: Comparison between different solutions

Table 4: Comparison of the second-order approximate solution with exact solution.

x	u_E	u_2	Error
0	1.0000000000	1.0000000000	0.00000E + 000
0.1	0.9946538262	0.9946577580	3.93180E – 006
0.2	0.9771222064	0.9771357780	1.35716E – 005
0.3	0.9449011656	0.9449268900	2.57244E – 005
0.4	0.8950948188	0.8951321100	3.72912E – 005
0.5	0.8243606355	0.8244058800	4.52445E – 005
0.6	0.7288475200	0.7288945300	4.70100E – 005
0.7	0.6041258121	0.6041666500	4.08379E – 005
0.8	0.4451081856	0.4451352800	2.70944E – 005
0.9	0.2459603111	0.2459701300	9.81890E – 006
1.0	0.0000000000	0.0000000000	0.00000E + 000

The exact solution for this problem is

$$u(x) = x^5 - 2x^4 + 2x^2. \tag{3.25}$$

According to (2.5), the iteration formulation is written as follows:

$$u_{n+1}(x) = u_n(x) + \int_0^x \frac{1}{6}(\tau - x)^3 \left(u_n^{(4)}(\tau) - u_n^2(\tau) - g(\tau)\right) d\tau. \tag{3.26}$$

Now it is assumed that an initial approximation has the form

$$u_0(x) = ax^3 + bx^2 + cx + d, \tag{3.27}$$

where a, b, c, and d are unknown constants to be further determined.

Table 5: Comparison of the first-order approximate solution with exact solution

x	u_E	u_1	Error
0	0.0000000000	0.0000000000	0.0000000E + 000
0.1	0.0198100000	0.0198624243	5.2424300E – 005
0.2	0.0771200000	0.0773022107	1.8221070E – 004
0.3	0.1662300000	0.1665781379	3.4813790E – 004
0.4	0.2790400000	0.2795490972	5.0909720E – 004
0.5	0.4062500000	0.4068747265	6.2472650E – 004
0.6	0.5385600000	0.5392178270	6.5782700E – 004
0.7	0.6678700000	0.6684511385	5.8113850E – 004
0.8	0.7884800000	0.7888727023	3.9270230E – 004
0.9	0.8982900000	0.8984356964	1.4569640E – 004
1.0	1.0000000000	1.0000000000	0.0000000E + 000

By the iteration formula (3.26), the following first-order approximation is obtained:

$$\begin{aligned} u_1(x) &= u_0(x) + \int_0^x \frac{1}{6}(\tau - x)^3 (u_0^{(4)}(\tau) - u_0^2(\tau) + \tau^{10} - 4\tau^9 + 4\tau^8 + 4\tau^7 - 8\tau^6 + 4\tau^4 - 120\tau + 48) d\tau \\ &= -\frac{1}{24024}x^{14} + \frac{1}{4290}x^{13} - \frac{1}{2970}x^{12} - \frac{1}{1980}x^{11} + \left(\frac{1}{5040}a^2 + \frac{1}{630}\right)x^{10} \\ &\quad + \frac{1}{1512}abx^9 + \left(-\frac{1}{420} + \frac{1}{1680}b^2 + \frac{1}{840}ac\right)x^8 + \left(\frac{1}{420}bc + \frac{1}{420}ad\right)x^7 \\ &\quad + \left(\frac{1}{180}bd + \frac{1}{360}c^2\right)x^6 + \left(1 + \frac{1}{60}cd\right)x^5 + \left(\frac{1}{24}d^2 - 2\right)x^4 + ax^3 + bx^2 + cx + d. \end{aligned} \tag{3.28}$$

Incorporating the boundary conditions (3.23), into $u_1(x)$, results in the following values: a

$$a = -0.006871650809; \quad b = 2.005929593; \quad c = 0, \quad d = 0. \tag{3.29}$$

The following first-order approximate solution is then achieved:

$$\begin{aligned} u_1(x) = &-4.162504162 \times 10^{-5}x^{14} + 2.331002331 \times 10^{-4}x^{13} \\ &- 3.367003367 \times 10^{-4}x^{12} - 5.050505050 \times 10^{-4}x^{11} \\ &+ 1.587310956 \times 10^{-3}x^{10} - 9.116433669 \times 10^{-6}x^{9} \\ &+1.4139007 \times 10^{-5}x^{8} + x^{5} - 2x^{4} - 6.871650809 \times 10^{-3}x^{3} \\ &+2.005929593x^{2}. \end{aligned} \tag{3.30}$$

Comparison of the first-order approximate solution with exact solution is tabulated in Table 5, which once again shows an excellent agreement.

Similarly, the following second-order approximation may be written:

$$u_2(x) = u_1(x) + \int_0^x \frac{1}{6}(\tau - x)^3 \left(u_1^{(4)}(\tau) - u_1^2(\tau) + \tau^{10} - 4\tau^9 + 4\tau^8 + 4\tau^7 - 8\tau^6 + 4\tau^4 - 120\tau + 48\right)d\tau. \tag{3.31}$$

Table 6: Comparison of the second-order approximate solution with exact solution.

x	U_E	U_2	Error
0	0.0000000000	0.0000000000	$0.000E+000$
0.1	0.0198100000	0.0198100068	$6.800E-009$
0.2	0.0771200000	0.0771200239	$2.390E-008$
0.3	0.1662300000	0.1662300464	$4.640E-008$
0.4	0.2790400000	0.2790400692	$6.920E-008$
0.5	0.4062500000	0.4062500874	$8.740E-008$
0.6	0.5385600000	0.5385600961	$9.610E-008$
0.7	0.6678700000	0.6678700906	$9.060E-008$
0.8	0.7884800000	0.7884800670	$6.700E-008$
0.9	0.8982900000	0.8982900292	$2.920E-008$
1.0	1.0000000000	1.0000000012	$1.200E-009$

Incorporating the boundary conditions, (3.23), into $u_2(x)$, yields

$$a = -8.269548014E-7; \quad b = 2.000000763; \quad c = 0, \quad d = 0. \tag{3.32}$$

The following second-order approximate solution is obtained:

$$\begin{aligned} u_2(x) = &-1.093855974 \times 10^{-9}x^{9} + 1.817 \times 10^{-9}x^{8} - 2x^{4} - 1.117934793 \times 10^{-8}x^{21} \\ &+ 1.463705892 \times 10^{-9}x^{20} + 6.586694874 \times 10^{-8}x^{19} + 2.000000763x^{2} \\ &- 8.269548014 \times 10^{-7}x^{3} - 1.047931585 \times 10^{-7}x^{18} - 3.536760165 \times 10^{-8}x^{17} \\ &+ 1.453571773 \times 10^{-7}x^{16} - 5.173598972 \times 10^{-13}x^{28} - 2.569735395 \times 10^{-14}x^{31} \\ &+ 1.252296566 \times 10^{-13}x^{30} - 2.016131906 \times 10^{-13}x^{29} + 2.007605778 \times 10^{-15}x^{32} \\ &+ 2.564345160 \times 10^{-12}x^{27} + 3.603899741 \times 10^{-9}x^{22} + 3.025 \times 10^{-13}x^{14} \\ &- 1.392179800 \times 10^{-10}x^{12} + 6.103539401 \times 10^{-10}x^{11} + x^{5} + 9.879565106 \times 10^{-12}x^{24} \\ &- 2.268156651 \times 10^{-12}x^{26} - 5.281071651 \times 10^{-12}x^{25} - 3.917282540 \times 10^{-10}x^{23} \\ &- 1.335600908 \times 10^{-13}x^{15} - 6.059998643 \times 10^{-10}x^{10}. \end{aligned} \tag{3.33}$$

The obtained solution is once again of remarkable accuracy, as shown in Table 6 and Figure 4.

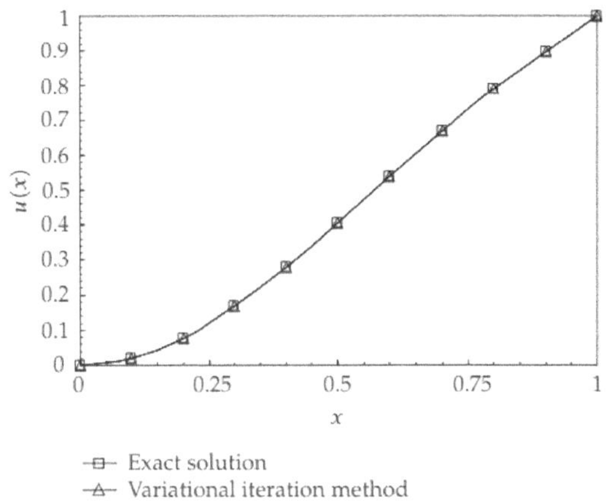

Figure 4: Comparison between different solutions.

CONCLUSION

This study showed that the variational iteration method is remarkably effective for solving boundary value problems. A fourth-order differential equation with particular engineering applications was solved using the VIM in order to prove its effectiveness. Different forms of the equation having boundary conditions of physical significance were considered. Comparison between the approximate and exact solutions showed that one iteration is enough to reach the exact solution. Therefore the VIM is able to solve partial differential equations using a minimum calculation process. This method is a very promoting method, which promises to find wide applications in engineering problems.

REFERENCES

1. R. A. Usmani, "On the numerical integration of a boundary value problem involving a fourth order linear differential equation," BIT, vol. 17, no. 2, pp. 227–234, 1977.
2. R. A. Usmani, "An O(h6) finite difference analogue for the solution of some differential equations occurring in plate deflection theory," Journal of the Institute of Mathematics and Its Applications, vol. 20, no. 3, pp. 331–333, 1977.
3. S. M. Momani, Some problems in non-Newtonian fluid mechanics, Ph.D. thesis, Wales University, Wales, UK, 1991.
4. T. F. Ma and J. da Silva, "Iterative solutions for a beam equation with nonlinear boundary conditions of third order," Applied Mathematics and Computation, vol. 159, no. 1, pp. 11–18, 2004.
5. J.-H. He, "New interpretation of homotopy perturbation method," International Journal of Modern Physics B, vol. 20, no. 18, pp. 2561–2568, 2006.
6. D. D. Ganji and A. Sadighi, "Application of He's homotopy-perturbation method to nonlinear coupled systems of reaction-diffusion equations," International Journal of Nonlinear Sciences and Numerical Simulation, vol. 7, no. 4, pp. 411–418, 2006.
7. M. Rafei and D. D. Ganji, "Explicit solutions of Helmholtz equation and fifth-order Kdv equation using homotopy-perturbation method," International Journal of Nonlinear Sciences and Numerical Simulation, vol. 7, no. 3, pp. 321–328, 2006.
8. N. H. Sweilam and M. M. Khader, "Variational iteration method for one dimensional nonlinear thermoelasticity," Chaos, Solitons & Fractals, vol. 32, no. 1, pp. 145–149, 2007.
9. S. M. Momani and Z. Odibat, "Numerical comparison of methods for solving linear differential equations of fractional order," Chaos, Solitons & Fractals, vol. 31, no. 5, pp. 1248–1255, 2007.
10. S. M. Momani and S. Abuasad, "Application of He's variational iteration method to Helmholtz equation,"Chaos, Solitons & Fractals, vol. 27, no. 5, pp. 1119–1123, 2006.
11. N. Bildik and A. Konuralp, "The use of variational iteration method, differential transform method and adomian decomposition method for solving different types of nonlinear partial differential equations,"International Journal of Nonlinear Sciences and Numerical Simulation, vol. 7, no. 1, pp. 65–70, 2006.
12. Barari, M. Omidvar, S. Gholitabar, and D. D. Ganji, "Variational iteration method and homotopy-perturbation method for solving second-order nonlinear wave equation," in Proceedings of the International Conference of Numerical Analysis and Applied Mathematics (ICNAAM '07), vol. 936, pp. 81–85, Corfu, Greece, September 2007.
13. D. D. Ganji, M. Jannatabadi, and E. Mohseni, "Application of He's variational iteration method to nonlinear Jaulent-Miodek equations and comparing it with ADM," Journal of Computational and Applied Mathematics, vol. 207, no. 1, pp. 35–45, 2007.
14. J.-H. He, "Variational iteration method—a kind of nonlinear analytical technique: some examples,"International Journal of Non-Linear Mechanics, vol. 34, no. 4, pp. 699–708, 1999.
15. Z. Odibat and S. Momani, "Application of variational iteration method to nonlinear differential equations of fractional order," International Journal of Nonlinear Sciences and Numerical Simulation, vol. 7, no. 1, pp. 27–34, 2006.
16. H. Tari, D. D. Ganji, and M. Rostamian, "Approximate solutions of K (2.2), KdV and modified KdV equations by variational iteration method, homotopy perturbation method and homotopy analysis method," International Journal of Nonlinear Sciences and Numerical Simulation, vol. 8, no. 2, pp. 203–210, 2007.
17. E. Yusufoglu, "Variational iteration method for construction of some compact and non-compact structures of Klein-Gordon equations," International Journal of Nonlinear Sciences and Numerical Simulation, vol. 8, no. 2, pp. 152–158, 2007.

CHAPTER 9

SOLVING THE PROBLEM OF MULTIPLE-CRITERIA BUILDING DESIGN DECISIONS WITH RESPECT TO THE FIRE SAFETY OF OCCUPANTS: AN APPROACH BASED ON PROBABILISTIC MODELLING

Egidijus Rytas Vaidogas and Jurgita Šakėnaitė

INTRODUCTION

Architectural and structural design of buildings face many problems and these problems include a selection among alternative plans of houses, individual floors, alternative use of floors in one building, alternative structural solutions, and structural materials. The selection problem is highly important in the early design stage. In the literature, attempts are described to provide some theoretical base for the selection among alternative architectural and structural solutions [1–3].

The presence of several alternative floor plans or similar alternative architectural solutions generates a problem of a multiple-criteria selection with respect to economic and architectural/structural characteristics (criteria) of each alternative. As fire prevails among hazards present in buildings, criteria related to fire safety should be included in the selection problem.

If considered from the standpoint of fire safety, alternative architectural and structural solutions mean different possibilities of fire spread, evacuation, and firefighting. A choice of a specific solution among a number of alternatives should take into account a number of economic, functional, technical, aesthetical criteria. Methods of multiple-criteria decision-making (MCDM) can be used for such a choice [4]. Until now, a building-related MCDM did not use criteria expressing fire safety [5–7]. Therefore, it makes sense to include criteria expressing fire safety into building-related MCDM problems.

In the field of fire safety, MCDM was applied to ranking fire safety criteria [8, 9]. MCDM was also used for the choice among alternative buildings with respect to fire safety and selection of fire safety systems [10, 11].

The present study proposes a building-related MCDM which takes into account fire risk to occupants. The conventional application of MCDM to build property is expanded by applying estimates of the fire risk as MCDM criteria. They are expressed in terms of probabilities that occupants will be exposed to untenable conditions which can build up on escape routes. Criteria expressed through time to untenable conditions and rescue time are applied to ranking alternative building design solutions. The proposed procedure allows solving MCDM problems when some elements of a decision-matrix are random. An application of the procedure is illustrated by a choice among different architectural solutions of a medical building.

Citation: Egidijus Rytas Vaidogas and Jurgita Šakėnaitė, "Solving the Problem of Multiple-Criteria Building Design Decisions with respect to the Fire Safety of Occupants: An Approach Based on Probabilistic Modelling," Mathematical Problems in Engineering, vol. 2015, Article ID 792658, 18 pages, 2015. doi:10.1155/2015/792658.

THE PROBLEM OF CHOOSING AMONG ALTERNATIVE DESIGN SOLUTIONS OF A BUILDING

Development of Alternative Designs

A generation of alternative architectural and structural designs (or, briefly, alternatives) is one of the main tasks arising in the early stage of a building creation. The process of developing different alternatives is to a large margin intuitive and is not easily amenable to a mathematical formalisation. This process will be highly influenced and sometimes constrained by regulations and client wishes. However, the designer (architect or structural engineer) will always have certain freedom in specifying alternatives and choosing among them, especially, in the early stage of building design. At the same time, an improper rating of alternatives and an erroneous choice of one as seemingly the best solution can protract the entire process of design and construction and encumber exploitation. Consequently, an application of mathematical means that facilitating selection of the best alternative among a set of developed alternatives will allow reducing probability of a wrong choice.

A generation of alternatives can be interpreted as a discretisation of a continuous process. In principle, small changes in the design will allow generating an almost endless set of alternatives. However, in many cases the designer will face the problem of a selection among a limited set of alternatives which represent principal or fairly different design solutions.

The set of n alternatives denoted, say, by $a_1, a_2, \ldots, a_i, \ldots, a_n$ can be generated at different levels of development (LODs) in the process of building design. Five LODs range from "conceptual" to "as-built" and offer different possibilities for specifying, comparing and rating the alternatives a_i (Table 1). On the conceptual level, a_i can represent different shapes or floor plans of a building under design (Figure 1). On the lower levels of schematic and detailed design, the alternatives ai can be developed within a given shape or floor plan of a building (Figure 2). Finally, the alternatives ai can be generated for a building with a specific configuration and floor plans. These alternatives can represent different structural systems and materials of the main structure and nonload bearing elements, alternative internal and external finishes, and different solutions of building utilities

Table 1: Definition of levels of development (LOD) in a building design process [12, 13].

Level	Brief description	Definition
LOD 100	Conceptual design	Vague information consisting of nongeometric data or line work, areas, volumes, zones, and so forth
LOD 200	Schematic design	Generic elements shown in three dimensions
LOD 300	Detailed design	Specific elements with object geometry (dimensions, capacities, and connections)
LOD 400	Construction	Shop drawing/fabrication
LOD 500	As-built (actual)	The project as it has been constructed. The model and associated data is suitable for maintenance and operations of the facility

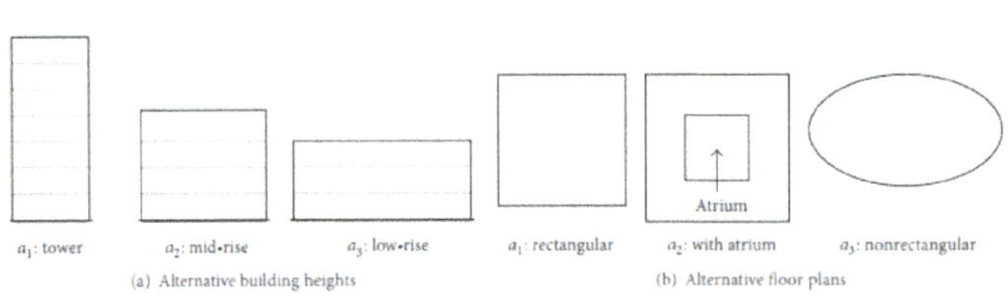

Figure 1: Alternative shapes and floor plans of a building specified on the conceptual level of design process.

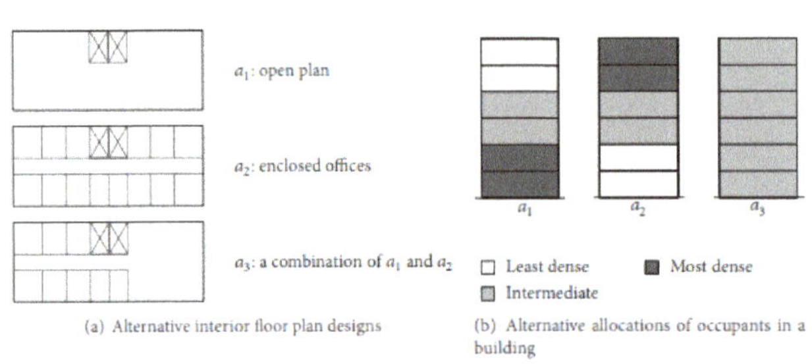

Figure 2: Alternative design solutions of a building within a given geometrical configuration.

A formal framework for comparison and ranking of the alternatives a_i is the methodology of MCDM. It allows interalternative comparisons of ai by applying a set of criteria c_1, c_2,...,,...,$_m$. A problem statement of MCDM is n-by- m decision-making matrixC. Its element c_{ij} represents a value of the criterion c_j related to the alternative a_i. A wide variety of MCDM methods allows ranking the alternatives a_i by applying both quantitative and qualitative criteria c_j. MCDM can accommodate subjective and objective, fixed, fuzzy, and random criteria c_j [14].

In the context of building design organised as a successive passing of LODs, a comparison of the alternatives a_i will make sense if ai will represent the same LOD. In principle, branching of the design process into the alternatives a_i is possible on the conceptual level LOD 100, as long as it is possible to characterise the vague information expressed by conceptual designs by some qualitative or quantitative criteria c_j (Figure 3(a)). Most of them will express subjective opinions of architect and client, because very little can be measured quantitatively and objectively at this stage of design

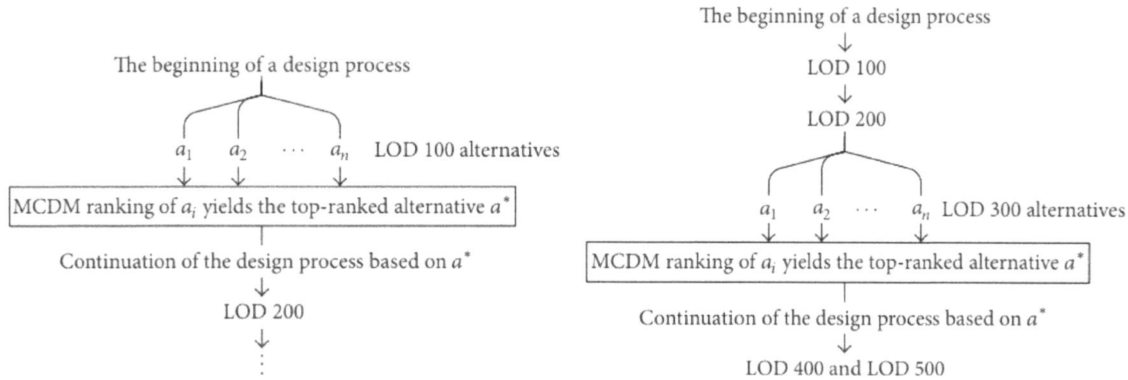

(a) Comparison of conceptual designs

(b) Comparison of designs on higher level of detail

Figure 3: A generation of the alternative building designs a_i at different levels of development.

Higher levels of detail represented by LODs from 200 to 400 give better opportunity to formulate and solve a MCDM problem. Branching of the design process into the alternatives a_i can be done on each of these levels (Figure 3(b)). At the same time, a development of a large number n of the alternatives ai and characterisation of each of them by the criteria c_j can encumber the design process. In essence, a solution of a MCDM problem will require preparing n different designs of building related to the LOD reached in the design process. On the other hand, a successful choice of the best alternative design (top-ranked alternative, say, *) may allow compensating for that extra effort, due to the effect of a multiple-criteria optimisation.

Inclusion of Criteria Related to Building Fire Safety

The prevailing hazard in most of nonindustrial occupations and many industrial buildings is the fire hazard. Its converse, a fire safety, is usually decomposed into safety of occupants, safety of fire-fighting operations, and safety of structures subjected to fire effects. A certain number ms of quantitative or qualitative measures expressing the three aforementioned aspects of safety can be included into an m-dimensional vector of MCDM criteria, c. For instance, the vector c can be composed of two subvectors:

$$c = (c' \mid c'') = (c_1, c_2, \ldots, c_{m-m_s} \mid c_{m-m_s+1}, c_{m-m_s+2}, \ldots, c_m), \quad (1)$$

where c; and c'' are vectors including criteria of general nature and safety-related criteria, respectively. To compose the decision-making matrixC, it is necessary to specify values of c' and c'' for each alternative a_i. The fire safety can be influenced by decisions made on all LODs. For instance, early decisions at LOD 100 or LOD 200 may pose the following selection problems related to the fire safety:(1)An inclusion of a large volume space like an atrium into the building will generate the need to deal with increased risk to the life of occupants. Atria allow smoke and heat to travel throughout all floors of a building (e.g., [15, 16]). The opposite alternative of a solid floor plan will not pose this problem.(2)A decision to erect a high-rise tower instead of a low-rise, longitudinal building will automatically pose the problems related to fire safety of tall buildings (e.g., [17]).(3)An allocation of people predominantly in the lower floors of an office tower will increase the chance of a successful evacuation in comparison to other occupant distributions (Figure 2(b)). Occupant loads depend on human behaviour which is sometimes difficult to control. However, alternative allocations can be assessed and compared by means of probabilistic measures [18].

Although LOD 100 and LOD 200 designs present a vague information, some criteria c_j belonging to the subvector c'' can be specified and used for solving a MCDM problem. Let us compare a building with an atrium (design concept a_1) and a solid plan building (design concept a_2) (Figure 1(b)). The criterion c_j can be a number of scores subjectively assigned by an expert to fire safety level of a_1 and a_2 (e.g., a_1 earns 6 and a_2 earns 8 in a 10-point scale). A component of c'' can also be an average cost (in C/m^2 , say) of fire safety systems which can or must be installed in a_1 and a_2. The concept $a1$ presupposes a provision of smoke and heat control equipment in the atrium. Therefore, fire safety system of a_1 may cost in average more than fire safety system of a_2. Even if a_1 is preferable from the architectural standpoint, higher average fire safety costs of a_1 and possibly higher risk to occupants can make a_1 less appealing than a_2. The cost of build-in fire protection is high (e.g., [19]). A trade-off between cost of fire safety and fire risk to occupants may be necessary in many design situations [20]. The need of such a trade-off is capable of creating a mutual relation between architectural design and fire protection at LOD 100 or LOD 200 (Figure 4(a)).

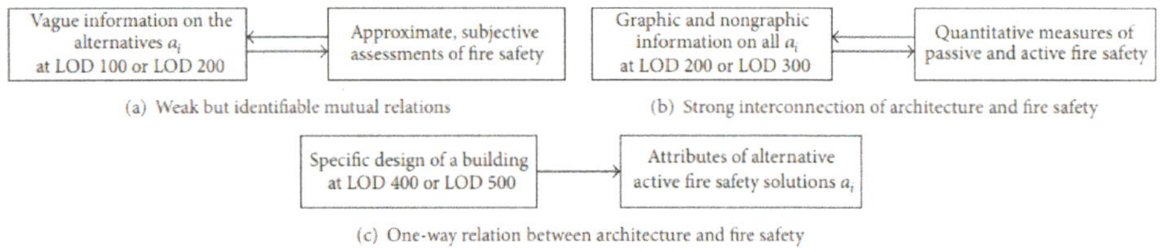

Figure 4: Three levels of an interconnection of building design and assurance of fire safety.

The best platform for a comparison of the alternatives a_i by means of formal MCDM methods is LOD 200 and LOD 300 designs (Figure 4(b)). The knowledge of at least approximate geometry and building materials specified at these development levels will allow applying a computer simulation of fire and evacuation processes. Many computer codes used for such a simulation require transferring geometry and material data between a CAD system and often coupled systems of fire and evacuation simulation (e.g., FDS and Evac codes [22, 23]). The computer simulation allows obtaining estimates of fire and evacuation processes. These estimates can be used as fire-related criteria c_j belonging to the subvector c''. The criteria c_j will be able to reflect both passive and active fire safety measures. For instance, different compartmentation and internal lining can be used to develop the alternatives a_i.

The alternatives a_i can also be developed for a unique architectural and structural solution of the building at LOD 400 or LOD 500, in which passive fire protection measures are unchangeable. However, alternative technical measures can be provided as an active fire protection [11, 12]. The components c_j of the subvector c'' will have to reflect a fire safety level achieved by a combined provision of passive and active measures.

MCDM methods allow ranking the alternatives a_i with deterministic, random, and fuzzy criteria c_j. Despite the fact that the praxis of fire and evacuation assessment is predominantly deterministic, all key characteristics related to a building fire safety are random [24, 25]. For instance, random quantities are

(1) Evacuation times related to the paths that will be available during a certain time after a fire outbreak,

(2) Times to untenable conditions in available evacuation paths,

(3) Times to structural failures which can endanger firefighting operations,

(4) Potential numbers of victims among occupants and fire fighters.

This randomness should be taken into account in specifying the components c_j of the subvector c''. The criteria c_j can be uncertain quantities or quantities expressing uncertainties (e.g., probabilities).

MATHEMATICAL SPECIFICATION OF DECISION CRITERIA

The criteria c_j related to fire protection of a building can be very diverse and reflect economic and technical characteristics of build-in fire safety [11, 12]. It is reasonable to state that the most important criteria c_j should measure the life safety of building occupants in a preflashover period of fire development.

In previous decades, the fire safety of buildings was quantified by means of nonprobabilistic fire risk indices [26]. They reflect in certain way the life safety and, technically speaking, can be used as c_j. However, the indices are relatively insensible to architectural and structural changes and so are not very useful for interalternative comparisons [27].

The most informative and integral criterion cj expressing the life safety is the probability of a successful evacuation from a building fire:

$$P_{\text{se}} = P\left(T_u \geq T_e\right) = P\left(T_u \geq T_d + T_r + T_{\text{tr}}\right), \qquad (2)$$

where T_u is the time to untenable conditions (the time gaseous combustion products take to travel from the fire room and produce untenable conditions on an escape route); T_e is the total evacuation time; T_d is the discovery time (the time period from ignition to discovery of a fire, known also as the perception time); T_r is the reaction time (the time period from fire discovery to the start of escape action, known also as action or recognition time or gathering phase);T_{tr} is the travel time (the time taken to move to the place of safety). Needless to say, the times in (2) are always subject to uncertainties and must be modelled as random variables.

The chance of fire injuries will increase with the increasing duration of exposure to untenable conditions, expressed by the difference T_e–T_u. The information behind the random times T_e and T_u can be utilised for a MCDM at least in two ways.

If it is possible to obtain accurate and "cheap" estimates of the probabilities ($T_u \geq T_e$), they can be used as MCDM criteria

$$c_{ij} = P\left(T_{uij} \geq T_{eij}\right), \qquad (3)$$

where T_{uij} and T_{eij} are random times related to the alternative ai and available evacuation path j. The criteria defined by (3) are in essence deterministic and can be easily included into the decision-making matrix C. However, scarcity of hard data on the times T_{uij} and T_{eij} may generate uncertainty related to the "true" values of the probabilities ($T_{uij} \geq T_{eij}$). In terms of the theory known as the quantitative risk assessment such uncertainty is called epistemic (state-of-knowledge) uncertainty and quantified by applying methods of Bayesian statistical theory (e.g., [28]). A result of a Bayesian estimation of ($T_{uij} \geq T_{eij}$) will be an epistemic random variable $\tilde{c}_{ij}$. It can be used as an element of the decision-making matrix. However, in the latter case a MCDM problem will be formulated as a decision-making matrix $\tilde{C}$, some elements of which are uncertain in the epistemic sense. Simply stated, the decision problem will be expressed by a random matrix $\tilde{C}$.

The second way of utilising information behind T_e and T_u is to use the random criteria

$$\tilde{c}_{ij} = T_{uij} - T_{eij}. \quad (4)$$

They will allow carrying out interalternative comparisons without estimating the probabilities ($T_{uij} \geq T_{eij}$). At the same time, information used to assess probability distributions of T_{uij} and T_{eij} will be in the main part the same as information used to estimate ($T_{uij} \geq T_{eij}$). With the criteria $\tilde{cij}$ defined by (4), a MCDM problem will be formulated through a random decision-making matrix $\tilde{C}$.

The direct data allowing us to fit the probability distributions of T_{uij} and T_{eij} will hardly be available in most design situations. However, distributions of $Tuij$ and $Teij$ and so the probabilities ($T_{uij} \geq T_{eij}$) can be estimated by means of a stochastic (Monte Carlo) simulation of fire and evacuation processes. Such a simulation became an intensive field of fire safety investigation in recent years [29–31]. In context of MCDM, the simulation can yield generated samples of T_{uij} and $T_{ei}j$. These samples can be used for either choosing the probability distributions of T_{uij} and T_{eij} or estimating ($T_{uij} \geq T_{eij}$).

A specification of probabilistic input information for a computer-aided fire simulation can be gained from various sources. There exist large collections of data on fires in general [8]. Databases related to fires are collected and maintained in such particular areas as safety of nuclear power plants [32]. Unfortunately, the data situation in some specific areas of fire safety (e.g., sprinklers and fire alarms) is not very encouraging [33]. Consequently, specification of input information for assessing such values as T_{uij} and T_{eij} will have to rely on Bayesian analysis widely used for a probabilistic risk assessment.

The distributions (estimates) of the times T_{uij} and T_{eij} can be applied to solving a MCDM problem with either deterministic or stochastic decision-making matrix, C or $\tilde{C}$. Several MCDM methods were proposed to solve the selection problem with the random matrix $\tilde{C}$ [34–36]. The following case study illustrates dealing with $\tilde{C}$ by means of a MCDM method based on stochastic simulation.

AN APPLICATION TO A RECONSTRUCTION OF A MEDICAL BUILDING

Alternatives under Comparison and Safety-Related MCDM Criteria

The present case study considers an existing two-storey wing of a hospital building in Lithuania. The first floor of the wing accommodates a haemodialysis unit and the second floor is used for a catering department (Figures5 and 6). Smoke detectors in all rooms of the building will activate an alarm system in case of fire. Fresh air supply fans installed in the building are supposed to automatically shut down when the fire alarm is activated.

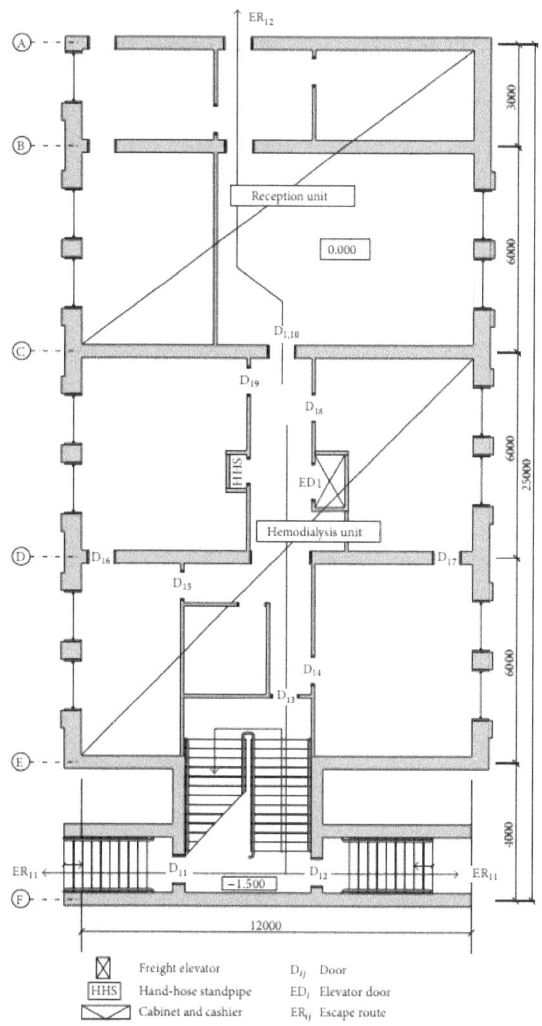

Figure 5: Plan of the haemodialysis unit in the first floor of the building.

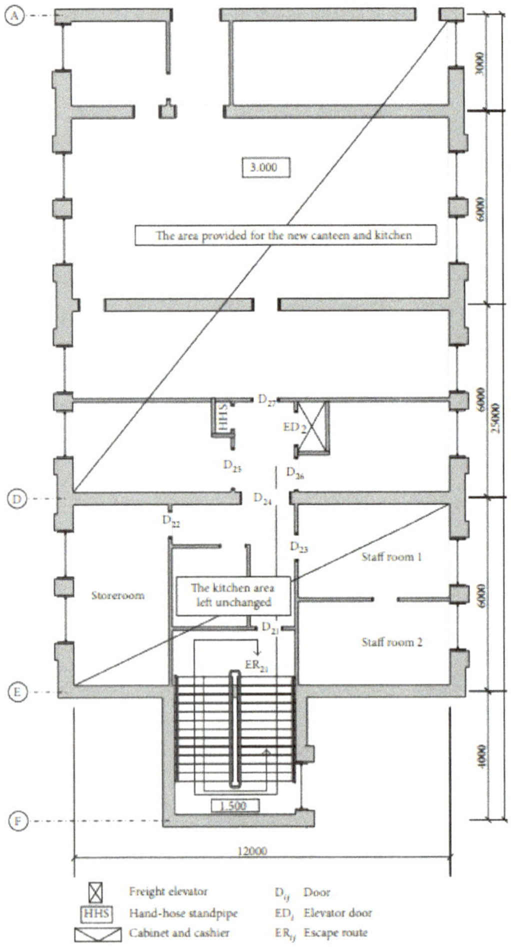

Figure 6: Existing wall plan of the catering department in the second floor of the building.

A hospital administration is going to reconstruct the catering department and to open a canteen in the second floor. The administration wants to create the most functional and efficient floor plan for running the canteen. At the same time, the hospital administration knows that cooking is the leading cause of fires in healthcare facilities.

Three new floor plans are considered in the design (Figures 7 to 9). They differ in wall plans and the position of the potential room of fire initiation (the room with stoves and ovens). The three floor plans will constitute the alternatives of a MCDM problem:

(i) a^1 is a floor plan with single-room seating area (66.1 m^2) and large service area (31.4 m^2) (Figure 7), a relatively large number of openings allowing a convenient movement of kitchen staff and products; the kitchen area is the smallest among the alternative floor plans (Table 2).

(ii) a_2 is a two-room seating area (51.9 m^2 and 26.3 m^2), relatively small service area (11 m^2) (Figure 8); among the three alternative plans, the room of potential fire initiation in the plan a^2 is at the largest distance from freight elevator, stairwell, and hospital units in the first floor.

(iii) $a^3$is a two-room seating area (50 m^2 and 18 m^2), relatively small service area (12.3 m^2) (Figure 9); the room of potential fire initiation is relatively well isolated from other rooms by compartmentation; the kitchen area is the largest among the alternative floor plans (Table 2).

Table 2: Characteristics of the floor plans a_1, a_2, and a_3 shown in respective Figures 7, 8, and 9 (the bold rows indicate characteristics of the alternatives which will be used as MCDM criteria).

Floor plan characteristic	Plan a_1	Plan a_2	Plan a_3
Seating area, m^2	66.1	78.2	68.0
Kitchen area, m^2	**42.7**	**49.1**	**57.2**
Service area, m^2	31.4	11.0	12.3
Sum of the above three areas, m^2	140	138	132
Floor reconstruction cost, thousands of €	**458.0**	**298.5**	**315.6**
Estimated reconstruction time, months	**3.5**	**2.1**	**2.0**

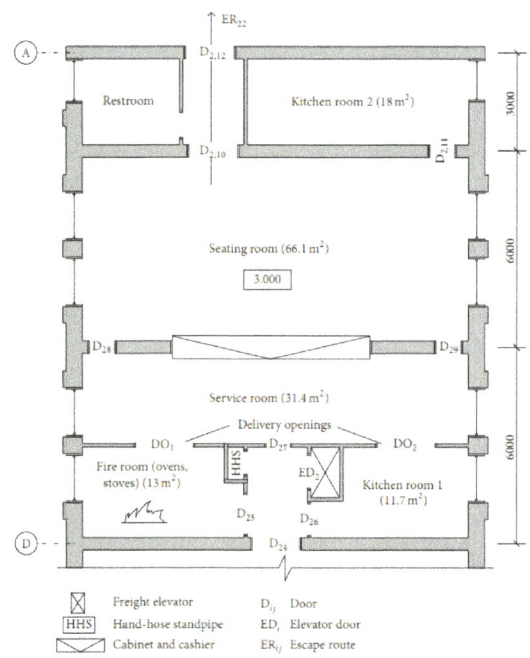

Figure 7: New wall plan a_1 for the canteen.

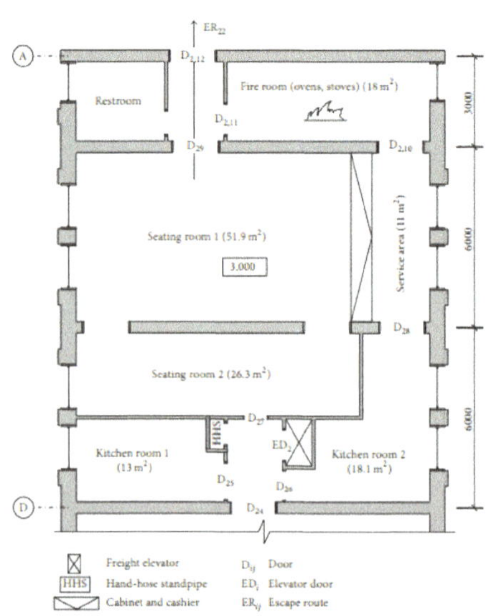

Figure 8: New wall plan a_2 for the canteen

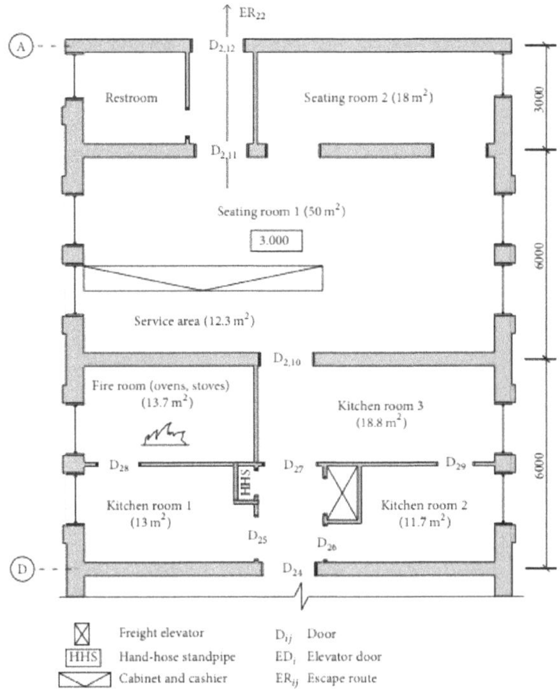

Figure 9: New wall plan a_3 for the canteen.

In terms of fire safety, the alternatives a_1, a_2, and a_3 can be characterised as follows:

(i) a_1 is the floor plan with the least effective compartmentation; the room of potential fire initiation is in front of the freight elevator and is relatively close to the stairwell; gaseous combustion products may flow from the fire room to the inside through two horizontal vents (door D_{25} and delivery opening DO_1, Figure 7); in case of fire, the combustion products will first affect kitchen staff and then visitors of the canteen; the horizontal evacuation is possible through the escape routes ending with doors D_{21} and $D_{2,12}$ (Figures 6 and 7).

(ii) a_2 is the floor plan with the most remote position of the fire room with respect to the freight elevator and stairwell, an average compartmentation in comparison to plans a_1 and a_3; there are two horizontal vents for the flow of gaseous combustion products to the inside (doors $D_{2,10}$ and $D_{2,11}$, Figure 8); in case of fire, the staff in the fire room and visitors of the canteen will be affected first by combustion products; the horizontal evacuation is possible through the escape route ending with door D_{21}, whereas the evacuation through $D_{2,12}$ can be problematic (Figures 6 and 8).

(iii) a_3 is the floor plan with a relatively effective compartmentation with respect to the position of the fire room; gaseous combustion products may flow from the fire room to the inside only through door D_{28} (Figure 9); the possible paths of horizontal evacuation ending with doors D_{21} and $D_{2,12}$ are similar to those in the plan a_1 (Figures 6 and 9).

The alternatives a_1 to a_3 will differ in the economic sense and, to a certain degree, in the necessary operations of reconstruction. For instance, the floor plan a_1 requires providing a large opening in the load-bearing masonry wall for cabinet and cashier (Figure 7). This opening will be 6 m wide and must be covered by a lintel with a relatively wide span. It is clear that construction of such an opening and lintel will increase the duration and cost of reconstruction works required by the floor plan a_1 with respect to the plans a_2 and a_3. On the other hand, the floor plans a_2 and a_3 will require constructing a larger number of partitions to form new rooms than the floor plan a_1.

The differences in the floor plans will lead to differences in the risk posed by a potential fire to patients, medical staff, and visitors of the canteen. A selection among the alternatives a_1 to a_3 must include a MCDM criterion which reflects the risk to lives of these three categories of occupants.

The alternative floor plans a_1, a_2, and a_3 can be compared in terms of the probability P_{se}. However, this probability is relatively difficult to be estimated and the estimation will involve information common to all three alternatives:

(1) The discovery time T_d may be considered equal for all three alternatives (the existing part of the building is and new part will be equipped with automatic detectors and alarm which reduce T_d to approximately one minute in the whole building (e.g., [8]).

(2) The reaction time T_r may also be assumed to be equal for all three alternatives because the types and numbers of occupants (haemodialysis patients, medical staff, and kitchen staff as well as visitors of canteen and reception unit) will not depend on the wall plans in individual alternatives.

(3) The time needed for evacuation of the haemodialysis unit, T_{tr}, may be considered equal for all three alternatives because different floor plans in the second floor will not influence possibilities of escape from this unit.

The alternative floor plans will create different possibilities for gaseous combustion products to travel horizontally through the second floor and to leak into the haemodialysis unit. This may cause differences in the time to untenable conditions on potential escape routes, T_u, in both floors. The different wall plans and position of fire room will also provide different routes of evacuation and so differences in the travel times T_{tr} related to the individual alternatives. The idea was to compare the alternative floor plans by using the times T_u and T_{tr} and not the probabilities of a successful evacuation, P_{se}.

The time to the production of untenable conditions in the haemodialysis unit (say T_{u1}) can be taken as a candidate MCDM criterion related to the live safety of haemodialysis patients, because the total time of evacuation from this unit, T_e, may be considered the same for all three alternatives:

$$c_{ij} = T_{u,i1} \quad (i = 1, 2, 3)\,, \tag{5}$$

where the values of the subscript i refer to a_1, a_2, and a_3 and subscript "1" denotes the first floor.

The life safety of kitchen staff and visitors of the canteen will depend on times to untenable conditions (say T_{u2}) and travel times (say T_{tr2}) related to available escape routes. A candidate MCDM criterion expressing the life safety of occupants in the second floor can be expressed by the difference of T_{u2} and T_{tr2}:

$$c_{i,j+1} = T_{u,i2} - T_{tr,i2} \quad (i = 1, 2, 3)\,, \tag{6}$$

where subscript "2" refers to the second floor The times $T_{u,1}$, $T_{u,i2}$, and $T_{tr,i2}$ depend on the development of fire and movement of occupants on the second floor. An estimation of probability distributions of these times will require modelling a development of fire in the building under analysis. Consequently, a specification of a MCDM decision-making matrix $\tilde{C}$ used in the present case study will include a computer fire simulation.

Probabilistic Simulation of Burning Objects

The fire is assumed to be initiated in the room used to operate stoves and ovens or, briefly, the fire room (Figures 7 to 9). The fire will be initiated by an ignition of an electrical kitchen stove and a subsequent ignition of flue above the stove. The fire will be caused by a malfunction of a thermostat in the stove and ignition of a mixture of oil and grease in a tray.

The heat release rate (HRR) of the oil and grease pool fire may be assumed to be constant over the fire duration (0, t_o), where the subscript "o" stands for "oil" [37, 38]. The constant HRR of this fire, $\dot{q}_o$, will be modelled by the expression

$$\dot{q}_o = \bar{q}_o + \xi_o, \tag{7}$$

where $\bar{q}_o$ is the average value of $\dot{q}_o$ estimated with the data given in Figure 10 and ξ_o is the random variable expressing stochastic uncertainty in values of $\dot{q}_o$. This uncertainty results from uncertainties in the variables mentioned in Figure 10.

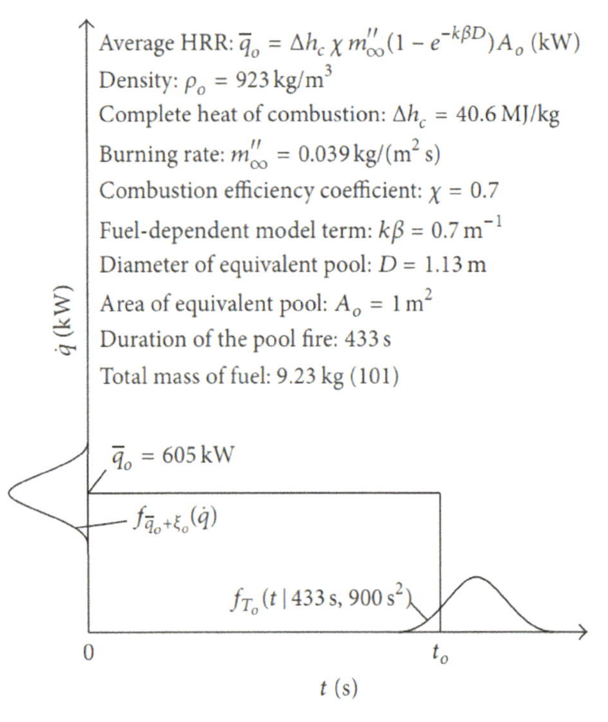

Figure 10: Mathematical model of the heat release rate (HRR) for the pool fire of oil and grease (the model of $\bar{q}_o$ was adopted from Karlsson and Quintiere [21]).

For the purpose of fire simulation, the value of $\bar{q}_o$ was estimated to be equal to 605 kW and the probability distribution of ξ_o was assumed to be normal with the mean of 0 kW and variance of 1200 (kW)^2. The duration of the pool fire, t_o, is assumed to be uncertain and modelled by the normal random variable T_o with the distribution N (433 s, 900 s^2). Densities of the random variables ξ_o and T_o are illustrated in Figure 10. The HRR of flue fire, which will be initiated by the pool fire of oil and grease at the time t_{f0}, will be simulated as a time history $\dot{q}(t)$ expressed as

$$\dot{q}_f(t) = \bar{q}_f(t) + \xi_f(t) \quad \text{with } t \in (t_{f0}, t_f), \tag{8}$$

where $\bar{q}_f(t)$ is the average value of $\dot{q}_f(t)$ and $\xi_f(t)$ is the random, "fluctuating part" of $\dot{q}_f(t)$ at the time t, and $t_{f0} - t_f$ is the duration of the flue fire. For the purpose of a computer fire simulation, the continuous process $\dot{q}(t)$ was replaced by the set of random variables

$$\Xi_f = \left\{\left(\bar{q}_f(t_l) + \xi_f(t_l)\right),\ l = 1, 2, \ldots, 17\right\}, \tag{9}$$

where t_l are time moments obtained by assuming that the flue fire will last 400 s and dividing the fire duration $t_f - t_{f0}$ into 16 intervals, each lasting 25 seconds (Figure 11); $\bar{q}_f(t_l)$ is the average HRR at the moment t_l; and $\xi_f(t_l)$ is a normally distributed random variable with the mean 0 kW and standard deviation equal to 0.07 × $\bar{q}_f(t_l)$. The random variables (t_l) were assumed to be correlated and the correlation coefficients were calculated by means of the following model [39]:

$$\rho_{lk} \equiv \rho\left(\xi_f(t_l), \xi_f(t_k)\right) = \exp\{-\kappa |t_l - t_k|\} \tag{10}$$

with κ = 0.02.

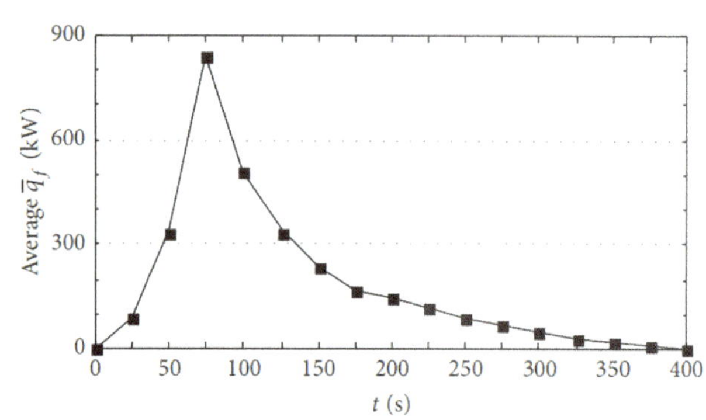

Figure 11: Time history of the average heat release rate of the flue fire, $\bar{q}_f(t_l)$, drawn for t_{f0} = 0 s and t_f = 400 s.

The discretisation of the fire duration (t_{f0}, t_f) follows from the measurement techniques of the so-called free burn experiments used for determining HRR values at given time moments t_l (e.g., [40]). A certain number of individual time histories (signals) measured in repeated experiments allow estimating the mean values $\bar{q}_f(t_l)$ at given t_l and fitting probability distributions of the random variables (t_l).

The stove and flue fire will spread to a set of cupboards in the surrounding of the stove. The cupboards will ignite when the upper gas layer reaches the temperature of 200∘ C. The HRR history of the cupboards was modelled similarly to the one of the flue fire, that is, by a set of random variables

$$\Xi_c = \left\{\left(\bar{q}_c(t_l) + \xi_c(t_l)\right),\ l = 1, 2, \ldots, 17\right\}, \tag{11}$$

where the subscript "c" stands for "cupboard," $\bar{q}_f(t_l)$ is mean values of HRR at the moment t_l, and $\xi_c(t_l)$ is the "fluctuating" part of HRR at t_l. The sequence of the mean values $\bar{q}_f(t_l)$ was adopted from Babrauskas [40] and is shown in Figure 12. The duration of the cupboard fire is assumed to be equal to 800 s and thus $t_1 = 0$ s, $t_{17} = 800$ s, and $t_{i+1} - t_i = 50$ s (Figure 12). The fluctuating parts (t_l) were modelled as correlated random variables with mean values equal to 0 kW and standard deviations equal to $0.05 \times \bar{q}_f(t_l)$. The coefficients of correlation $((t_l), \xi_c(t_k))$ were calculated with the model given by (10) with $\kappa = 0.01$.

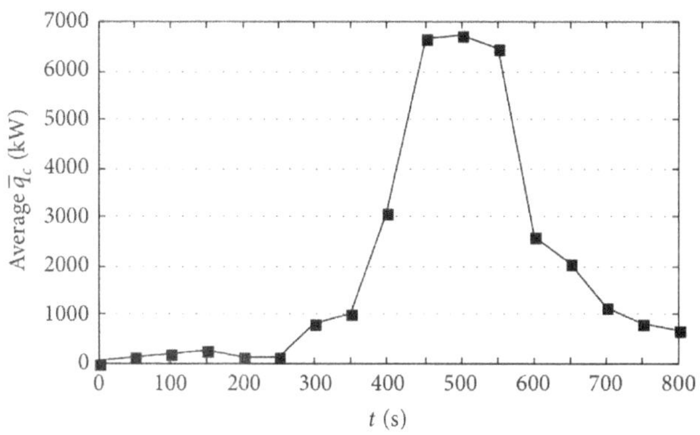

Figure 12: A sequence of average heat release rates of the cupboard fire $\bar{q}_f(t_l)$.

The window in the kitchen room is assumed to break when the temperature reaches a sufficiently high temperature. In the present study, the breaking temperature and fraction of broken glass area will be modelled by two respective random variables $\xi_w t$ and ξ_{wa}. The following probability distributions were assumed for these variables: $\xi_{wt} \sim N$ (250∘ C, 155(∘ C)^2) (normal distribution) and $\xi_{wa} \sim$ Be (6, 21) (a beta distribution with the mode of 0.2).

The variables ξ_{wt} and ξ_{wa} are assumed to be uncorrelated because any data, which can substantiate presence or absence of a stochastic dependence between ξ_{wt} and ξ_{wa}, is not known to us. Characteristics of the random variables ξ_{wt} and ξ_{wa}, along with ones of other random variables used in the problem, are summarized in Table 3.

The scenario of fire initiation is the same in all three alternatives. The position of stove and cupboards with respect to the window in the fire room is also assumed to be the same in all alternative floor plans.

Table 3: Probability distributions of the random variables used for the computer fire simulation in the hospital building considered in a MCDM problem.

Variable	Symbol	Mean	Coeff. of var.	Distribution
The fluctuation of the oil fire HRR$^{(1)}$ around the mean value $\bar{q}_o$ (Figure 10)	ξ_o	605 kW	0.05	Normal
The random duration of oil fire (Figure 10)	T_o	433 s	0.07	Normal
The fluctuation of the flue fire HRR$^{(1)}$ around the mean value $\bar{q}_f(t_l)$ at t_l (Figure 11)	$\xi_f(t_l)^{(3)}$	0 kW	0.07	Lognormal
The fluctuation of the cupboard fire HRR around the mean value $\bar{q}_c(t_l)$ at t_o (Figure 12)	$\xi_c(t_o)^{(2)}$	0 kW	0.05	Lognormal
The temperature of wind glass breaking	ξ_{wt}	250°C	0.05	Normal
The fraction of the broken glass area	ξ_{wa}	0.222	0.354	Beta$^{(3)}$

$^{(1)}$HRR: heat release rate.
$^{(2)}$The variables $\xi_f(t_l)$ and $\xi_c(t_l)$ $(l = 1, 2, \ldots, 17)$ are assumed to be correlated with the correlation coefficients calculated by (10).
$^{(3)}$Beta distribution with the parameters 6 and 21 and the mode of 0.2.

Fire Development and Scenarios

Haemodialysis unit and catering department are operated only in the daytime and so lives of patients and staff cannot be threatened if the fire will break out in the night time. If the fire breaks out, the people staying in the second floor will be threatened by heat radiation, toxic gases in the smoke, and impaired visibility which can complicate the evacuation.

People staying in the first floor will be endangered by toxic combustion gasses. They can penetrate from the floor of fire origin into the rooms where patients and staff stay. The most relevant criterion of reaching untenable conditions is the smoke interface height. In the present case study, this height is assumed to be equal to 2 m above the floor in all rooms of the building.

The combustion gasses can spread from the second floor to the first floor by two ways:(1)through the vertical shaft of the freight elevator if the elevator doors ED_1 and ED_2 are left open due to negligence or when the fire outbreaks during loading and unloading operations (Figures 5 and 7 to 9);(2) through the stairwell in the case that doors to the stairwell, D_{21} and D_{13}, are open in both first and second floor (Figures 5 and 6). Leakage of combustion products between the second and first floors through other paths is considered to be negligible. The ventilations located in the first and second floors are not interconnected. In case of fire, these systems will be shut down by fire alarm.

Occupants in haemodialysis unit, kitchen, and canteen have four escape routes:(1)Escape route ER_{11} (first floor, horizontal evacuation): from the haemodialysis unit to the stairwell through the door D_{13} and then to the outside through D_{11} and D_{12} (Figure 5).(2)Escape route ER_{12} (first floor, horizontal evacuation): from the haemodialysis unit to the reception unit and then to the main building of the hospital (Figure 5).(3)Escape route ER_{21} (second floor, horizontal, and vertical evacuation): from the kitchen and canteen to the stairwell through the door D_{21} and then to the outside through D_{11} and D_{12} (Figures 5 and 6).(4)Escape route ER_{22} (second floor, horizontal evacuation): from the kitchen and canteen to the main building of the hospital through the door $D_{2,12}$ (Figures 7 to 9). Two fire scenarios leading to a maximum growth in combustion product concentration (toxicity and limitation of visibility) in the first and second floor will be considered. In the first scenario, the fire is confined to the second floor because the elevator door ED_2 and the door to the stairwell, D_{21}, are closed (Figure 6). This fire scenario involves different event sequences for the alternative floor plans a_1, a_2, and a_3:

(i) a_1: the staff rooms 1 and 2 and the storeroom are unoccupied when the fire outbreaks (Figure 6); the combustion products will block the escape route ER_{21} in a relatively short time and the evacuation will occur through the escape route ER_{22} (Figure 7); the untenable conditions on ER_{22} will first be reached in the service room at the time $T_{u,12}$.

(ii) a_2: untenable conditions on the escape route ER22 will be first reached in the room including service area and seating room at the time $T_{u,22}$ (Figure 8). Evacuation of staff and visitors of the canteen will be possible through the escape route ER_{21} (Figure 6).

(iii) a_3: untenable conditions will be reached first in the fire room and kitchen room 1 (Figure 9); however, the kitchen staff will leave these rooms relatively quickly; the kitchen room 2 can be evacuated through the kitchen room 3; consequently, the escape route ER_{22} will be blocked for kitchen staff staying in room 2 when the untenable conditions will occur in the kitchen room 3 at the time $T_{u,32}$; the evacuation of kitchen staff in the service area and visitors in the seating areas will be possible through the escape route ER_{22}.

In the second scenario, the combustion products will spread to the haemodialysis unit in the first floor through the shaft of the freight elevator. This scenario will take place when the elevator doors in the first and second floors, ED_1 and ED_2, are left open in the course of fire (Figures 5 and 6). The travel of the combustion products from the first to the second floor through the stairwell is considered to be small because the spring doors $D_{13}$and D_{21} close automatically after evacuees pass them. In the second scenario, the times to untenable conditions in the second floor will not be considered because the leakage through elevator shaft will decrease the concentrations of combustion products with respect to concentrations reached in the first scenario.

The differences between the position of the fire room in a_1, a_2, and a_3 as well as in the wall plans of these alternatives will lead to different times to untenable conditions in the first floor, $T_{u,i1}$ $(i = 1, 2, 3)$.

Fitting Probability Distributions of the Criteria

The times to untenable conditions $T_{u,1}$ and $T_{u,i2}$ can be estimated by means of a computer fire simulation. Input information for this simulation can be generated by means of a stochastic simulation. In the present study, the fire was simulated by means of the computer model CFAST for each of a_1, a_2, and a_3. A visualisation of the CFAST model developed for the alternative $a1$ is given in Figure 13.

For each MCDM alternative, the fire simulation was carried out by embedding the CFAST algorithm in the loop of a stochastic simulation. The loop was repeated 100 times. The number of repetitions will be denoted by N_k (i.e., $N_k = 100$). The following four-step procedure was repeated each time (the time k):

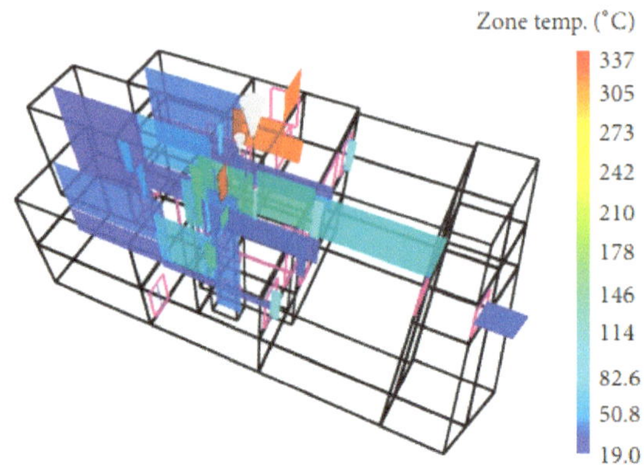

Figure 13: A visualisation of the CFAST model developed for the second fire scenario and the alternative floor plan a_1 (Figure 7)

(1) The values ξ_{ok} to $\xi_{wa,}$ of the random variables ξ_o to ξ_{wa} summarised in Table 3 were generated by means of a stochastic simulation.

(2) The kth value of the heat release rate of the pool fire, $\dot{q}_{ok} = \bar{q}_o + \xi_{ok}$, was calculated and the pool fire with constant HRR $\dot{q}_{ok}$ and duration t_{ok} was uploaded into the CFAST model.

(3) The kth values Ξ_{fk} and Ξ_{ck} of the HRR histories Ξ_f and Ξ_c defined by (9) and (11) were computed and uploaded into the CFAST model. These values were composed of the respective sums $\bar{q}_f(t_l) + \xi_{fk}(t_l)$ and $\bar{q}_c(t_l)+\xi_{ck}(t_l)$ $(l = 1, 2, \ldots, 17)$. The flue fire succeeded the pool fire after the time $t_{f0} = 50$ s.

(4) The computer fire simulation was carried out with the input information specified as indicated above and the kth values of the times to untenable conditions, $T_{u,1}$ or $T_{u,i2}$, were obtained from the output information of CFAST. These times will be denoted by $t_{u,1k}$ or $t_{u,i2k}$.

The simulation of the first fire scenario, in which the fire is initiated in the second floor and can spread to the first floor, produced three samples of the random times to untenable conditions, $T_{u,i1}$ $(i = 1, 2, 3)$. The samples are denoted by the symbols τ_{11}, τ_{21}, and τ_{31} and explained in Table 4. These samples were used to fit the probability distributions of the respective random variables $T_{u,1}$ $(i = 1, 2, 3)$ used as MCDM criteria and expressed by (5). Results of the distribution fitting are given in Table 5.

Table 4: Simulated samples of the times to untenable conditions.

Random time	Simulated sample*	Escape route blocked by smoke	Available escape route
	Alternative a_1, Figure 7 ($i = 1$)		
$T_{u,11}$	$\boldsymbol{\tau}_{11} = \{t_{u,11k},\ k = 1, 2, \ldots, N_k\}^{**}$	ER_{12}, Figure 5	ER_{11}, Figure 5
$T_{u,12}$	$\boldsymbol{\tau}_{12} = \{t_{u,12k},\ k = 1, 2, \ldots, N_k\}$	ER_{21}, Figure 6	ER_{22}, Figure 7
	Alternative a_2, Figure 8 ($i = 2$)		
$T_{u,21}$	$\boldsymbol{\tau}_{21} = \{t_{u,21k},\ k = 1, 2, \ldots, N_k\}$	ER_{12}, Figure 5	ER_{11}, Figure 5
$T_{u,22}$	$\boldsymbol{\tau}_{22} = \{t_{u,22k},\ k = 1, 2, \ldots, N_k\}$	ER_{22}, Figure 8	ER_{21}, Figure 6
	Alternative a_3, Figure 9 ($i = 3$)		
$T_{u,31}$	$\boldsymbol{\tau}_{31} = \{t_{u,31k},\ k = 1, 2, \ldots, N_k\}$	ER_{12}, Figure 5	ER_{11}, Figure 5
$T_{u,32}$	$\boldsymbol{\tau}_{32} = \{t_{u,32k},\ k = 1, 2, \ldots, N_k\}$	ER_{21}, Figure 6	ER_{22}, Figure 9

*Sample elements $t_{u,i2k}$ ($i = 1, 2, 3$) are values of the respective random times $T_{u,i2}$.
** $N_k = 100$ in all computer fire simulations.

Table 5: Descriptive measures of the samples of the times to untenable $T_{u,i1}$ related to the first fire scenario and the three alternative floor plans in the hospital building under reconstruction ($i = 1, 2, 3$).

Sample statistic/distribution fitting results	Sample of $t_{u,11k}$	Sample of $t_{u,21k}$	Sample of $t_{u,31k}$
Sample size	100	100	100
Mean, s	126.3	236.1	57.5
Median, s	126.0	236.0	57.25
Coeff. of variation, %	2.02	1.66	4.26
Standardised skewness	3.66	1.17	2.34
Standardised kurtosis	2.19	0.37	1.75
Minimum, s	14.7	223.0	51.75
Maximum, s	29.0	245.0	65.75
KS-DN for normal distribution/p value	0.1015/0.2555	0.071/0.6938	0.0912/0.3805
KS-DN for lognormal distribution/p value	0.1039/0.2306	0.0738/0.6471	0.0826/0.5156
KS-DN for Gumbel distribution/p value	0.1584/0.0132	0.1179/0.1242	0.0886/0.4175
Fitted distribution	Normal	Normal	Lognormal
Estimate of the 1st parameter*	123.33 (mean)	236.1 (mean)	4.05 (scale)
Estimate of the 2nd parameter*	2.56 (std. dev.)	3.91 (std. dev.)	0.04 (shape)

*Parameter of the fitted distribution.

The simulation of the second fire scenario, in which the fire is confined to the second floor, yielded three samples of the random times to untenable conditions, $T_{u,i2}$ ($i = 1, 2, 3$). These samples are denoted by the respective symbols τ_{12}, τ_{22}, and τ_{32}, (Table 4). The samples τ_{12}, τ_{22}, and τ_{32} were used to fit probability distributions for the random variables $T_{u,2}$ ($i = 1, 2, 3$). The fitted distributions are described in Table 6. These variables appear in the MCDM criterion expressed by (6).

Table 6: Descriptive measures of the samples of the times to untenable $T_{u,2}$ related to the second fire scenario and the three alternative floor plans ($i = 1, 2, 3$).

Sample statistic/distribution fitting results	Sample of $t_{u,12k}$	Sample of $t_{u,22k}$	Sample of $t_{u,32k}$
Sample size	100	100	100
Mean, s	143.09	123.63	74.03
Median, s	144	123	74
Coeff. of variation, %	3.98	4.33	3.51
Standardised skewness	0.19	2.03	1.01
Standardised kurtosis	−1.05	0.9	−0.55
Minimum, s	132.0	110.0	68.5
Maximum, s	156.0	139.0	80.5
KS-DN/p value (normal distr.)	0.0914/0.3771	0.1125/0.1594	0.1075/0.1981
KS-DN/p value (lognormal d.)	0.0979/0.2944	0.1038/0.2322	0.1008/0.2629
KS-DN/p value (Gumbel distr.)	0.1294/0.0702	0.0868/0.4456	0.1314/0.0632
Fitted distribution	Normal	Gumbel	Lognormal
Estimate of the 1st parameter*	143.09 (mean)	121.09 (mode)	4.30 (scale)
Estimate of the 2nd parameter*	5.69 (std. dev.)	4.89 (scale)	0.04 (shape)

*Parameter of the fitted distribution.

The travel times $T_{tr,2}$ were estimated by means of the computer model SIMULEX. The computer simulation was repeated 100 times and produced three samples of $T_{tr,2}$ values $\{t_{tr,i2k}, k = 1, 2, \ldots, 100\}$ ($i = 1, 2, 3$). Elements of these samples, $t_{tr,2k}$, were obtained with different simulated numbers of visitors in the canteen, ξ_{Vk}. In addition, locations of visitors and kitchen staff were randomly distributed for each k and i. The random numbers of the visitors, ξ_{Vk}, were sampled from a binomial distribution by assuming that the maximum number of seats in the canteen is equal to 30: $\xi_V \sim (0.3, 30)$ (binomial distribution with the mean of 9).

In the simulation of evacuation, the number of kitchen staff members remained constant and equal to six persons in all simulations. Figure 14 shows the distribution of nine visitors ($\xi_{Vk} = 9$) in the canteen and six kitchen staff members at the commencement of evacuation

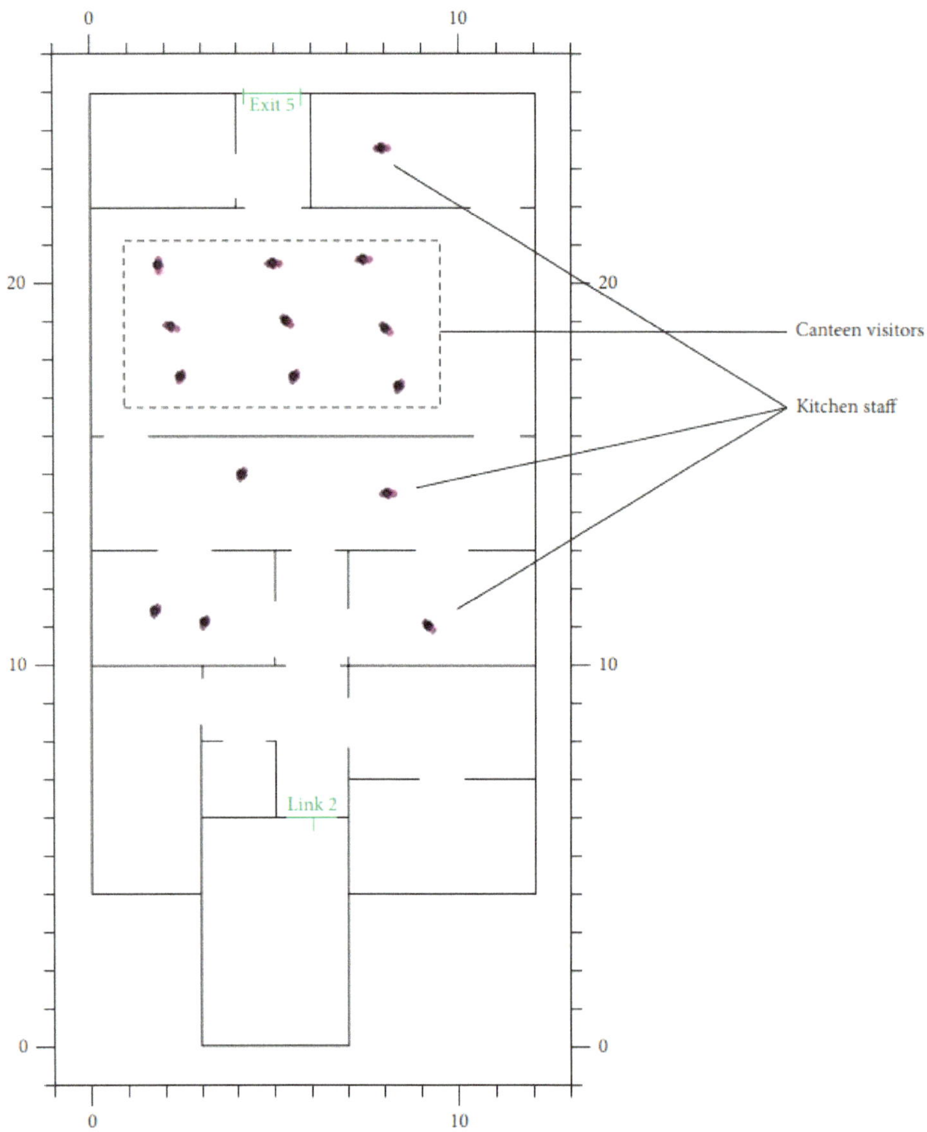

Figure 14: Random distribution of the kitchen staff and nine visitors of the canteen before commencement of evacuation in the floor a_1 shown in Figure 7 (SIMULEX model).

Descriptive measures of the samples $\{t_{tr,2k}, k = 1, 2, \ldots, 100\}$ $(i = 1, 2, 3)$ as well as probability distributions fitted to these samples are presented in Table 7.

Table 7: Descriptive measures of the samples of travel times $T_{tr,2}$ elapsed during the evacuation from the second floor $(i = 1, 2, 3)$

Sample statistic/distribution fitting results	Sample of $t_{tr,12k}$	Sample of $t_{tr,22k}$	Sample of $t_{tr,32k}$
Sample size	100	100	100
Mean, s	19.72	18.47	22.0
Median, s	19.5	18.35	21.6
Coeff. of variation, %	0.134	0.18	0.14
Standardised skewness	3.6553	1.0029	0.6941
Standardised kurtosis	2.189	−1.721	−1.0195
Minimum, s	14.7	12.6	15.8
Maximum, s	29.0	25.7	28.9
KS-DN/*p* value (normal distr.)	0.1118/0.1646	0.0631/0.8213	0.0658/0.7790
KS-DN/*p* value (lognormal d.)	0.0890/0.4118	0.0650/0.7925	0.0516/0.9529
KS-DN/*p* value (Gumbel distr.)	0.0687/0.7335	0.0744/0.6379	0.0723/0.6728
Fitted distribution	Gumbel	Normal	Lognormal
Estimate of the 1st parameter*	18.51 (mode)	18.47 (mean)	3.08 (scale)
Estimate of the 2nd parameter*	2.148 (scale)	3.32 (std. dev.)	0.14 (shape)

*Parameter of the fitted distribution.

The Random Decision-Making Matrix and Selection of the Best Alternative

A MCDM analysis was carried out by taking into account five criteria *c*1 to *c*5 explained in Table 8. The criteria related to the fire safety, c_1 and c_2, were the respective random times$T_{u,1}$ and$T_{tr,i2}-T_{u,i2}$, selected as MCDM criteria in Section 4.1 (see (5) and (6)). Hereby, Table 8 contains random and nonrandom components of the decision-making matrix $\tilde{C}$. The kitchen area *c*3 was included among the criteria because the kitchen will still perform the catering function in the hospital even after the reconstruction of the second floor. The larger the kitchen area is, the better the conditions will be to perform this function. Floor reconstruction cost *c*4 and estimated reconstruction time c_5 are natural criteria of a building-related MCDM and they do not need further explanation. The weights *wj* given in Table 8 mean that the greatest significance was assigned to the criteria associated with the life safety ($w_1 + w_2 = 0.6$). The floor reconstruction cost is also among the significant criteria ($w_4 = 0.25$).

Table 8: The transposed random decision-making matrix $\tilde{C}^T$ (bold cells) composed for the choice among the fire alternative floor plans in a hospital building shown in Figures 7 to 9.

Criteria	Unit of c_j	Preference	Weights, w_j	Plan a_1	Plan a_2	Plan a_3
Random time to untenable conditions, c_1	Sec.	Min	0.35	$\mathbf{T_{u,11}}$	$\mathbf{T_{u,21}}$	$\mathbf{T_{u,31}}$
Random time to untenable conditions minus random travel time, c_2	Sec.	Max	0.25	$\mathbf{T_{u,12} - T_{tr,12}}$	$\mathbf{T_{u,22} - T_{tr,22}}$	$\mathbf{T_{u,32} - T_{tr,32}}$
Kitchen area, c_3	m^2	Max	0.1	42.7	49.1	52.2
Floor reconstruction cost, c_4	Thous. of €	Min	0.25	458.0	298.5	315.6
Estimated reconstruction time, c_5	Months	Min	0.05	3.5	2.1	2.0

The alternatives a_1, a_2, and a_3 were ranked by applying a simulation-based MCDM procedure proposed by Vaidogas and Zavadskas [37]. Six deterministic MCDM methods developed in the game theory and described in the book [38] (criteria K_1 to K_6) were embedded in a simulation loop. All criteria were applied to the matrix of estimates, C, obtained from the matrix of dimensionless criteria, C, with the criterion weights w_j given in Table 8. The matrix C was computed with the vector-norm normalisation method (e.g., [37, 41]).

A total of a million simulation steps were applied to propagate the uncertainty modelled by the times $T_{u,1}$ and $T_{tr,i2} - T_{u,i2}$. In the *l*th step, the criteria K_1 to K_6 were applied to find a^* by using the sampled decision-making matrix $\tilde{C}$. The frequencies of choosing the a_1, a_2, and a_3 as the best ones, f_{r1}, f_{r2}, and f_{r3}, are summarised in Table 9.

Table 9: Results of the choice among the alternative floor plans a_1, a_2, and a_3 designed for a hospital building and shown in Figures 7 to 9.

MCDM criterion	Frequency[(1)] of choosing a_i as a^*		
	fr_1	$\mathbf{fr_2}$	fr_3
Wald's K_1	0.4639	**0.5361**	0
Savage's K_2	0	**1**	0
Bernoulli-Laplace K_3	**0.9988**	0.1156×10^{-2}	0
Hurwicz's K_4 ($\lambda = 0.5$)	0	**1**	0
Bayes's criterion K_5	0.331×10^{-3}	**0.9997**	0
Hodges-Lehman's ($\delta = 0.5$)	0.1827×10^{-1}	**0.9817**	0

(1)The frequencies fr_1, fr_2, and fr_3 were computed with $N_l = 1 \times 10^6$.

CONCLUSIONS

The building design that takes into account alternative architectural and structural solutions has been considered. A universal methodology known as MCDM can be applied to rank available alternative designs. MCDM can be used at different levels of development of a building project. Each alternative design is characterised by a number of MCDM criteria which are juggled simultaneously.

Fire is prevailing hazard in most buildings and a built-in fire protection is an important part of each building project. Methods of MCDM allow including criteria related to building fire safety. The most important criteria express life safety of occupants. The main finding of this paper is that two criteria can be used for MCDM (1)probability of a successful evacuation of occupants from a building in fire and (2) difference between evacuation time and time to untenable conditions along available evacuation paths. These criteria are in general uncertain quantities. The probability can be uncertain in the epistemic sense and the difference in times will be uncertain in the aleatory (stochastic) sense. Probability distributions of these criteria can be estimated by a Monte Carlo simulation of fire and evacuation processes.

Problem with uncertain safety-related criteria can be solved by means of stochastic MCDM methods. An application of such a method was illustrated by an example, in which alternative architectural floor plans of a hospital building were compared.

CONFLICT OF INTERESTS

The authors declare that there is no conflict of interests regarding the publication of this paper.

REFERENCES

1. R. Binnekamp, Preference—Based Design in Architecture, IOS Press, Amsterdam, The Netherlands, 2010.
2. D. Kim, S. Lee, and S. A. Kim, "A visualization system for the comfort analysis of modular architecture: a case study," in Cooperative Design, Visualization, and Engineering, vol. 7467 of Lecture Notes in Computer Science, pp. 247–254, Springer, 2012.
3. V. Granadeiro, J. P. Duarte, J. R. Correia, and V. M. S. Leal, "Building envelope shape design in early stages of the design process: integrating architectural design systems and energy simulation," Automation in Construction, vol. 32, pp. 196–209, 2013.
4. M. Ehrgott, J. R. Figueira, and S. Greco, Trends in Multiple Criteria Decision Analysis, Springer, New York, NY, USA, 2010.
5. J. Wong, H. Li, and J. Lai, "Evaluating the system intelligence of the intelligent building systems. Part 1: development of key intelligent indicators and conceptual analytical framework," Automation in Construction, vol. 17, no. 3, pp. 284–302, 2008.
6. R. Mora, G. Bitsuamlak, and M. Horvat, "Integrated life-cycle design of building enclosures," Building and Environment, vol. 46, no. 7, pp. 1469–1479, 2011.
7. K. Mela, T. Tiainen, and M. Heinisuo, "Comparative study of multiple criteria decision making methods for building design," Advanced Engineering Informatics, vol. 26, no. 4, pp. 716–726, 2012.
8. D. J. Rasbash, G. Ramachandran, B. Kandola, J. M. Watts, and M. Law, Evaluation of Fire Safety, John Wiley & Sons, Chichester, UK, 2004.
9. C. M. Zhao, S. M. Lo, J. A. Lu, and Z. Fang, "A simulation approach for ranking of fire safety attributes of existing buildings," Fire Safety Journal, vol. 39, no. 7, pp. 557–579, 2004.
10. E. R. Vaidogas and J. Šakėnaitė, "Multi-attribute decision-making in economics of fire protection,"Engineering Economics, vol. 22, no. 3, pp. 262–270, 2011.
11. E. R. Vaidogas and J. Šakenaite, "Protecting built property against fire disasters: Multi-atribute decisio n making with respect to fire risk," International Journal of Strategic Property Management, vol. 14, no. 4, pp. 391–407, 2010.
12. J. Choi, H. Kim, and I. Kim, "Open BIM-based quantity take-off system for schematic estimation of building frame in early design stage," Journal of Computational Design and Engineering, vol. 2, no. 1, pp. 16–25, 2015.
13. J. Wood, K. Panuwatwanich, and J.-H. Doh, "Using LOD in structural cost estimation during building design stage: pilot study," Procedia Engineering, vol. 85, no. 5, pp. 543–552, 2014.
14. C. Kahraman, Ed., Fuzzy Multi-Criteria Decision Making. Theory and Applications with Recent Developments, vol. 16 of Springer Optimization and Its Applications, Springer, New York, NY, USA, 2008.
15. P. Ayala, A. Cantizano, C. Gutiérrez-Montes, and G. Rein, "Influence of atrium roof geometries on the numerical predictions of fire tests under natural ventilation conditions," Energy and Buildings, vol. 65, pp. 382–390, 2013.
16. R. M. Doheim, Y. G. Yohanis, A. Nadjai, and H. Elkadi, "The impact of atrium shape on natural smoke ventilation," Fire Safety Journal, vol. 63, pp. 9–16, 2014.
17. G.-Y. Wu and H.-Ch. Huang, "Modeling the emergency evacuation of the high rise building based on the control volume model," Safety Science, vol. 73, no. 5, pp. 62–72, 2015.
18. G. de Sanctis, J. Kohler, and M. Fontana, "Probabilistic assessment of the occupant load density in retail buildings," Fire Safety Journal, vol. 69, pp. 1–11, 2014.
19. J. R. Hall, The Total Cost of Fire in the United States, NFPA, Quincy, Mass, USA, 2014.
20. J. Xin and C. Huang, "Fire risk analysis of residential buildings based on scenario clusters and its application in fire risk management," Fire Safety Journal, vol. 62, pp. 72–78, 2013.
21. B. Karlsson and J. G. Quintiere, Enclosure Fire Dynamics, CRC Press, Boca Raton, Fla, USA, 2000.
22. T. Korhonen and S. Hostikka, Fire Dynamics Simulator with Evacuation: FDS+Evac, VTT Technical Research Centre of Finland, Otaniemi, Finland, 2009.
23. R. Lovreglio, E. Ronchi, and D. Borri, "The validation of evacuation simulation models through the analysis of behavioural uncertainty," Reliability Engineering and System Safety, vol. 131, no. 11, pp. 166–174, 2014.
24. M. Hasofer, V. R. Beck, and I. D. Bennetts, Risk Analysis in Building Fire Safety Engineering, Butterworth & Heinermann, Amsterdam, The Netherlands, 2007.
25. D. Yung, Principles of Fire Risk Assessment in Buildings, Wiley, Chichester, UK, 2008.
26. J. M. Watts, "Fire risk indexing," in SFPE Handbook of Fire Protection Engineering, pp. 5-125–5-142, NFPA & SFPE, Quincy, Mass, USA, 3rd edition, 2002.
27. J. Šakėnaitė, Combined application of multi-attribute selection and risk analysis to the assessment of building fire safety [Doctoral disertation], Technika, Vilnius, Lithuania, 2012.
28. E. K. Zavadskas and E. R. Vaidogas, "Multiattribute selection from alternative designs of infrastructure components for accidental situations," Computer-Aided Civil and Infrastructure Engineering, vol. 24, no. 5, pp. 346–358, 2009.
29. S. Hostikka and O. Keski-Rahkonen, "Probabilistic simulation of fire scenarios," Nuclear Engineering and Design, vol. 224, no. 3, pp. 301–311, 2003.
30. S. K. Au, Z.-H. Wang, and S.-M. Lo, "Compartment fire risk analysis by advanced Monte Carlo simulation," Engineering Structures, vol. 29, no. 9, pp. 2381–2390, 2007.
31. X. Zhang, X. Li, and G. Hadjisophocleous, "A probabilistic occupant evacuation model for fire emergencies using Monte Carlo methods," Fire Safety Journal, vol. 58, no. 5, pp. 15–24, 2013.
32. L. C. Cadwallader and S. A. Eide, "Component failure

rate data sources for probabilistic safety and reliability," Process Safety Progress, vol. 29, no. 3, pp. 236–241, 2010.

33. E. R. Vaidogas and J. Šakenaite, "A brief look at data on the reliability of sprinklers used in conventional buildings," Journal of Civil Engineering and Management, vol. 17, no. 1, pp. 115–125, 2011.

34. M. Nowak, "Aspiration level approach in stochastic MCDM problems," European Journal of Operational Research, vol. 177, no. 3, pp. 1626–1640, 2007.

35. Y. Liu, Z.-P. Fan, and Y. Zhang, "A method for stochastic multiple criteria decision making based on dominance degrees," Information Sciences, vol. 181, no. 19, pp. 4139–4153, 2011.

36. Z.-P. Fan, Y. Liu, and B. Feng, "A method for stochastic multiple criteria decision making based on pairwise comparisons of alternatives with random evaluations," European Journal of Operational Research, vol. 207, no. 2, pp. 906–915, 2010.

37. E. R. Vaidogas and E. K. Zavadskas, "Introducing reliability measures into multi-criteria decision-making," International Journal of Management and Decision Making, vol. 8, no. 5-6, pp. 475–496, 2007.

38. S. French, Decision Theory: An Introduction to the Mathematics of Rationality, Ellis Harwood, Chichester, UK; Wiley, New York, NY, USA, 1988.

39. H. Bottenbruch, H. J. Pradlwarter, and G. I. Schuëller, "The influence of spatial correlation of concrete strength on the failure probabilities of reinforced concrete chinneys," Materials and Structures, vol. 22, no. 4, pp. 255–263, 1989.

40. V. Babrauskas, "Heat release rates," in SFPE Handbook of Fire Protection Engineering, pp. 3-1–3-37, NFPA & SFPE, Quincy, Mass, USA, 3rd edition, 2002.

41. E. R. Vaidogas and L. Linkute, "Sitting the barrier aimed at protecting roadside property from accidental fires and explosions on road: a pre-optimisation stage," The Baltic Journal of Road and Bridge Engineering, vol. 7, no. 4, pp. 277–287, 2012.

CHAPTER 10 FREE VIBRATION ANALYSIS OF SYMMETRICALLY LAMINATED FOLDED PLATE STRUCTURES USING AN ELEMENT-FREE GALERKIN METHOD

L. X. Peng

INTRODUCTION

Because of high strength/weight ratio, easy forming, and low cost, folded plate structures have been widely used in many engineering branches, such as roofs, corrugated-cores, and cooling towers. They have much higher load carrying capacity compared to flat plates. Before the invention of fiber-reinforced material, folded plates were often made of medal or timber. The application of fiber-reinforced material to folded plate structures was a remarkable advance in engineering, which combined the advantages of fiber-reinforced material and folded plate structure directly and made the structure even lighter and stiffer.

The study of isotropic folded plates had a quite long history, and a variety of methods had emerged. In early days, researchers were short of powerful numerical tools and tried to analyze the structures with various approximations. The beam method and the theory that ignores relative joint displacement were introduced [1]. Although the methods were weak in dealing with generalized folded plate problems, they were simple and fulfilled the demand of fast and easy computation in engineering. Therefore, they are still used in some design environments, where accurate analysis is not the first concern. Researchers such as Gaafar [2], Yitzhaki [3], Yitzhaki and Reiss [4], and Whitney et al. [5] were the first to consider the relative joint displacement of the structures in their methods, which led to more precise analysis results. Goldberg and Leve [6] used the two-dimensional theory of elasticity and the two-way slab theory to analyze folded plates. In their method, both the simultaneous bending and the membrane action of a folded plate were taken into account, and the degree of freedom (DOF) of each point along the joint of the folded plate was chosen to be four (three components of translation and one rotation). Niyogi et al. [7] considered this method as the first to give an exact static solution for folded plates. Yitzhaki and Reiss [4] took the moments along the joints of folded plates as unknown and applied the slope deflection method to the analysis of the folded plates. Bar-Yoseph and Hersckovitz [8] proposed an approximated method for folded plates based on Vlasov's theory of thin-wall beams. Their method considered a folded plate as a monolithic structure composed of longitudinal beams, which can give good results for long folded plates. Bandyopadhyay and Laad [9] compared two classical methods for folded plates and studied the suitability of these methods for the preliminary analysis of folded plate structures. Lai et al. [10] gave an equation of the middle surface of a simply supported cross V-shaped folded plate roof by using the inclined coordinate system and generalized functions, sign function and step function, and carried out a nonlinear analysis for the folded plate.

The development of computation techniques and computers has aroused research interest in numerical methods

Citation: L. X. Peng, "Free Vibration Analysis of Symmetrically Laminated Folded Plate Structures Using an Element-Free Galerkin Method," Mathematical Problems in Engineering, vol. 2015, Article ID 124296, 13 pages, 2015. doi:10.1155/2015/124296.

for folded plates. A number of methods, such as the finite strip methods (Cheung [11], Golley and Grice [12], Eterovic and Godoy [13]), the combined boundary element-transfer matrix method (Ohga et al. [14]), and the finite element methods (FEM) (Liu and Huang [15], Perry et al. [16], Niyogi et al. [7], and Duan and Miyamoto [17]), have been introduced to solve folded plate problems. Among these methods, the FEMs are the most successful. They are very versatile as they can deal with the problems with complicated geometry, boundary conditions, or loadings easily. However, FEM also has disadvantages. Their solution of a problem is based on the meshes that discretize the problem domain, and any dramatic change of the problem domain will lead to remeshing of the domain, which results in programming complexity, diminished accuracy, and long computation time. Regarding the disadvantage, some researchers proposed the element-free, meshfree, or meshless methods [18–22]. As alternatives to FEMs, the meshfree methods construct their approximated solution of a problem completely in terms of a set of ordered or scattered points that discretize the problem domain; that is, their solution relies on the points other than meshes. No element is required. Without the limit of meshes, the meshfree methods are more applicable than the FEMs and avoid the aforementioned difficulties caused by remeshing in the FEMs.

Bui et al. [23], Bui and Nguyen [24], and Somireddy and Rajagopal [25] have introduced the meshfree methods for vibration analysis of laminated plates. However, few studies on folded laminated plates have been found. There are only Niyongi et al. [7] and Lee et al.'s [26] work on vibration and the author's work on bending with a meshfree method, which is also the motive for this paper.

The objective of this paper is to introduce an element-free Galerkin method based on the first-order shear deformation theory (FSDT) [27, 28] for the free vibration analysis of folded laminated plates. A symmetrical folded laminated plate is regarded as a composite structure composed of symmetric laminates. The analysis process includes (a) deriving the stiffness and mass matrices of the symmetric laminates that make up a folded plate by the element-free Galerkin method; (b) considering the laminates as super elements and superposing their stiffness and mass matrices to obtain the global stiffness and mass matrices of the folded plate. Some numerical examples are used to demonstrate the convergence and accuracy of the proposed method. The calculated results are compared with the results from the finite element analytical software ANSYS. The proposed method may be used as a potential meshfree tool for the analysis of laminated shell structures.

MOVING LEAST-SQUARES APPROXIMATION

In the moving least-squares approximation (MLS) [18], a function V(x) in a domain Ω can be approximated by $V^d(x)$ in the subdomain Ω_x and

$$v^d(\mathbf{x}) = \sum_{i=1}^{m} q_i(\mathbf{x}) b_i(\mathbf{x}) = \mathbf{q}^T(\mathbf{x})\mathbf{b}(\mathbf{x}) \tag{1}$$

where $q_i(x)$ are the monomial basis functions, $b_i(x)$ are the corresponding coefficients, d is a factor that measures the domain of influence (or the support) of the nodes, and m is the number of basis functions. In this paper, the quadratic basis $q^T = [1, x, y, x^2, xy, y^2]$ (m=6) are used for the laminates. The unknown coefficients $b_i(x)$ are obtained by the minimization of a weighted discrete L_2 norm

$$\Gamma = \sum_{I=1}^{n} \varpi(\mathbf{x}-\mathbf{x}_I)\left[\mathbf{q}(\mathbf{x}_I)^T \mathbf{b}(\mathbf{x}) - v_I\right]^2, \tag{2}$$

where $\varpi(x-x_I)$ or $\varpi_I(x)$ is the weight function that is associated with node I, $\varpi_I(x)$=0 outside Ω_x, n is the number of nodes in Ω_x that make the weight function $\varpi_I(x)$>0, and V_I are the nodal parameters. The minimization of Γ in (2)with respect to b(x)

$$\frac{\partial \Gamma}{\partial \mathbf{b}(\mathbf{x})} = 0 \tag{3}$$

leads to a set of linear equations

$$\mathbf{B}(\mathbf{x})\mathbf{b}(\mathbf{x}) = \mathbf{A}(\mathbf{x})\mathbf{v}, \tag{4}$$

where

$$\mathbf{B}(\mathbf{x}) = \sum_{I=1}^{n} \varpi(\mathbf{x}-\mathbf{x}_I)\mathbf{q}(\mathbf{x}_I)\mathbf{q}^T(\mathbf{x}_I), \tag{5}$$

$$\mathbf{A}(\mathbf{x}) = \left[\varpi(\mathbf{x}-\mathbf{x}_1)\mathbf{q}(\mathbf{x}_1), \ldots, \varpi(\mathbf{x}-\mathbf{x}_n)\mathbf{q}(\mathbf{x}_n)\right]. \tag{6}$$

The coefficients b(x) are then derived from (4):

$$\mathbf{b}(\mathbf{x}) = \mathbf{B}^{-1}(\mathbf{x})\mathbf{A}(\mathbf{x})\mathbf{v}. \tag{7}$$

By substituting (7) into (1), the approximation $V^d(x)$ is expressed in a standard form as

$$v^d(\mathbf{x}) = \sum_{I=1}^{n} N_I(\mathbf{x}) v_I, \tag{8}$$

where the shape function $N_I(x)$ is given by

$$N_I(\mathbf{x}) = \mathbf{q}^T(\mathbf{x})\mathbf{B}^{-1}(\mathbf{x})\mathbf{A}_I(\mathbf{x}). \tag{9}$$

From (6), we obtain

$$\mathbf{A}_I(\mathbf{x}) = \varpi(\mathbf{x}-\mathbf{x}_I)\mathbf{q}(\mathbf{x}_I), \tag{10}$$

and thus (9) can be rewritten as

$$N_I(\mathbf{x}) = \mathbf{q}^T(\mathbf{x})\mathbf{B}^{-1}(\mathbf{x})\mathbf{q}(\mathbf{x}_I)\varpi(\mathbf{x}-\mathbf{x}_I). \tag{11}$$

MESHLESS MODEL OF A LAMINATE

The first step in our analysis is to obtain the stiffness and mass matrices of the laminates that make up a folded plate. The meshless model for a laminate in the local coordinate, as shown in Figure 1, is prescribed with a set of nodes. The DOF of every node is (u_0, V_0, w, φ_x, φ_y), where u_0, V_0, and w are the nodal translations of the laminate in the x-direction, y-direction, and z-direction, respectively. φ_x and φ_y are the rotation about the y-axis and the x-axis, respectively. The laminate is assumed to have N layers, and the thickness of each layer is z_i (i = 1, . . . , N). Therefore, the thickness of kth layer is $h_k = z_{k+1} - z_k$.

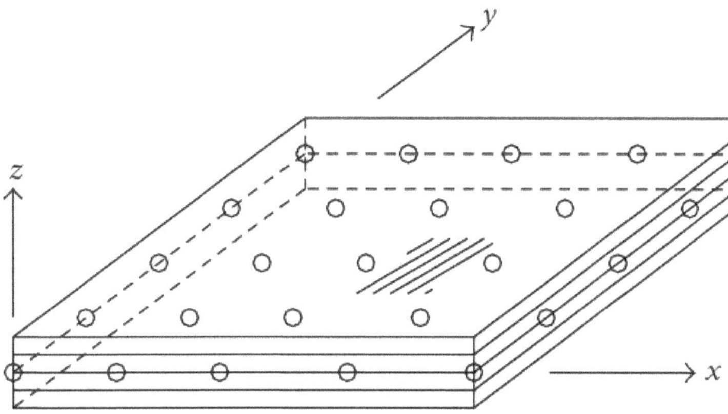

Figure 1: Meshfree model of a laminate.

Displacement Approximation

Based on the FSDT and the MLS approximation, the displacements of the laminate can be approximated by

$$u(x,y,z,t) = u_0(x,y,t) - z\varphi_x(x,y,t)$$
$$= \sum_{I=1}^{n} H_I(x,y)\,u_{0I}(t) - z\sum_{I=1}^{n} H_I(x,y)\,\varphi_{xI}(t),$$
$$v(x,y,z,t) = v_0(x,y,t) - z\varphi_y(x,y,t)$$
$$= \sum_{I=1}^{n} H_I(x,y)\,v_{0I}(t) - z\sum_{I=1}^{n} H_I(x,y)\,\varphi_{yI}(t),$$
$$w(x,y,t) = \sum_{I=1}^{n} H_I(x,y)\,w_I(t), \tag{12}$$

where $\{u_{0I}(t), V_{0I}(t), wI(t)\ \varphi_{xI}(t), \varphi_{yI}(t)\}^{T} = \delta I$ are the nodal parameters of the Ith node of the laminate, n is the number of nodes of the laminate, and φx and φy are independent of w. The shape functions (x, y) are obtained from (9), and the cubic spline function

$$\varpi(s) = \begin{cases} \frac{2}{3} - 4s^2 + 4s^3, & s \le \frac{1}{2}, \\ \frac{4}{3} - 4s + 4s^2 - \frac{4}{3}s^3, & \frac{1}{2} < s \le 1, \\ 0, & s > 1 \end{cases} \tag{13}$$

is used as the weight function. Equation (12) can be written in a matrix form as

$$\mathbf{U} = \begin{Bmatrix} u \\ v \\ w \end{Bmatrix} = \sum_{I=1}^{n} \begin{bmatrix} H_I(x,y) & 0 & 0 & -zH_I(x,y) & 0 \\ 0 & H_I(x,y) & 0 & 0 & -zH_I(x,y) \\ 0 & 0 & H_I(x,y) & 0 & 0 \end{bmatrix}$$
$$\times \begin{Bmatrix} u_{0I}(t) \\ v_{0I}(t) \\ w_I(t) \\ \varphi_{xI}(t) \\ \varphi_{yI}(t) \end{Bmatrix}. \tag{14}$$

The strains of the laminate are defined as

$$\boldsymbol{\kappa} = \begin{Bmatrix} \varepsilon_x \\ \varepsilon_y \\ \gamma_{xy} \end{Bmatrix} = \begin{bmatrix} u_{0,x} - z\varphi_{x,x} \\ v_{0,y} - z\varphi_{y,y} \\ u_{0,y} + v_{0,x} - z\left(\varphi_{x,y} + \varphi_{y,x}\right) \end{bmatrix} = \sum_{I=1}^{n} \mathbf{B}_I^b \boldsymbol{\delta}_I,$$
$$\boldsymbol{\gamma} = \begin{Bmatrix} \gamma_{xz} \\ \gamma_{yz} \end{Bmatrix} = \begin{bmatrix} w_{,x} - \varphi_x \\ w_{,y} - \varphi_y \end{bmatrix} = \sum_{I=1}^{n} \mathbf{B}_I^s \boldsymbol{\delta}_I, \tag{15}$$

Where

$$\mathbf{B}_I^b = \left[\mathbf{B}_{0I}\ -z\mathbf{B}_{1I}\right] = \begin{bmatrix} H_{I,x} & 0 & 0 & -zH_{I,x} & 0 \\ 0 & H_{I,y} & 0 & 0 & -zH_{I,y} \\ H_{I,y} & H_{I,x} & 0 & -zH_{I,y} & -zH_{I,x} \end{bmatrix},$$
$$\mathbf{B}_{0I} = \begin{bmatrix} H_{I,x} & 0 \\ 0 & H_{I,y} \\ H_{I,y} & H_{I,x} \end{bmatrix}, \qquad \mathbf{B}_{1I} = \begin{bmatrix} 0 & H_{I,x} & 0 \\ 0 & 0 & H_{I,y} \\ 0 & H_{I,y} & H_{I,x} \end{bmatrix},$$
$$\mathbf{B}_I^s = \left[\mathbf{0}\ \mathbf{B}_{2I}\right] = \begin{bmatrix} 0 & 0 & H_{I,x} & -H_I & 0 \\ 0 & 0 & H_{I,y} & 0 & -H_I \end{bmatrix},$$
$$\mathbf{B}_{2I} = \begin{bmatrix} H_{I,x} & -H_I & 0 \\ H_{I,y} & 0 & -H_I \end{bmatrix}, \tag{16}$$

"$_x$" refers to the derivatives of x, and "," refers to the derivatives of y.

Governing Equation

In free vibration, the strain energy and kinetic energy of the laminate are, respectively

$$\Pi = \frac{1}{2}\iint\int_{-h/2}^{h/2} \boldsymbol{\kappa}^{\mathrm{T}}\mathbf{D}\boldsymbol{\kappa}\,\mathrm{d}z\,\mathrm{d}x\,\mathrm{d}y + \frac{1}{2}\iint \boldsymbol{\gamma}^{\mathrm{T}}\mathbf{A}_s\boldsymbol{\gamma}\,\mathrm{d}x\,\mathrm{d}y,$$
$$\mathrm{T}_0 = \frac{1}{2}\iint\int_{-h/2}^{h/2} \dot{\mathrm{U}}^{\mathrm{T}}\rho\dot{\mathrm{U}}\,\mathrm{d}z\,\mathrm{d}x\,\mathrm{d}y, \tag{17}$$

where

$$\mathbf{D} = \begin{bmatrix} \overline{Q}_{11}^{(k)} & \overline{Q}_{12}^{(k)} & \overline{Q}_{16}^{(k)} \\ \overline{Q}_{21}^{(k)} & \overline{Q}_{22}^{(k)} & \overline{Q}_{26}^{(k)} \\ \overline{Q}_{61}^{(k)} & \overline{Q}_{62}^{(k)} & \overline{Q}_{66}^{(k)} \end{bmatrix},$$
$$\widetilde{\mathbf{A}}_s = \begin{bmatrix} \widetilde{A}_{55} & \widetilde{A}_{45} \\ \widetilde{A}_{45} & \widetilde{A}_{44} \end{bmatrix},$$
$$\widetilde{A}_{ij} = k_c\int_{-h/2}^{h/2} \overline{Q}_{ij}^{(k)}\,dz = k_c\sum_{k=1}^{N} \overline{Q}_{ij}^{(k)}\left(z_{k+1} - z_k\right),$$
$$(i, j = 4, 5), \tag{18}$$

$\overline{Q}_{ij}\ (i, j = 1,2,6,4,5)$ are the material stiffness that are defined in [28], $k_c = 5/6$ is the shear correction factor, and h is the thickness of the laminate. ρ is the density of the material.

According to Hamilton's principle

$$\delta \int_{t_1}^{t_2} (T_0 - \Pi)\, dt = 0. \tag{19}$$

The substitution of (14) to (17) into (19) gives

$$\mathbf{K}\delta + \mathbf{M}\ddot{\delta} = \mathbf{0}, \tag{20}$$

Where

$$\boldsymbol{\delta} = \{\boldsymbol{\delta}_1^{\mathrm{T}}\ \boldsymbol{\delta}_2^{\mathrm{T}} \cdots \boldsymbol{\delta}_n^{\mathrm{T}}\}^{\mathrm{T}} = \{u_{01}, v_{01}, w_1, \varphi_{x1}, \varphi_{y1}, \ldots, u_{0n}, v_{0n}, w_n, \varphi_{xn}, \varphi_{yn}\}^{\mathrm{T}},$$

$$[\mathbf{M}]_{IJ} = \iint \begin{bmatrix} h & & & & \\ & h & 0 & & \\ & & h & & \\ & 0 & & \frac{h^3}{12} & \\ & & & & \frac{h^3}{12} \end{bmatrix} \rho H_I H_J \,dx\,dy,$$

$$\begin{aligned}\mathbf{K}_{IJ} &= \iint \int_{-h/2}^{h/2} \begin{bmatrix} \mathbf{B}_{0I}^{\mathrm{T}} \\ -z\mathbf{B}_{1I}^{\mathrm{T}} \end{bmatrix} \mathbf{D} \,[\mathbf{B}_{0J}\ \ -z\mathbf{B}_{1J}]\, dz\,dx\,dy \\ &+ \iint \begin{bmatrix} \mathbf{0} \\ \mathbf{B}_{2I}^{\mathrm{T}} \end{bmatrix} \widetilde{\mathbf{A}}_s \,[\mathbf{0}\ \ \mathbf{B}_{2J}]\, dx\,dy \\ &= \iint \int_{-h/2}^{h/2} \begin{bmatrix} \mathbf{B}_{0I}^{\mathrm{T}}\mathbf{D}\mathbf{B}_{0J} & -z\mathbf{B}_{0I}^{\mathrm{T}}\mathbf{D}\mathbf{B}_{1J} \\ -z\mathbf{B}_{1I}^{\mathrm{T}}\mathbf{D}\mathbf{B}_{0J} & z^2\mathbf{B}_{1I}^{\mathrm{T}}\mathbf{D}\mathbf{B}_{1J} \end{bmatrix} dz\,dx\,dy \\ &+ \iint \begin{bmatrix} \mathbf{0} & \mathbf{0} \\ \mathbf{0} & \mathbf{B}_{2I}^{\mathrm{T}}\widetilde{\mathbf{A}}_s\mathbf{B}_{2J} \end{bmatrix} dx\,dy. \end{aligned} \tag{21}$$

If $(\widetilde{\mathbf{A}}, \widetilde{\mathbf{B}}, \widetilde{\mathbf{H}}) = \int_{-h/2}^{h/2} \mathbf{D}(1, z, z^2)\, dz$, we obtain

If $(\widetilde{\mathbf{A}}, \widetilde{\mathbf{B}}, \widetilde{\mathbf{H}}) = \int_{-h/2}^{h/2} \mathbf{D}(1, z, z^2)\, dz$, we obtain

$$\mathbf{K}_{IJ} = \iint \begin{bmatrix} \mathbf{B}_{0I}^{\mathrm{T}}\widetilde{\mathbf{A}}\mathbf{B}_{0J} & -\mathbf{B}_{0I}^{\mathrm{T}}\widetilde{\mathbf{B}}\mathbf{B}_{1J} \\ -\mathbf{B}_{1I}^{\mathrm{T}}\widetilde{\mathbf{B}}\mathbf{B}_{0J} & \mathbf{B}_{1I}^{\mathrm{T}}\widetilde{\mathbf{H}}\mathbf{B}_{1J} + \mathbf{B}_{2I}^{\mathrm{T}}\widetilde{\mathbf{A}}_s\mathbf{B}_{2J} \end{bmatrix} dx\,dy. \tag{22}$$

For the symmetric laminates $\widetilde{\mathbf{B}} = 0$.

FORMULATION FOR FOLDED LAMINATED PLATES

In the paper, a folded laminated plate is regarded as a composite structure composed of laminates. We have obtained the stiffness and mass matrices of a single laminate. Therefore, the next step is to take each laminate of the composite structure as a super element, to superpose their stiffness and mass matrices by applying the displacement compatibility conditions along the joints between the laminates, and to give the governing equation of the entire folded laminated plate (Figure 2).

Nevertheless, as pointed out by the author in [20], due to a lack of Kronecker delta properties in the meshfree shape functions given by (9), and that δ of (20) are nodal parameters other than actual nodal displacements, the stiffness and mass matrices cannot be directly superposed. The full transformation method that was first introduced by Chen et al. [19] to enforce the essential boundary conditions is extended by the author in the paper to modify the stiffness and mass matrices before a superposition. After the modification, the essential boundary conditions can be implemented as those in FEMs.

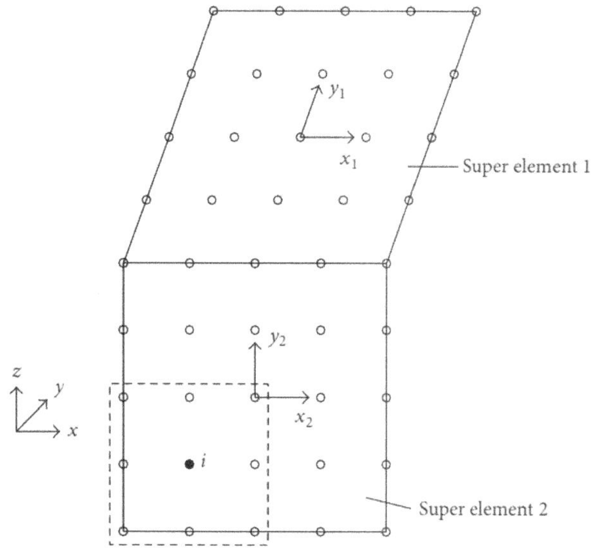

Figure 2: A meshfree model of a folded laminated plate that is made up of two super elements (laminates). The domain of influence of node i is indicated by the dashed line.

Modification of Stiffness and Mass Matrices

From (1), the actual displacement of the nodes, $\bar{\mathbf{v}}(\mathbf{x})$,, can be approximated by $\mathbf{v}^d(\mathbf{x})$

$$\bar{\mathbf{v}}(\mathbf{x}) \approx \mathbf{v}^d(\mathbf{x}) = \sum_{I=1}^{n} N_I(\mathbf{x})\, v_I = \boldsymbol{\Phi}\mathbf{v}, \tag{23}$$

Where

$$\boldsymbol{\Phi} = \begin{bmatrix} N_1(\mathbf{x}_1) & N_2(\mathbf{x}_1) & \ldots & N_n(\mathbf{x}_1) \\ N_1(\mathbf{x}_2) & N_2(\mathbf{x}_2) & \ldots & N_n(\mathbf{x}_2) \\ \vdots & \vdots & \vdots & \vdots \\ N_1(\mathbf{x}_n) & N_2(\mathbf{x}_n) & \ldots & N_n(\mathbf{x}_n) \end{bmatrix} \tag{24}$$

and V_I are the nodal parameters. Therefore, the nodal parameters can be expressed in terms of the actual displacement

$$\mathbf{v} = \Lambda\bar{\mathbf{v}}, \tag{25}$$

where $\Lambda = \Phi^{-1}$

For the folded laminated plate, we accordingly have

$$\mathbf{u}_0 = \Lambda\bar{\mathbf{u}}_0, \quad \mathbf{v}_0 = \Lambda\bar{\mathbf{v}}_0, \quad \mathbf{w} = \Lambda\bar{\mathbf{w}}, \quad \boldsymbol{\varphi}_x = \Lambda\bar{\boldsymbol{\varphi}}_x, \quad \boldsymbol{\varphi}_y = \Lambda\bar{\boldsymbol{\varphi}}_y, \tag{26}$$

where $\bar{\mathbf{u}}_0 = \{\bar{u}_0(\mathbf{x}_1), \bar{u}_0(\mathbf{x}_2), \ldots, \bar{u}_0(\mathbf{x}_n)\}^{\mathrm{T}} = \{\bar{u}_{01}, \bar{u}_{02}, \ldots, \bar{u}_{0n}\}^{\mathrm{T}}$ are the actual nodal displacement (translation) in the x direction and $\mathbf{u}_0 = \{u_{01}, u_{02}, \ldots, u_{0n}\}^{\mathrm{T}}$ are the corresponding nodal parameters. $\mathbf{v}_0, \bar{\mathbf{v}}_0, \mathbf{w}, \bar{\mathbf{w}}, \boldsymbol{\varphi}_x, \bar{\boldsymbol{\varphi}}_x, \boldsymbol{\varphi}_y$, and $\bar{\boldsymbol{\varphi}}_y$ have similar definition. Equation (26) can be written as

$$\boldsymbol{\delta} = \overline{\Lambda\boldsymbol{\delta}}, \tag{27}$$

where $\bar{\boldsymbol{\delta}} = \{\bar{\boldsymbol{\delta}}_1^{\mathrm{T}}\ \bar{\boldsymbol{\delta}}_2^{\mathrm{T}} \cdots \bar{\boldsymbol{\delta}}_n^{\mathrm{T}}\}^{\mathrm{T}} = \{\bar{u}_{01}, \bar{v}_{01}, \bar{w}_1, \bar{\varphi}_{x1}, \bar{\varphi}_{y1}, \ldots, \bar{u}_{0n}, \bar{v}_{0n}, \bar{w}_n, \bar{\varphi}_{xn}, \bar{\varphi}_{yn}\}^{\mathrm{T}}$ are the actual nodal displacement of all nodes. $\bar{\Lambda}$ is a $5n \times 5n$ modification matrix which combines five Λ. Substituting (27) into (20) and premultiplying both sides of the equation with $\bar{\Lambda}^{\mathrm{T}}$, we obtain

$$\overline{\Lambda}^{\mathrm{T}}\overline{\mathbf{K}}\overline{\Lambda}\overline{\boldsymbol{\delta}} + \overline{\Lambda}^{\mathrm{T}}\overline{\mathbf{M}}\overline{\Lambda}\ddot{\overline{\boldsymbol{\delta}}} = 0. \tag{28}$$

Assuming that $\overline{\mathbf{K}} = \overline{\Lambda}^{\mathrm{T}}\mathbf{K}\overline{\Lambda}$ and $\overline{\mathbf{M}} = \overline{\Lambda}^{\mathrm{T}}\mathbf{M}\overline{\Lambda}$, we have

$$\overline{\mathbf{K}}\overline{\boldsymbol{\delta}} + \overline{\mathbf{M}}\ddot{\overline{\boldsymbol{\delta}}} = 0, \tag{29}$$

the modified governing equation.

The Governing Equation of the Folded Laminated Plate

Because the stiffness and mass matrices of a laminate in (20) were established in a local coordinate attached to the laminate (Figure 1), the matrices in (29) need to be transformed to the global coordinates before the superposition. The stiffness and mass matrices and the nodal displacement in the global coordinates are

$$\widetilde{\mathbf{K}} = \mathbf{T}\overline{\mathbf{K}}\mathbf{T}^{\mathrm{T}}, \qquad \widetilde{\mathbf{M}} = \mathbf{T}\overline{\mathbf{M}}\mathbf{T}^{\mathrm{T}}, \qquad \widetilde{\boldsymbol{\delta}} = \mathbf{T}\overline{\boldsymbol{\delta}}, \tag{30}$$

where T is the $6n \times 6n$ coordinate transformation matrix derived in [20] (note: a drilling degree of freedom φz has been added to $\overline{\boldsymbol{\delta}}$, and $\overline{\mathbf{K}}$ and $\overline{\mathbf{M}}$ must be expanded accordingly by inserting some zero elements). If there are J (no. 1 to J) coincident nodes along the joint between super elements 1 and 2 (Figure 2), we have

$$\widetilde{\boldsymbol{\delta}}_b^1 = \widetilde{\boldsymbol{\delta}}_b^2, \qquad \ddot{\widetilde{\boldsymbol{\delta}}}_b^1 = \ddot{\widetilde{\boldsymbol{\delta}}}_b^2, \tag{31}$$

where $\widetilde{\boldsymbol{\delta}}_b^1 = \{\widetilde{u}_{01}^1, \widetilde{v}_{01}^1, \widetilde{w}_1^1, \widetilde{\varphi}_{x1}^1, \widetilde{\varphi}_{y1}^1, \widetilde{\varphi}_{z1}^1, \ldots, \widetilde{u}_{0J}^1, \widetilde{v}_{0J}^1, \widetilde{w}_J^1, \widetilde{\varphi}_{xJ}^1, \widetilde{\varphi}_{yJ}^1, \widetilde{\varphi}_{zJ}^1\}^{\mathrm{T}}$ are the actual displacement of the nodes of super element 1 along the joint between super element 1 and super element 2 and $\widetilde{\boldsymbol{\delta}}_b^2 = \{\widetilde{u}_{01}^2, \widetilde{v}_{01}^2, \widetilde{w}_1^2, \widetilde{\varphi}_{x1}^2, \widetilde{\varphi}_{y1}^2, \widetilde{\varphi}_{z1}^2, \ldots, \widetilde{u}_{0J}^2, \widetilde{v}_{0J}^2, \widetilde{w}_J^2, \widetilde{\varphi}_{xJ}^2, \widetilde{\varphi}_{yJ}^2, \widetilde{\varphi}_{zJ}^2\}^{\mathrm{T}}$ are the actual displacement of the nodes of super element 2 along the joint between super element 1 and super element 2. After necessary elementary transformation, the governing equation of super elements 1 and 2 can be written in block forms

$$\begin{bmatrix} \widetilde{\mathbf{k}}_{ii}^1 & \widetilde{\mathbf{k}}_{ib}^1 \\ \widetilde{\mathbf{k}}_{bi}^1 & \widetilde{\mathbf{k}}_{bb}^1 \end{bmatrix} \begin{Bmatrix} \widetilde{\boldsymbol{\delta}}_i^1 \\ \widetilde{\boldsymbol{\delta}}_b^1 \end{Bmatrix} + \begin{bmatrix} \widetilde{\mathbf{m}}_{ii}^1 & \widetilde{\mathbf{m}}_{ib}^1 \\ \widetilde{\mathbf{m}}_{bi}^1 & \widetilde{\mathbf{m}}_{bb}^1 \end{bmatrix} \begin{Bmatrix} \ddot{\widetilde{\boldsymbol{\delta}}}_i^1 \\ \ddot{\widetilde{\boldsymbol{\delta}}}_b^1 \end{Bmatrix} = 0,$$

$$\begin{bmatrix} \widetilde{\mathbf{k}}_{bb}^2 & \widetilde{\mathbf{k}}_{bi}^2 \\ \widetilde{\mathbf{k}}_{ib}^2 & \widetilde{\mathbf{k}}_{ii}^2 \end{bmatrix} \begin{Bmatrix} \widetilde{\boldsymbol{\delta}}_b^2 \\ \widetilde{\boldsymbol{\delta}}_i^2 \end{Bmatrix} + \begin{bmatrix} \widetilde{\mathbf{m}}_{bb}^2 & \widetilde{\mathbf{m}}_{bi}^2 \\ \widetilde{\mathbf{m}}_{ib}^2 & \widetilde{\mathbf{m}}_{ii}^2 \end{bmatrix} \begin{Bmatrix} \ddot{\widetilde{\boldsymbol{\delta}}}_b^2 \\ \ddot{\widetilde{\boldsymbol{\delta}}}_i^2 \end{Bmatrix} = 0, \tag{32}$$

where $\widetilde{\boldsymbol{\delta}}_i^1$ and $\widetilde{\boldsymbol{\delta}}_i^2$ are the actual displacement of the nodes of super elements 1 and 2 that are not along the joint, respectively. Equation (32) are supposed to give the equation governing the dynamic behaviors of the entire structure

$$\begin{bmatrix} \widetilde{\mathbf{k}}_{ii}^1 & \widetilde{\mathbf{k}}_{ib}^1 & 0 \\ \widetilde{\mathbf{k}}_{bi}^1 & \widetilde{\mathbf{k}}_{bb}^1 + \widetilde{\mathbf{k}}_{bb}^2 & \widetilde{\mathbf{k}}_{bi}^2 \\ 0 & \widetilde{\mathbf{k}}_{ib}^2 & \widetilde{\mathbf{k}}_{ii}^2 \end{bmatrix} \begin{Bmatrix} \widetilde{\boldsymbol{\delta}}_i^1 \\ \widetilde{\boldsymbol{\delta}}_b^1 \\ \widetilde{\boldsymbol{\delta}}_i^2 \end{Bmatrix} + \begin{bmatrix} \widetilde{\mathbf{m}}_{ii}^1 & \widetilde{\mathbf{m}}_{ib}^1 & 0 \\ \widetilde{\mathbf{m}}_{bi}^1 & \widetilde{\mathbf{m}}_{bb}^1 + \widetilde{\mathbf{m}}_{bb}^2 & \widetilde{\mathbf{m}}_{bi}^2 \\ 0 & \widetilde{\mathbf{m}}_{ib}^2 & \widetilde{\mathbf{m}}_{ii}^2 \end{bmatrix} \begin{Bmatrix} \ddot{\widetilde{\boldsymbol{\delta}}}_i^1 \\ \ddot{\widetilde{\boldsymbol{\delta}}}_b^1 \\ \ddot{\widetilde{\boldsymbol{\delta}}}_i^2 \end{Bmatrix} = 0. \tag{33}$$

Taking

$$\mathbf{K}_G = \begin{bmatrix} \widetilde{\mathbf{k}}_{ii}^1 & \widetilde{\mathbf{k}}_{ib}^1 & 0 \\ \widetilde{\mathbf{k}}_{bi}^1 & \widetilde{\mathbf{k}}_{bb}^1 + \widetilde{\mathbf{k}}_{bb}^2 & \widetilde{\mathbf{k}}_{bi}^2 \\ 0 & \widetilde{\mathbf{k}}_{ib}^2 & \widetilde{\mathbf{k}}_{ii}^2 \end{bmatrix},$$

$$\mathbf{M}_G = \begin{bmatrix} \widetilde{\mathbf{m}}_{ii}^1 & \widetilde{\mathbf{m}}_{ib}^1 & 0 \\ \widetilde{\mathbf{m}}_{bi}^1 & \widetilde{\mathbf{m}}_{bb}^1 + \widetilde{\mathbf{m}}_{bb}^2 & \widetilde{\mathbf{m}}_{bi}^2 \\ 0 & \widetilde{\mathbf{m}}_{ib}^2 & \widetilde{\mathbf{m}}_{ii}^2 \end{bmatrix},$$

$$\boldsymbol{\delta}_G = \begin{Bmatrix} \widetilde{\boldsymbol{\delta}}_i^1 \\ \widetilde{\boldsymbol{\delta}}_b^1 \\ \widetilde{\boldsymbol{\delta}}_i^2 \end{Bmatrix}, \tag{34}$$

we have

$$\mathbf{K}_G\boldsymbol{\delta}_G + \mathbf{M}_G\ddot{\boldsymbol{\delta}}_G = 0. \tag{35}$$

The solution of the corresponding eigenvalue problem

$$\left(\mathbf{K}_G - \omega^2\mathbf{M}_G\right)\boldsymbol{\delta}_0 = \mathbf{0} \tag{36}$$

gives us the free vibration frequencies of the folded laminated plate.

RESULTS AND DISCUSSION

In order to show the convergence and accuracy of the proposed method, several numerical examples are calculated with the method and the finite element software ANSYS. For all the laminates in the examples, the plies are assumed to have the same thickness and material properties $E_1 = 2.5 \times 10^7$ Pa, $E_2 = 1 \times 10^6$ Pa, $G_{12} = G_{13} = 5 \times 10^5$ Pa, $G_{23} = 2 \times 10^5$ Pa, $\mu_{12} = 0.25$, and $\rho = 2823$ kg/m^3 .. Unless otherwise specified, for each example a total of two cases of symmetric laminates, cross-ply or angle-ply, which make up the folded plates, are studied: (0°/90°/90°/0°) and $(-45°/45°/-45°/45°/-45°\ldots)_{10}$. In ANSYS, the folded laminated plates in the examples are all modeled as shells, and the linear layered structural shell element SHELL99 [29] is used to discretise the folded plates.

Validation Studies

To carry out the validation studies, a cantilever square laminate with a lamination scheme of (−45°/45°/45°/−45°) is considered

(Figure 3). The width of the plate is 1.8 m, and the thickness is 0.018 m. The solution from ANSYS (5000 elements to discretise the laminate) is taken to be the exact solution. The validation studies consist of a convergence study and a study on the effect of the size of support and the completeness order of the basis functions on the convergence of solutions.

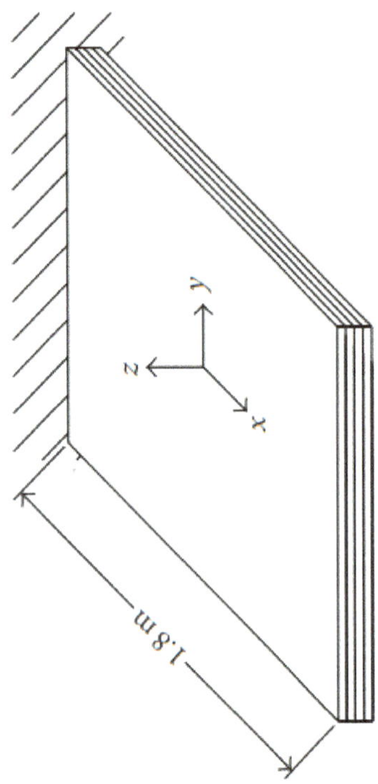

Figure 3: A cantilevered laminate.

Firstly, we choose a certain meshless scheme (11 × 11 nodes for the laminate) and let the scaling factor β and the completeness order N_c of the basis function vary. β defines the size of the support of nodes. In this paper, a rectangular support is used and

$$h_x = \beta \cdot I_x, \quad h_y = \beta \cdot I_y, \tag{37}$$

where h_x, h_y are the lengths of the support in the x and y directions, respectively, and I_x, I_y are the distances between two neighbouring nodes in the x- and y-directions, respectively. The dimensionless fundamental frequency of the laminate as calculated by the proposed method under different values of β and N_c is shown in Figure 4 and is compared with the solution that is given by ANSYS. The dimensionless frequency is defined as

$$\overline{\omega} = \omega\left(\frac{L^2}{h}\right)\sqrt{\left(\frac{\rho}{E_2}\right)}, \tag{38}$$

where L is the width of the laminate and ω is the vibration frequency. From Figure 4, it can be observed that for a certain meshless scheme (in this case 11 × 11 nodes), all of the solutions for different completeness orders (N_c) of the basis functions converge when the support size (β) is larger than 5. Higher completeness orders (N_c) need a larger support size to make the solution converge (the solution under $N_c = 2$ converges at β=4 while the solution under $N_c = 5$ converges at β=5).

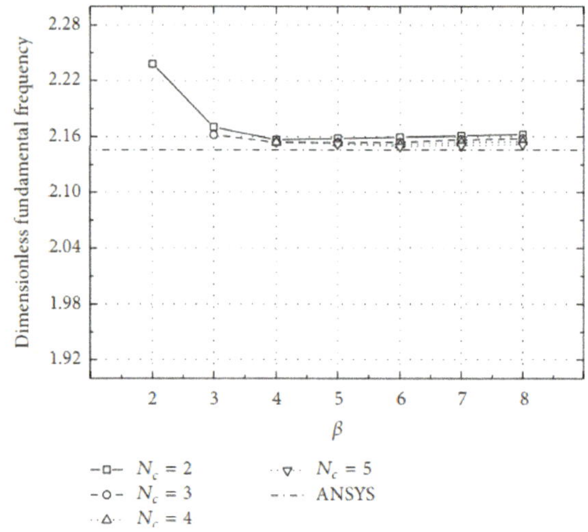

Figure 4: Variation of dimensionless fundamental frequency of the laminate under different β and N_c.

Secondly, we vary the meshless scheme and obtain the variations of the dimensionless fundamental frequency under certain completeness order of the basis functions (N_c), which are shown in Figures 5, 6, 7, and 8, respectively. The solution that is given by ANSYS is also in the figures for comparison. Figures 5 to 8 indicate that for certain β, the solution converges when the number of nodes increases. For an N_c, the solutions for larger support sizes (β) converge before those for smaller support sizes do.

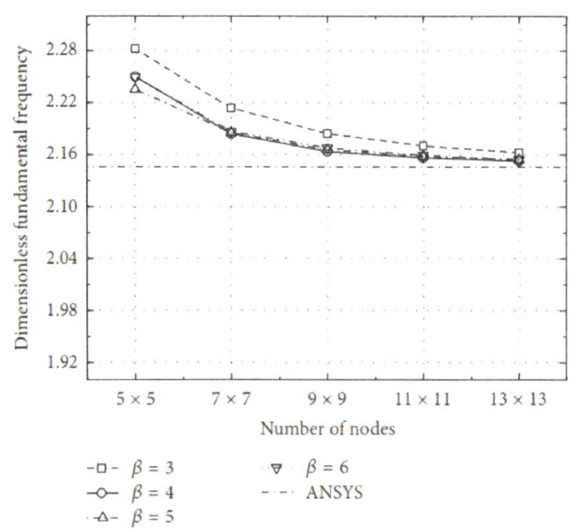

Figure 5: Variation of dimensionless fundamental frequency of the laminate, $N_c = 2$.

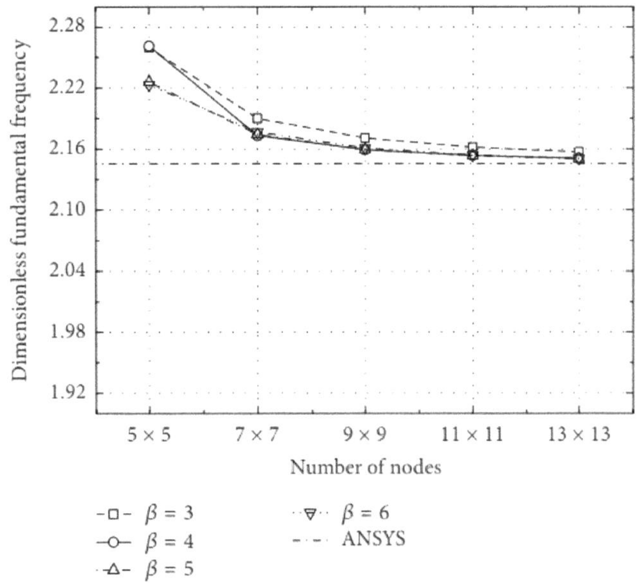

Figure 6: Variation of dimensionless fundamental frequency of the laminate, $N_c = 3$.

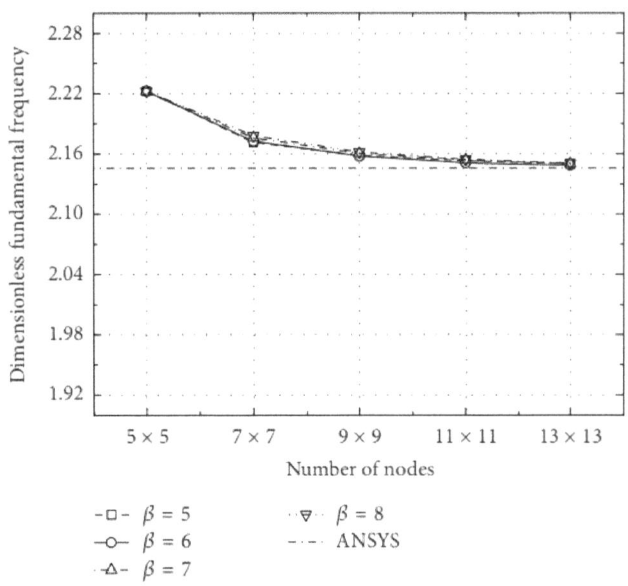

Figure 7: Variation of dimensionless fundamental frequency of the laminate, $N_c = 4$.

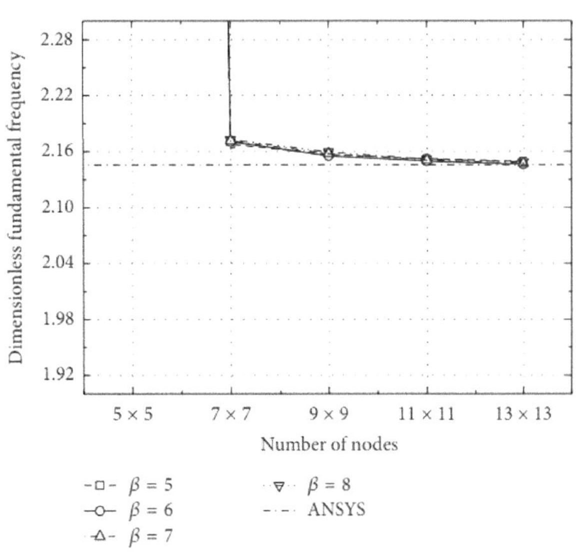

Figure 8: Variation of dimensionless fundamental frequency of the laminate, $N_c = 5$.

From the studies, we find that when the order of basis functions $N_c = 2$ and the support size β=4 for the laminate, the solutions are precise enough with a relatively lower computational cost. Therefore, all of the following examples are calculated with $N_c = 2$, β=4.

A Folded Plate That Is Made Up of Two Laminates

A clamped laminated folded plate that is made up of two identical square laminates is studied (Figure 9). The width of each laminate is L=1 m, and the thickness h = 0.012 m. The dimensionless frequencies of the first five vibration modes of the folded plate that are obtained by the proposed method (11 × 11 nodes for each laminate) are listed in Tables 1 and 2 alongside the results that are given by ANSYS (3200 elements to discretise the folded plate) for comparison, and the first five mode shapes of vibration of the folded plate are also plotted graphically in Figures 10 and 11. Different crank angles of the folded plate, $\alpha = 90^\circ$ and $\alpha = 150^\circ$, are considered. The results of the two methods are very close.

Table 1: Dimensionless free vibration frequencies of the onefold laminated plate with the lamination scheme (0°/90°/90°/0°).

Crank angle	Modes	Present results	ANSYS results	Relative errors
	1	8.9128	8.89037	0.25%
	2	10.0977	10.06576	0.32%
$\alpha = 90^\circ$	3	12.9276	12.87695	0.39%
	4	13.8224	13.77692	0.33%
	5	28.836	28.70447	0.46%
	1	8.91325	8.89121	0.25%
	2	10.1036	10.06659	0.37%
$\alpha = 150^\circ$	3	12.9008	12.8508	0.39%
	4	13.7894	13.74465	0.33%
	5	28.841	28.71003	0.46%

Table 2: Dimensionless free vibration frequencies of the onefold laminated plate with the lamination scheme $(-45°/45°/-45°/45°/-45°\ldots)_{10}$.

Crank angle	Modes	Present results	ANSYS results	Relative errors
	1	10.5522	10.48862	0.61%
	2	15.7311	15.63222	0.63%
$\alpha = 90^\circ$	3	19.2923	19.11888	0.91%
	4	22.6269	22.45392	0.77%
	5	35.233	34.9528	0.80%
	1	10.5784	10.49001	0.84%
	2	15.4292	15.32843	0.66%
$\alpha = 150^\circ$	3	19.5467	19.09968	2.34%
	4	21.6468	21.4844	0.76%
	5	35.3159	34.97227	0.98%

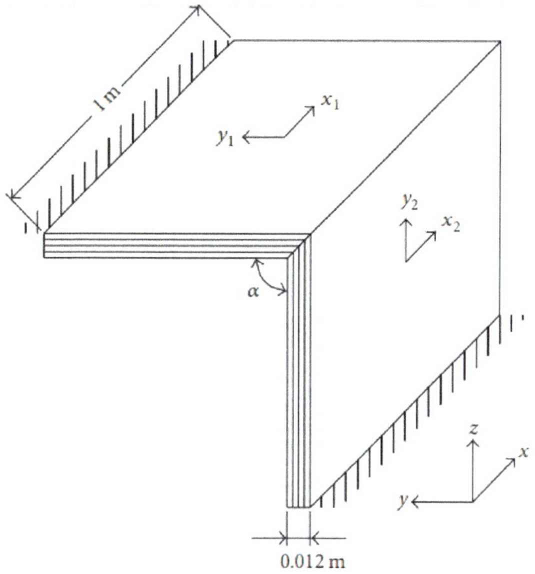

Figure 9: The onefold laminated plate with two sides fixed.

Figure 10: First five vibration mode shapes of the clamped folded plate ($\alpha = 90°$)

Figure 11: First five vibration mode shapes of the clamped folded plate ($\alpha = 150°$)

When the folded plate is clamped at one side, which makes it a cantilevered folded plate (Figure 12), the dimensionless frequencies of the first five vibration modes of the structure are listed in Tables 3 and 4, and the five vibration mode shapes are shown in Figures 13 and 14.

Table 3: Dimensionless free vibration frequencies of the cantilevered onefold laminated plate with the lamination scheme (0°/90°/90°/0°).

Crank angle	Modes	Present results	ANSYS results	Relative errors
$\alpha = 90^\circ$	1	4.97836	4.97752	0.02%
	2	5.47994	5.47883	0.02%
	3	10.9922	10.97741	0.13%
	4	14.3614	14.32497	0.25%
	5	29.6419	29.62531	0.06%
$\alpha = 150^\circ$	1	4.97887	4.97752	0.03%
	2	5.47507	5.47410	0.02%
	3	11.0002	10.98326	0.15%
	4	14.0093	13.97611	0.24%
	5	29.6493	29.63643	0.04%

Figure 13: First five vibration mode shapes of the cantilevered folded plate ($\alpha = 90°$)

Table 4: Dimensionless free vibration frequencies of the cantilevered onefold laminated plate with the lamination scheme $(-45°/45°/-45°/45°/-45°\ldots)_{10}$.

Crank angle	Modes	Present results	ANSYS results	Relative errors
$\alpha = 90^\circ$	1	5.19091	5.16586	0.48%
	2	6.37998	6.35766	0.35%
	3	15.5232	15.42413	0.64%
	4	18.0588	17.93153	0.71%
	5	20.3733	20.27757	0.47%
$\alpha = 150^\circ$	1	5.23049	5.16336	1.30%
	2	6.30941	6.27921	0.48%
	3	15.6594	15.4258	1.51%
	4	17.6398	17.50867	0.75%
	5	20.2856	20.15211	0.66%

Figure 14: First five vibration mode shapes of the cantilevered folded plate ($\alpha = 150°$)

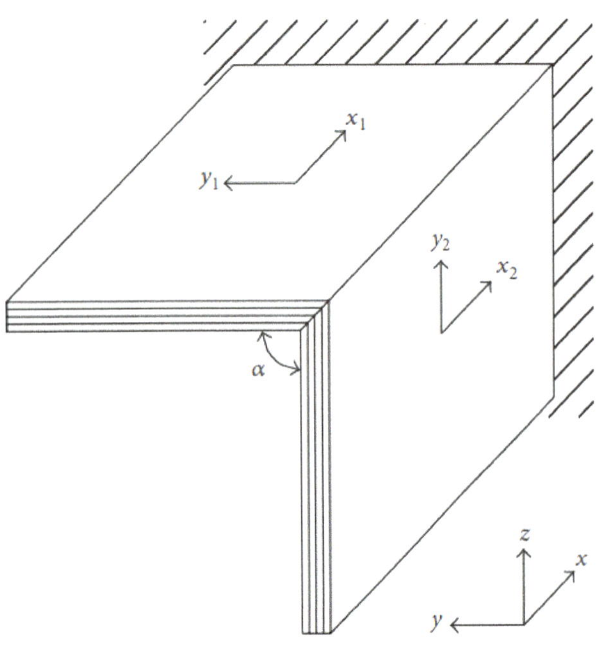

Figure 12: The cantilevered onefold laminated plate.

A Folded Plate That Is Made Up of Three Laminates

A folded plate that is made up of three identical square laminates and clamped at one side is studied (Figure 15, α = 90o). The width of each laminate is L=1 m, and the thickness h = 0.01 m. The dimensionless frequencies of the first five vibration modes of the folded plate, which are calculated with the proposed method and ANSYS, are listed in Tables 5 and 6, and the first five mode shapes of vibration of the folded plate are also plotted graphically in Figures 16 and 17. In ANSYS, 4800 elements are used to discretise the structure. The agreement between the two sets of results is good.

Table 5: Dimensionless free vibration frequencies of the cantilevered twofold laminated plate with the lamination scheme (0°/90°/90°/0°).

Crank angle	Modes	Present results	ANSYS results	Relative errors
$\alpha = 90^\circ$	1	5.15051	5.14912	0.03%
	2	5.27945	5.27765	0.03%
	3	9.42541	9.40889	0.18%
	4	12.4088	12.37671	0.26%
	5	14.3156	14.25922	0.40%
$\alpha = 150^\circ$	1	5.15063	5.14912	0.03%
	2	5.27544	5.27331	0.04%
	3	9.41916	9.40088	0.19%
	4	12.3053	12.27255	0.27%
	5	14.162	14.10833	0.38%

Table 6: Dimensionless free vibration frequencies of the cantilevered twofold laminated plate with the lamination scheme $(-45°/45°/-45°/45°/-45°\ldots)_{10}$.

Crank angle	Modes	Present results	ANSYS results	Relative errors
$\alpha = 90°$	1	5.55864	5.53036	0.51%
	2	6.01129	5.98305	0.47%
	3	13.3613	13.2764	0.64%
	4	16.8974	16.755	0.85%
	5	18.1847	18.02859	0.87%
$\alpha = 150°$	1	5.55885	5.51233	0.84%
	2	5.9971	5.94399	0.89%
	3	13.4186	13.21197	1.56%
	4	16.8014	16.59041	1.27%
	5	17.9625	17.78489	1.00%

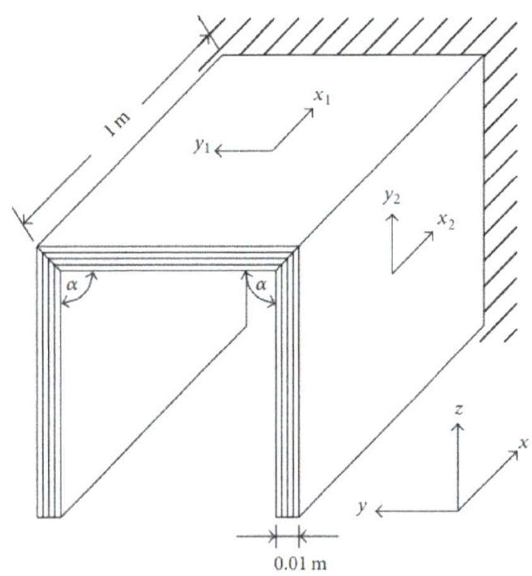

Figure 15: A cantilevered folded plate that is made up of three identical square laminates.

Figure 16: First five vibration mode shapes of the cantilevered folded plate ($\alpha = 90°$) that is made up of three identical square laminates.

Figure 17: First five vibration mode shapes of the cantilevered folded plate ($\alpha = 150°$) that is made up of three identical square laminates.

When the crank angle$\alpha = 60°$ and the laminates are assumed to be connected with one another, we obtain a tub structure with three folds (Figure 18). The dimensionless frequencies of the first five vibration modes of the structure are listed in Tables 7 and 8, and the first five mode shapes are shown in Figure 19.

Table 7: Dimensionless free vibration frequencies of the cantilevered tub structure with the lamination scheme (0°/90°/90°/0°).

Modes	Present results	ANSYS results	Relative errors
1	9.14727	9.126463	0.23%
2	9.14957	9.126797	0.25%
3	14.1263	14.06026	0.47%
4	23.6554	23.53825	0.50%
5	31.4647	31.43285	0.10%

Table 8: Dimensionless free vibration frequencies of the cantilevered tub structure with the lamination scheme $(-45°/45°/-45°/45°/-45°\ldots)_{10}$.

Modes	Present results	ANSYS results	Relative errors
1	12.6681	12.56399	0.83%
2	12.7663	12.56399	1.61%
3	18.6702	18.49329	0.96%
4	31.407	31.05495	1.13%
5	33.1907	32.96984	0.67%

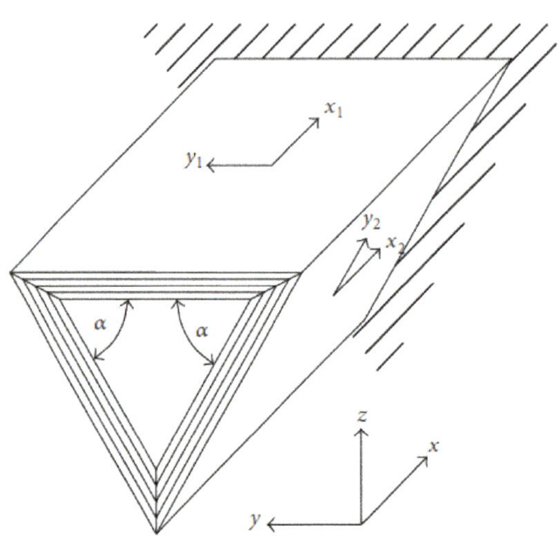

Figure 18: A tub structure with three folds.

Figure 19: First five vibration mode shapes of the tub structure.

A Laminated Shell

If the three laminates in Section 5.3 are joined with each other vertically, we obtain a laminated shell (or half of a box structure) (Figure 20). The structure is pinned at points A, B, and C. All of the DOFs, except the rotations of these points, are set to zero.

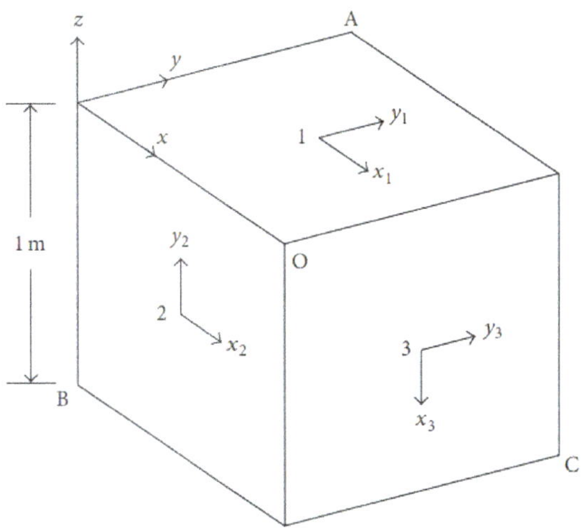

Figure 20: A laminated shell structure.

Three lamination schemes are considered. Case 1: laminate 1 is taken to be (45°/−45°/−45°/45°) and laminates 2 and 3 to be (−45°/45°/45°/−45°), as is demonstrated in Figure 21(a); Case 2: laminate 1 is taken to be (−45°/45°/45°/−45°) and laminates 2 and 3 to be (45°/−45°/−45°/45°), as is demonstrated in Figure 21(b); and Case 3: laminate 1 is taken to be (90°/0°/0°/90°) and laminates 2 and 3 to be (0°/90°/90°/0°), as is shown in Figure 21(c).

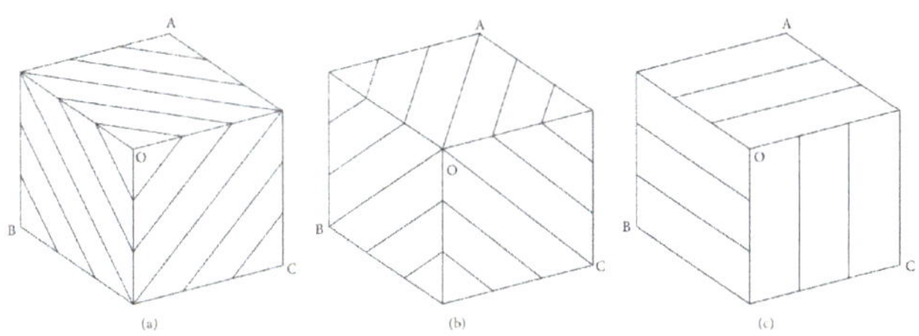

Figure 21: Lamination schemes of the laminated shell.

The dimensionless frequencies of the first five vibration modes of the structures are computed by both the proposed method and ANSYS and listed in Tables 9, 10, and11, and the first five mode shapes are shown in Figures 22, 23, and 24. In ANSYS, 4800 elements are used to discretise the structure. The agreement of the two sets of results is good.

Table 9: Dimensionless free vibration frequencies of the laminated shell (Figure 21(a)).

Modes	Present results	ANSYS results	Relative errors
1	8.10352	8.01745	1.1%
2	8.10352	8.01745	1.1%
3	10.01294	9.86091	1.5%
4	13.70969	13.49473	1.6%
5	15.17441	14.93191	1.6%

Table 10: Dimensionless free vibration frequencies of the laminated shell (Figure 21(b)).

Modes	Present results	ANSYS results	Relative errors
1	9.18846	9.08674	1.1%
2	9.18846	9.08674	1.1%
3	9.65333	9.58148	0.7%
4	13.41558	13.23601	1.4%
5	14.99186	14.81606	1.2%

Table 11: Dimensionless free vibration frequencies of the laminated shell (Figure 21(c)).

Modes	Present results	ANSYS results	Relative errors
1	7.09973	7.07069	0.4%
2	7.09973	7.07069	0.4%
3	9.75645	9.71035	0.5%
4	15.16616	15.11418	0.3%
5	18.02388	17.94379	0.4%

Figure 22: First five vibration mode shapes of the laminated shell (Case 1).

Figure 23: First five vibration mode shapes of the laminated shell (Case 2).

Figure 24: First five vibration mode shapes of the laminated shell (Case 3).

CONCLUSIONS

An element-free Galerkin method that is based on the FSDT is proposed for the free vibration analysis of folded symmetrically laminated plate structures. A folded laminated plate is considered to be a composite structure of flat symmetrical laminates. The global stiffness and mass matrices of the folded plate are formed by superposing the stiffness and mass matrices of the laminates that are derived with the meshfree method. In order to ensure the success of the superposition, a treatment initially developed for the enforcement of essential boundary conditions is extended to modify the stiffness and mass matrices before the superposition, which has overcome the difficulties that the EFG handles displacement compatibility and improved the applicability of EFG to composite structures. The proposed method does not rely on meshes; therefore, mesh disorder due to the large deformation of problem domain is avoided. The convergence and accuracy of the proposed method are demonstrated by a comparison of the solutions of several examples with those that are given by ANSYS. Good agreement between the two sets of results is observed. The proposed method used a relatively small number of nodes to obtain the calculated results close to the solutions given by ANSYS with a large number of nodes, and the linear analysis by the proposed meshless method in the paper can be the basis for future nonlinear analysis. The treatment introduced with the proposed method has provided a clue for EFG to be applied to composite structures.

CONFLICT OF INTERESTS

The author declares that there is no conflict of interests regarding the publication of this paper.

ACKNOWLEDGMENTS

The work that is described in this paper has been supported by the grants awarded by the National Natural Science Foundation of China (Projects nos. 11102044, 51168003, and 11562001), the Systematic Project of Guangxi Key Laboratory of Disaster Prevention and Structural Safety (Project no. 2012ZDX07), and Key Project of Guangxi Science and Technology Lab Center (Fund no. LGZX201101).

REFERENCES

1. Task Committee, "Phase I report on folded plate construction. Report of the Task Committee on folded plate construction. Committee on Masonry and Reinforced Concrete of the Structural Division," Journal of Structural Division, vol. 89, pp. 365–406, 1963.
2. Gaafar, "Hipped plate analysis considering joint displacement," Transaction of ASCE, vol. 119, pp. 743–784, 1954.
3. D. Yitzhaki, The Design of Prismatic and Cylindrical Shell Roofs, Haifa Science, Haifa, Israel, 1958.
4. D. Yitzhaki and M. Reiss, "Analysis of folded plates," ASCE Journal of the Structural Division, vol. 88, pp. 107–142, 1962.
5. C. S. Whitney, B. G. Anderson, and H. Birnbaum, "Reinforced concrete folded plate construction," ASCE Journal of the Structural Division, vol. 85, pp. 15–43, 1959.
6. J. E. Goldberg and H. L. Leve, "Theory of prismatic folded plate structures,"International Association for Bridge and Structural Engineering, vol. 17, pp. 59–86, 1957.
7. G. Niyogi, M. K. Laha, and P. K. Sinha, "Finite element vibration analysis of laminated composite folded plate structures," Shock and Vibration, vol. 6, no. 5, pp. 273–283, 1999.
8. P. Bar-Yoseph and I. Hersckovitz, "Analysis of folded plate structures," Thin-Walled Structures, vol. 7, no. 2, pp. 139–158, 1989.
9. J. N. Bandyopadhyay and P. K. Laad, "Comparative analysis of folded plate structures," Computers & Structures, vol. 36, no. 2, pp. 291–296, 1990.
10. Y. Lai, Z. Wu, Y. Zhu, and A. Sun, "Geometrical non-linear analysis of a simply-supported cross V-shaped folded plate roof," Thin-Walled Structures, vol. 37, no. 3, pp. 259–275, 2000.
11. Y. K. Cheung, "Folded plate structures by finite strip method," Journal of the Structural Division, ASCE, vol. 95, pp. 2963–2979, 1969.
12. B. W. Golley and W. A. Grice, "Prismatic folded plate analysis using finite strip-elements," Computer Methods in Applied Mechanics and Engineering, vol. 76, no. 2, pp. 101–118, 1989.
13. L. Eterovic and L. A. Godoy, "An exact strip method for folded plate structures," Computers and Structures, vol. 32, no. 2, pp. 263–276, 1989
14. M. Ohga, T. Shigematsu, and S. Kohigashi, "Analysis of folded plate structures by a combined boundary element-transfer matrix method," Computers and Structures, vol. 41, no. 4, pp. 739–744, 1991.
15. W. H. Liu and C. C. Huang, "Vibration analysis of folded plates," Journal of Sound and Vibration, vol. 157, no. 1, pp. 123–137, 1992.
16. B. Perry, P. Bar-Yoseph, and G. Rosenhouse, "Rectangular hybrid shell element for analysing folded plate structures," Computers and Structures, vol. 44, no. 1-2, pp. 177–185, 1992.

17. M. Duan and Y. Miyamoto, "Effective hybrid/mixed finite elements for folded-plate structures," Journal of Engineering Mechanics, vol. 128, no. 2, pp. 202–208, 2002.

18. T. Belytschko, Y. Y. Lu, and L. Gu, "Element-free Galerkin methods,"International Journal for Numerical Methods in Engineering, vol. 37, no. 2, pp. 229–256, 1994.

19. J.-S. Chen, C. Pan, C.-T. Wu, and W. K. Liu, "Reproducing kernel particle methods for large deformation analysis of non-linear structures," Computer Methods in Applied Mechanics and Engineering, vol. 139, no. 1–4, pp. 195–227, 1996.

20. L. X. Peng, S. Kitipornchai, and K. M. Liew, "Bending analysis of folded plates by the FSDT meshless method," Thin-Walled Structures, vol. 44, no. 11, pp. 1138–1160, 2006.

21. L. X. Peng, S. Kitipornchai, and K. M. Liew, "Analysis of rectangular stiffened plates under uniform lateral load based on FSDT and element-free Galerkin method," International Journal of Mechanical Sciences, vol. 47, no. 2, pp. 251–276, 2005.

22. L. X. Peng, S. Kitipornchai, and K. M. Liew, "Free vibration analysis of folded plate structures by the FSDT mesh-free method," Computational Mechanics, vol. 39, no. 6, pp. 799–814, 2007.

23. T. Q. Bui, M. N. Nguyen, and C. Zhang, "An efficient meshfree method for vibration analysis of laminated composite plates," Computational Mechanics, vol. 48, no. 2, pp. 175–193, 2011.

24. T. Q. Bui and M. N. Nguyen, "Meshfree Galerkin Kriging model for bending and buckling analysis of simply supported laminated composite plates,"International Journal of Computational Methods, vol. 10, no. 3, Article ID 1350011, 2013.

25. M. Somireddy and A. Rajagopal, "Meshless natural neighbor Galerkin method for the bending and vibration analysis of composite plates," Composite Structures, vol. 111, no. 1, pp. 138–146, 2014.

26. S.-Y. Lee, S.-C. Wooh, and S.-S. Yhim, "Dynamic behavior of folded composite plates analyzed by the third order plate theory," International Journal of Solids and Structures, vol. 41, no. 7, pp. 1879–1892, 2004.

27. J. N. Reddy, Theory and Analysis of Elastic Plates, Taylor & Francis, London, UK, 1999.

28. J. N. Reddy and A. Miravete, Practical Analysis of Composite Laminates, CRC Press, Boca Raton, Fla, USA, 1995.

29. SHELL99, Element Reference, ANSYS Documentation

CHAPTER 11 SOLUTION OF THE POROUS MEDIA EQUATION BY A COMPACT FINITE DIFFERENCE METHOD

Murat Sari

INTRODUCTION

Most physical phenomena and processes encountered in various fields of science are governed by partial differential equations. The nonlinear heat equation describing various physical phenomena, for instance see references [1–4], is called the porous media equation

$$u_t = \frac{\partial}{\partial x}(u^m u_x) + bu \quad (1.1)$$

with the initial condition

$$u(x,0) = f(x) \quad (1.2)$$

and the boundary conditions

$$u(0,t) = g_1(t), \qquad u(1,t) = g_2(t). \quad (1.3)$$

Here m is a rational number while b is real constant. In (1.1)–(1.3), x and t denote derivatives with respect to space and time, respectively, when they are used as subscripts.

Computing the solutions that have a physical or biological interpretation for the nonlinear equations of the form (1.1) is of essential importance. This equation is often encountered in nonlinear problems of heat and mass transfer, combustion theory, and flows in porous media. For instance, it illustrates unsteady heat transfer in a quiescent medium with the heat diffusivity being a power-law function of temperature [5].

Equation (1.1) has also applications to many physical systems including the fluid dynamics of thin films [6]. How this model has been used to represent population pressure in biological systems was analyzed by Murray [7]. Equation (1.1) is known as a degenerate parabolic differential equation as the diffusion term $(u)=um$ does not satisfy the condition for classical diffusion equations, $D(u)>0$ [6].

For the motion of thin viscous films, (1.1) with $m=3$ can be obtained from the Navier-Stokes equations. Lacking a physical law to describe the complex behavior in a system, an appropriate value for the parameter $m$can be determined by comparing known solutions with empirical data [6]. A number of analytical and numerical methods are available in the literature for the investigation of the solution of the corresponding equation [4, 8–13].

In this work, it is aimed to effectively employ a combination of a sixth-order compact finite difference (CFD6) scheme in space [14–16] and a low-storage total variation diminishing third-order Runge-Kutta (TVD-RK3) scheme in time [17, 18] to obtain accurate solutions of (1.1). Since the TVD-RK3 is an explicit time integration scheme, there is no need for linearization of the nonlinear terms.

Citation: Murat Sari, "Solution of the Porous Media Equation by a Compact Finite Difference Method," Mathematical Problems in Engineering, vol. 2009, Article ID 912541, 13 pages, 2009. doi:10.1155/2009/912541.

THE COMPACT FINITE DIFFERENCE SCHEME

The compact finite difference schemes can be dealt within two essential categories: explicit compact and implicit compact approaches. Whilst the first category computes the numerical derivatives directly at each grid by using large stencils, the second one obtains all the numerical derivatives along a grid line using smaller stencils and solving a linear system of equations. Because of the reasons given previously, the present work uses the second approach. To attain the solutions of the equation, discretizations are needed in both space and time.

Discretization

Spatial derivatives are evaluated by the compact finite difference scheme [14]. For any scalar pointwise value u, the derivatives of u can be obtained by solving a tridiagonal or pentadiagonal system. Much work was done in deriving such formulae [14, 19]. More details on the formulae can be found in [14, 19]. A uniform one-dimensional mesh is considered, consisting of N points: $x_1, _2, \ldots, x_i-1, x_i, x_i+1, \ldots, x_N$. The mesh size is denoted by $h=x_{i+1}-x_i$. The first derivatives can be given at interior nodes as follows [14, 19]:

$$\alpha u'_{i-1} + u'_i + \alpha u'_{i+1} = b\frac{u_{i+2} - u_{i-2}}{4h} + a\frac{u_{i+1} - u_{i-1}}{2h} \tag{2.1}$$

which gives rise to an α-family of fourth-order tridiagonal schemes with

$$a = \frac{2}{3}(\alpha + 2), \qquad b = \frac{1}{3}(4\alpha - 1), \tag{2.2}$$

where $\alpha=0$ leads to the explicit fourth-order scheme for the first derivative. A sixth-order tridiagonal scheme is obtained by $\alpha=1/3$,

$$\frac{1}{3}u'_{i-1} + u'_i + \frac{1}{3}u'_{i+1} = \frac{1}{9}\frac{u_{i+2} - u_{i-2}}{4h} + \frac{14}{9}\frac{u_{i+1} - u_{i-1}}{2h}. \tag{2.3}$$

For the nodes near the boundary, approximation formulae for the derivatives of nonperiodic problems can be derived with the consideration of one-sided schemes. More details on the derivations for the first- and second-order derivatives can be found in [14, 19]. The derived formulae at boundary points 1, 2, $N-1$, and N, respectively, are as follows:

$$\begin{aligned}
u'_i + 5u'_{i+1} &= \frac{1}{h}\left(-\frac{197}{60}u_i - \frac{5}{12}u_{i+1} + 5u_{i+2} - \frac{5}{3}u_{i+3} + \frac{5}{12}u_{i+4} - \frac{1}{20}u_{i+5}\right),\\
\frac{2}{11}u'_{i-1} + u'_i + \frac{2}{11}u'_{i+1} &= \frac{1}{h}\left(-\frac{20}{33}u_{i-1} - \frac{35}{132}u_i + \frac{34}{33}u_{i+1} - \frac{7}{33}u_{i+2} + \frac{2}{33}u_{i+3} - \frac{1}{132}u_{i+4}\right),\\
\frac{2}{11}u'_{i-1} + u'_i + \frac{2}{11}u'_{i+1} &= \frac{1}{h}\left(\frac{20}{33}u_{i+1} + \frac{35}{132}u_i - \frac{34}{33}u_{i-1} + \frac{7}{33}u_{i-2} - \frac{2}{33}u_{i-3} + \frac{1}{132}u_{i-4}\right),\\
5u'_{i-1} + u'_i &= \frac{1}{h}\left(\frac{197}{60}u_i + \frac{5}{12}u_{i-1} - 5u_{i-2} + \frac{5}{3}u_{i-3} - \frac{5}{12}u_{i-4} + \frac{1}{20}u_{i-5}\right).
\end{aligned} \tag{2.4}$$

In order to obtain the formulae presented above, the procedure of Gaitonde and Visbal [19] was followed. The formulae can be re-expressed as

$$\mathbf{B}U' = \mathbf{A}U, \tag{2.5}$$

where $U=(u_1,\ldots,u_n)^T$. The second-order derivative terms are obtained by applying the first-order operator twice, that is,

$$\mathbf{B}U'' = \mathbf{A}U', \tag{2.6}$$

Where

$$\mathbf{A} = \frac{1}{h}\begin{bmatrix}
\frac{-197}{60} & \frac{-5}{12} & 5 & \frac{-5}{3} & \frac{5}{12} & \frac{-1}{20} & & \\
\frac{-20}{33} & \frac{-35}{132} & \frac{34}{33} & \frac{-7}{33} & \frac{2}{33} & \frac{-1}{132} & & \\
\frac{-1}{36} & \frac{-7}{9} & 0 & \frac{7}{9} & \frac{1}{36} & & & \\
 & \ddots & \ddots & \ddots & \ddots & \ddots & & \\
 & & \frac{-1}{36} & \frac{-7}{9} & 0 & \frac{7}{9} & \frac{1}{36} \\
 & \frac{1}{132} & \frac{-2}{33} & \frac{7}{33} & \frac{-34}{33} & \frac{35}{132} & \frac{20}{33} \\
 & \frac{1}{20} & \frac{-5}{12} & \frac{5}{3} & -5 & \frac{5}{12} & \frac{197}{60}
\end{bmatrix}_{N\times N}, \quad
\mathbf{B} = \begin{bmatrix}
1 & 5 & & & & & \\
\frac{2}{11} & 1 & \frac{2}{11} & & & & \\
 & \frac{1}{3} & 1 & \frac{1}{3} & & & \\
 & & \ddots & \ddots & \ddots & & \\
 & & & \frac{1}{3} & 1 & \frac{1}{3} & \\
 & & & & \frac{2}{11} & 1 & \frac{2}{11} \\
 & & & & & 5 & 1
\end{bmatrix}_{N\times N}. \tag{2.7}$$

In the calculation of the numerical derivatives, a tridiagonal band matrix algorithm is encountered in both (2.5) and (2.6).

In the current work, the equations are integrated in time with consideration of the low-storage TVD-RK3 scheme [17]. Application of the CFD6 technique to (1.1) leads to the following first-order ordinary differential equation in time:

$$\frac{du_i}{dt} = Lu_i, \tag{2.8}$$

where L stands for the differential operator. The spatial nonlinear terms are approximated by the CFD6 scheme as follows:

$$Lu_i = \left(\frac{\partial}{\partial x}(u^m u_x) + bu\right)_i = \left(mu^{m-1}u_x^2 + u^m u_{xx} + bu\right)_i. \tag{2.9}$$

The choice of time integration is influenced by several factors such as desired accuracy, available memory, computer speed, and stability. A class of high-order TVD time discretization technique was developed by Gottlieb and Shu [17] for solving hyperbolic conservation laws with stable spatial discretizations. As was pointed out by them, unlike in the case of linear operators, there is no stability criterion for fully discrete methods where Lu is nonlinear. However, the TVD methods guarantee the stability properties expected of the forward Euler method [18]. The total variation (TV) of the numerical solution

$$\mathrm{TV}(u) = \sum_i |u_{i+1} - u_i| \tag{2.10}$$

does not increase in time, that is, the TVD property holds:

$$\mathrm{TV}(u^{k+1}) \le \mathrm{TV}(u^k). \tag{2.11}$$

More details can be found in [17, 18]. The scheme integrates equation (2.8) from time t_0 (step k) to $t_0+\Delta t$ (step$k+1$) through the operations [17, 18]

$$\begin{aligned}
u_i^{(1)} &= u_i^k + \Delta t\, Lu_i^k,\\
u_i^{(2)} &= \frac{3}{4}u_i^k + \frac{1}{4}u_i^{(1)} + \frac{1}{4}\Delta t\, Lu_i^{(1)},\\
u_i^{k+1} &= \frac{1}{3}u_i^k + \frac{2}{3}u_i^{(2)} + \frac{2}{3}\Delta t\, Lu_i^{(2)}.
\end{aligned} \tag{2.12}$$

The CFD6 scheme will be applied to physical models to illustrate the strength of the present method. Each spatial derivative on the right-hand side of (2.8) was computed with the use of the present method, and then the semidiscrete equation (2.8) was solved with the help of the low-storage explicit TVD-RK3 scheme. Thus, the solution is obtained without requiring neither linearization nor transformation. For the approximate [a,]solution of the problem (1.1) with the taken boundary and initial conditions using the current scheme, first the interval is discretized such that $a=x_1<x_2<\cdots<x_N=b$, where N is the number of grid points.

NUMERICAL ILLUSTRATIONS

In order to see numerically whether the present methodology leads to accurate solutions, the CFD6 solutions are evaluated for some examples of the porous media equations given above.

Now, numerical solutions of the porous media equation are obtained to validate the current numerical scheme. To verify the efficiency, measure its accuracy and the versatility of the present scheme for our problem in comparison with the exact solution, the absolute errors are reported which are defined by

$$\left|u^{\mathrm{num}}(x_i,t_j)-u(x_i,t_j)\right| \qquad (3.1)$$

in the point (x_i,t_j), where $u^{\mathrm{num}}(x_i,t_j)$ is the solution obtained by (2.8) solved by the combination suggested here, and $u(x_i,t_j)$ is the exact solution.

Consider the porous media equation in the form (1.1) with the initial condition (1.2) and boundary conditions (1.3). The results are compared with the analytical solutions. The numerical computations were performed using uniform grids. All computations were carried out using some codes produced in Visual Basic for Applications. All computations were performed using a laptop computer, Genuine Intel(R) T2500 2.00 GHz CPU, 997 MHz 1.00 GB RAM. All computations were carried out using double-length arithmetic. The differences between the computed solutions and the exact solutions are shown in Tables 1–5, 7, 8. A comparison between the CFD6 solution and the solutions produced by a noncompact fourth-order finite difference scheme (FD4) has also been made in Table 5. In Table 6, computational orders of the method were calculated. In Table 7, the absolute errors were computed for various values of x and N for t=0.05 and Δt=1.0E-06 in Example 3.2. In Table 8, computation of the absolute errors was carried out for various values of$x$, Δt and N at t=0.05 in Example 3.2.

Table 1: The absolute errors for various values of x and t for N=11, h=0.1, Δt=0.0001 in Example 3.1.

x	t	Absolute error
	0.01	1.07E-15
0.10	0.10	1.08E-14
	1.00	6.51E-14
	0.01	1.11E-15
0.50	0.10	1.10E-14
	1.00	1.32E-13
	0.01	1.33E-15
0.90	0.10	3.11E-15
	1.00	3.51E-14

Table 2: The absolute errors for various values of x and t for N=11, h=0.1, Δt=0.0001 in Example 3.2.

x	t	Absolute error
	0.01	1.09E-14
0.10	0.10	1.20E-14
	1.00	3.70E-14
	0.01	1.28E-15
0.50	0.10	2.32E-15
	1.00	4.80E-15
	0.01	1.35E-14
0.90	0.10	1.74E-14
	1.00	4.86E-14

Table 3: The absolute errors for various values of x and t for N=11, h=0.1, Δt=0.0001 in Example 3.3.

x	t	Absolute error
	0.01	7.05E-13
0.10	0.10	1.01E-12
	1.00	2.91E-11
	0.01	1.07E-13
0.50	0.10	2.09E-13
	1.00	3.72E-12
	0.01	1.26E-12
0.90	0.10	1.73E-12
	1.00	3.68E-11

Table 4: The absolute errors for various values of x and t for N=11, h=0.1, Δt=0.0001 in Example 3.4.

x	t	Absolute error
0.10	0.01	8.10E-07
	0.10	1.22E-07
	1.00	1.92E-10
0.50	0.01	6.49E-08
	0.10	4.75E-07
	1.00	1.54E-11
0.90	0.01	1.01E-06
	0.10	1.00E-06
	1.00	1.85E-10

Table 5: Comparison of the CFD6 with the FD4 for various values of x and t with Δt=0.0001, N=11, computational domain [1,2] in Example 3.5.

x	t	Exact	Numerical		Absolute error	
			FD4	CFD6	FD4	CFD6
1.1	0.10	1.110366	1.110365	1.110367	1.38E-06	1.31E-06
	1.00	6.717324	6.717289	6.717331	3.45E-05	7.52E-06
	5.00	20024.059813	20023.956641	20024.082234	1.03E-01	2.24E-02
1.5	0.10	0.814269	0.814235	0.814268	3.33E-05	7.68E-07
	1.00	4.926037	4.925791	4.926030	2.47E-04	6.95E-06
	5.00	14684.310530	14683.574752	14684.289675	7.36E-01	2.09E-02
1.9	0.10	0.642844	0.642830	0.642842	1.34E-05	1.70E-06
	1.00	3.888977	3.888881	3.888964	9.61E-05	1.25E-05
	5.00	11592.876734	11592.589375	11592.839334	2.87E-01	3.74E-02

Table 6: Convergence rate (CR) of the present scheme: Comparisons on various values of h, and Δt for 3.5.

t = 0.0001, Δt = 1.0E-06			t = 0.0003, Δt = 1.0E-06			t = 0.0003, Δt = 1.0E-05		
h	Average absolute error	CR	h	Average absolute error	CR	h	Average absolute error	CR
1/10	5.84E-09		1/18	6.82E-10		1/18	8.95E-10	
1/15	6.86E-10	5.28	1/20	3.84E-10	5.45	1/20	5.06E-10	5.42
1/20	1.48E-10	5.34	1/22	2.27E-10	5.54	1/22	2.99E-10	5.51
1/25	4.36E-11	5.46	1/24	1.39E-10	5.62	1/24	1.84E-10	5.60
1/30	1.58E-11	5.57	1/26	8.80E-11	5.71	1/26	1.17E-10	5.68
1/35	6.57E-12	5.69	1/28	5.72E-11	5.80	1/28	7.60E-11	5.77
1/40	3.03E-12	5.80	1/30	3.81E-11	5.89	1/30	5.07E-11	5.86
1/45	1.51E-12	5.91	1/32	2.59E-11	5.98	1/32	3.45E-11	5.95
1/50	8.02E-13	6.02	1/34	1.79E-11	6.07	1/34	2.39E-11	6.05
1/55	4.47E-13	6.14	1/36	1.26E-11	6.16	1/36	1.68E-11	6.15

Table 7: The absolute errors for various values of x and N for t=0.05 and Δt=1.0E-06 in Example 3.2.

x	N	Absolute error
0.1	11	4.77E-14
	21	3.04E-14
	51	2.97E-14
	81	3.00E-14
	101	3.00E-14
0.5	11	8.17E-14
	21	7.66E-14
	51	7.64E-14
	81	7.68E-14
	101	7.68E-14
0.9	11	1.15E-13
	21	2.28E-14
	51	2.25E-14
	81	2.33E-14
	101	2.31E-14

Table 8: The absolute errors for various values of x, Δt and N for t=0.05 in Example 3.2.

x	Δt	N = 11		N = 21	
		Absolute error	CPU (s)	Absolute error	CPU (s)
0.1	1.0E-04	1.11E-14	0.0	1.74E-04	0.0
	1.0E-05	7.09E-14	2.0	3.04E-15	3.0
	1.0E-06	4.77E-14	16.0	3.04E-14	30.0
	1.0E-07	2.49E-13	162.0	3.03E-13	295.0
	1.0E-08	3.25E-12	1640.0	3.27E-12	2964.0
0.5	1.0E-04	2.26E-15	0.0	8.19E-04	0.0
	1.0E-05	1.19E-14	2.0	7.61E-15	3.0
	1.0E-06	8.17E-14	16.0	7.66E-14	30.0
	1.0E-07	8.26E-13	162.0	7.84E-13	295.0
	1.0E-08	8.74E-12	1640.0	8.62E-12	2964.0
0.9	1.0E-04	1.65E-14	0.00	7.02E-04	0.0
	1.0E-05	8.23E-14	2.00	2.50E-15	3.0
	1.0E-06	1.15E-13	16.00	2.28E-14	30.0
	1.0E-07	3.66E-13	162.00	2.12E-13	295.0
	1.0E-08	3.07E-12	1640.00	2.76E-12	2964.0

As various problems of science were modeled by nonlinear partial differential equations, and since therefore the porous media equation is of high importance, the following examples [1, 4, 5, 10, 13, 20] have been considered. Computational domain $[a,]$ is taken to be [0,1] in Examples 3.1–3.4 and [1,2] in Example 3.5. In Examples 3.1–3.4 b=0, while b=2 in Example 3.5.

Example 3.1 (see [13]). Let us take m=1 and b=0 in (1.1). Thus the equation becomes

$$u_t = \frac{\partial}{\partial x}(uu_x) \tag{3.2}$$

with the initial condition

$$u(x,0) = x \tag{3.3}$$

subject to boundary conditions

$$u(0,t) = t, \qquad u(1,t) = 1 + t. \tag{3.4}$$

An exact solution of (1.1) is

$$u(x,t) = x + t. \tag{3.5}$$

In Table 1, the absolute errors have been shown for various values of x and t. A comparison has been made between the results of the present scheme and the exact results. It is seen from the results that the current scheme are to be powerful and efficient.

Example 3.2 (see [5]). For m=−1 and b=0, (1.1) becomes

$$u_t = \frac{\partial}{\partial x}(u^{-1}u_x). \tag{3.6}$$

In [5] the authors give an exact solution to (3.6) as follows:

$$u(x,t) = \frac{1}{c_1 x - c_1^2 t + c_2}, \tag{3.7}$$

where c_1 and c_2 are arbitrary constants. For the sake of simplicity, c_1=1 and c_2=10. With these choices, the solution becomes

$$u(x,t) = \frac{1}{x - t + 10}. \tag{3.8}$$

Now (3.6) is solved with the consideration of the initial condition

$$u(x,0) = \frac{1}{x + 10} \tag{3.9}$$

and the boundary conditions

$$u(0,t) = \frac{1}{10 - t}, \qquad u(1,t) = \frac{1}{9 - t}. \tag{3.10}$$

The absolute errors have been shown for various values of x and t in Table 2. Also, a comparison between the exact solution and the CFD6 solutions is given in Table 2. The obtained results are seen to be very accurate.

Example 3.3 (see [5]). When m=−4/3 and b=0, (1.1) becomes

$$u_t = \frac{\partial}{\partial x}(u^{-4/3}u_x). \tag{3.11}$$

An exact solution to (3.11) is given as follows [5]:

$$u(x,t) = \left(2c_1 x - 3c_1^2 t + c_2\right)^{-3/4}, \tag{3.12}$$

where c_1 and c_2 are arbitrary constants, c_1=1 and c_2=10. Thus, one has

$$u(x,t) = (2x - 3t + 10)^{-3/4}. \tag{3.13}$$

Now (3.11) is solved with the initial condition

$$u(x,0) = (2x + 10)^{-3/4} \tag{3.14}$$

subject to the boundary conditions

$$u(0,t) = \frac{1}{10 - 3t}, \qquad u(1,t) = \frac{1}{8 - 3t}. \tag{3.15}$$

The absolute errors have been shown for various values of x and t in Table 3. Also, a comparison between the exact solution and the CFD6 solution is given in Table 3. The computed results are seen to be very accurate.

Example 3.4 (see [1, 4, 10]). when m=−2 and b=0, (1.1) represents different nonlinear models and becomes

$$u_t = \frac{\partial}{\partial x}(u^{-2}u_x). \tag{3.16}$$

An exact solution of (3.16) is

$$u(x,t) = \left(x^2 + e^{2t}\right)^{-1/2}. \tag{3.17}$$

Now (3.16) is solved with the initial condition

$$u(x,0) = \left(x^2 + 1\right)^{-1/2} \tag{3.18}$$

subject to the boundary conditions

$$u(0,t) = e^{-t}, \qquad u(1,t) = (1 + e^{2t})^{-1/2} \tag{3.19}$$

The computed results for this example have been given for various values of x and t in Table 4. The results of the combination of the CFD6 method with the low-storage explicit TVD-RK3 have been presented in the table. Comparisons of the current results with the exact solution showed that the presented results are very accurate.

Example 3.5 (see [20]). When m=−2 and b=2, (1.1) represents nonlinear realistic models and becomes

$$u_t = \frac{\partial}{\partial x}(u^{-2}u_x) + 2u. \tag{3.20}$$

An exact solution of (3.20) is [18]

$$u(x,t) = x^{-1}e^{2t}. \quad (3.21)$$

Now (3.20) is solved with the initial condition

$$u(x,0) = x^{-1} \quad (3.22)$$

subject to the boundary conditions

$$u(1,t) = e^{2t}, \qquad u(2,t) = \frac{e^{2t}}{2}. \quad (3.23)$$

In Table 5, computation of the absolute errors was carried out for various values of x and t in Example 3.5. To see the effect of the CFD6 solution, Example 3.5 was also solved using the FD4. Note that the effect of the CFD6 solutions is very clear as seen in Table 5. Numerical rate of convergence (CR) has also been studied to know about the convergency of the scheme. To achieve this, the CR was calculated using

$$\mathrm{CR} = \frac{\log\left(E_1/E_2\right)}{\log\left(h_1/h_2\right)}, \quad (3.24)$$

where $E1$ and $E2$ are errors correspond to grids with mesh size $h1$ and $h2$, respectively, (see Table 6). Also computational orders in Table 7 show the high-order accuracy of the present method for solving such problems.

In this paper, it is shown that the approximate solutions of the porous media equation are very close to the exact solutions. The absolute errors have been calculated for m=1, m=−1, and m=−4/3, and m=−2, in Tables 1, 2, 3, and 4, respectively, in the point (x_i,). For all four cases b=0, but m=−2 is used twice, the second one with b=2 in Example 3.5. As seen from Tables 1–5 and 7-8, the absolute errors in all cases are very small. Very good approximations to the exact solutions are achieved.

The method was also applied to realistic problem sizes (see Table 7), and seen that the solution becomes stable for small number of grid points. Increasing in the number of grid points requires small time step size for satisfying numerical stability of the solutions. This situation makes necessary the use of excessive CPU time and gives rise to accumulation of numerical errors (see Tables 7 and 8). Compact high-order schemes are closer to spectral methods [14] and at the same time maintain the freedom to retain accuracy even though relatively small number of grid points is used. Note that the same error is obtained for all discretizations; the error is close to the level of machine accuracy, and this is believed to be a saturation effect (see Table 7).

CONCLUSIONS

In this paper, numerical simulations of the porous media equation were dealt with using a combination of the CFD6 scheme in space and a low-storage explicit TVD-RK3 scheme in time. The method successfully worked to give very reliable and accurate solutions to the equation. The method gives convergent approximations and handles nonlinear problems. In this method, there is no need for linearization of nonlinear terms. Nonlinear scientific models arise frequently in scientific problems for expressing nonlinear phenomena. For nonlinear problems, the present method is seen to be a very good choice to achieve a high degree of accuracy while dealing with the problems. The computed results justify the advantage of this method. The present method needs less use of storage space.

ACKNOWLEDGMENTS

The author would like to thank anonymous referees of the journal of MPE for their valuable comments and suggestions to improve most of this paper. The author is very grateful to G. Gürarslan (Department of Civil Engineering, Faculty of Engineering, Pamukkale University, Denizli, Turkey) for his comments and careful reading the paper.

REFERENCES

1. E. A. Saied, "The non-classical solution of the inhomogeneous non-linear diffusion equation," Applied Mathematics and Computation, vol. 98, no. 2-3, pp. 103–108, 1999.
2. E. A. Saied and M. M. Hussein, "New classes of similarity solutions of the inhomogeneous nonlinear diffusion equations," Journal of Physics A, vol. 27, no. 14, pp. 4867–4874, 1994.
3. L. Dresner, Similarity Solutions of Nonlinear Partial Differential Equations, vol. 88 of Research Notes in Mathematics, Pitman, Boston, Mass, USA, 1983.
4. Q. Changzheng, "Exact solutions to nonlinear diffusion equations obtained by a generalized conditional symmetry method," IMA Journal of Applied Mathematics, vol. 62, no. 3, pp. 283–302, 1999.
5. D. Polyanin and V. F. Zaitsev, Handbook of Nonlinear Partial Differential Equations, Chapman & Hall/CRC Press, Boca Raton, Fla, USA, 2004.
6. T. P. Witelski, "Segregation and mixing in degenerate diffusion in population dynamics," Journal of Mathematical Biology, vol. 35, no. 6, pp. 695–712, 1997.
7. J. D. Murray, Mathematical Biology, vol. 19 of Biomathematics, Springer, Berlin, Germany, 2nd edition, 1993.
8. J. R. King, "Exact results for the nonlinear diffusion equations ∂u/∂t=(∂/∂x)(u−4/3∂u/∂x) and ∂u/∂t=(∂/∂x)(u−2/3∂u/∂x)," Journal of Physics A, vol. 24, no. 24, pp. 5721–5745, 1991.
9. W. F. Ames, Nonlinear Partial Differential Equations in Engineering. Vol. II, Mathematics in Science and Engineering, Vol. 18, Academic Press, New York, NY, USA, 1972.
10. A.-M. Wazwaz, "Exact solutions to nonlinear diffusion equations obtained by the decomposition method," Applied Mathematics and Computation, vol. 123, no. 1, pp. 109–122, 2001.
11. Sadighi and D. D. Ganji, "Exact solutions of nonlinear diffusion equations by variational iteration method," Computers & Mathematics with Applications, vol. 54, no. 7-8, pp. 1112–1121, 2007.
12. A.-M. Wazwaz, "The variational iteration method: a powerful scheme for handling linear and nonlinear diffusion equations," Computers & Mathematics with Applications, vol. 54, no. 7-8, pp. 933–939, 2007.

13. S. Pamuk, "Solution of the porous media equation by Adomian's decomposition method," Physics Letters A, vol. 344, no. 2-4, pp. 184–188, 2005.

14. S. K. Lele, "Compact finite difference schemes with spectral-like resolution," Journal of Computational Physics, vol. 103, no. 1, pp. 16–42, 1992.

15. M. Sari and G. Gürarslan, "A sixth-order compact finite difference scheme to the numerical solutions of Burgers' equation," Applied Mathematics and Computation, vol. 208, no. 2, pp. 475–483, 2009.

16. M. Sari, G. Gürarslan, and İ. Dağ, "A compact finite difference method for the solution of the generalized Burgers-Fisher equation," Numerical Methods for Partial Differential Equations. In press.

17. S. Gottlieb and C.-W. Shu, "Total variation diminishing Runge-Kutta schemes," Mathematics of Computation, vol. 67, no. 221, pp. 73–85, 1998.

18. S. Gottlieb, C.-W. Shu, and E. Tadmor, "Strong stability-preserving high-order time discretization methods," SIAM Review, vol. 43, no. 1, pp. 89–112, 2001.

19. D. V. Gaitonde and M. R. Visbal, "High-order schemes for Navier–Stokes equations: algorithms and Implementation into FDL3DI," Air Force Research Laboratory, Wright-Patterson AFB, Dayton, Ohio, USA, 1998.

20. N. Pamuk, "Series solution for porous medium equation with a source term by Adomian›s decomposition method," Applied Mathematics and Computation, vol. 178, no. 2, pp. 480–485, 2006.

CHAPTER
12

ARCHITECTURE OF WIRELESS VEHICLE WEIGHT MEASUREMENT SYSTEM FOR STRUCTURAL HEALTH MONITORING IN CIVIL ENGINEERING APPLICATION

Artur Andrzejczak, PaweB Awczycki, MichaB Wojtera, Piotr Pietrzak, Bartosz PwkosBawski, and Andrzej Napieralski

INTRODUCTION

Wireless Sensor Networks are widely applied in structural health monitoring solutions [1]. Measurement data, for example, strain of concrete bridge, provide valuable feedback for civil engineers [2, 3]. Acquisition of different types of data and their correlation leads to a more thorough diagnosis of a monitored object. Wireless communication enables usage of such systems in places, where structural cabling is not advisable, for instance, in heritage buildings [4] or due to formal restrictions. Additionally, deployment time and cost are reduced. Location of measurement points can be easily changed.

The aim of the described project is a design of scalable, multipoint vehicle weight measurement system with wireless communication, short response time, and in-node processing. To achieve this goal, wireless mesh networking was applied. IPv6 connectivity simplifies integration with backbone network and remote data acquisition system.

The WVWMS cooperates with other subsystems, which are the part of TULCOEMPA project. Therefore, the following section describes them and the project objectives.

TULCOEMPA Project Description. The TULCOEMPA project is based on the Swiss-Polish cooperation between the Lodz University of Technology and Swiss Material Science Institute EMPA. From the Lodz University of Technology two departments are involved: DMCS (Department of Microelectronics and Computer Science) and KBB (Concrete Construction Department). Multipurpose program of the project covers civil engineering and structural health monitoring of buildings. Unanchored prestressed carbon fibre reinforced polymer laminates will be used for bridge strengthening. A bridge located in Szczercowska Wieś, central part of Poland, is used as a model one. To ensure effectiveness of this innovative solution, a long-term monitoring of the structure will be applied [5]. It will include multipoint measuring system composed of wireless sensors, sink base station, and computer system for collecting, processing, and data coding. Strain will be measured with low frequency and, additionally, with high frequency for short periods of time. This enables the behaviour assessment of the bridge during heavy loadings. Dynamic measurements will occur only when presence of a heavy vehicle is detected by TRS (Truck Recognition System). It will reduce WSS (Wireless Strain Sensor) devices power consumption. The Truck Recognition System is described below.

Results from structural health monitoring installation are required to correlate with weights of passing vehicles. A level of concrete lamellae strain paired with weight of a vehicle passing through the bridge provides valuable information about efficiency of lamella reinforcement solution. Furthermore, measurement results are required for verification of the simulation model of the bridge. What is more, vehicle weight data can be used to asses TRS effectiveness, and when needed, to support it. On the other hand, when TRS shows high accuracy,

Citation: Artur Andrzejczak, Paweł Łęczycki, Michał Wojtera, Piotr Pietrzak, Bartosz Pękosławski, and Andrzej Napieralski, "Architecture of Wireless Vehicle Weight Measurement System for Structural Health Monitoring in Civil Engineering Application," International Journal of Distributed Sensor Networks, vol. 2015, Article ID 202545, 10 pages, 2015. doi:10.1155/2015/202545.

it can be used to activate weight measurement. Avoidance of continuous weight measurement can significantly increase node battery life. Due to mains power supply unavailability, all subsystems are powered by the Hybrid Power System. As a result, low energy consumption is necessary. To meet the described requirements, the authors propose the Wireless Vehicle Weight Measurement System, described in Section 2.

All subsystems are deployed on the bridge as shown in Figure 1 and relation between them in Figure 2.

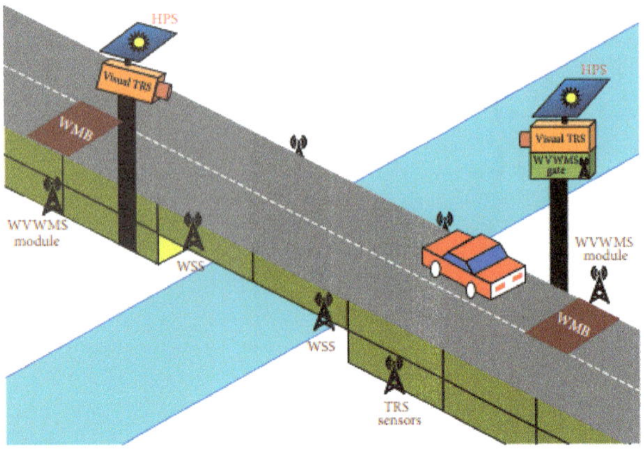

Figure 1: TULCOEMPA subsystems deployed on the bridge: Wireless Strain Sensors, Truck Recognition System, and Wireless Vehicle Weight Measurement System.

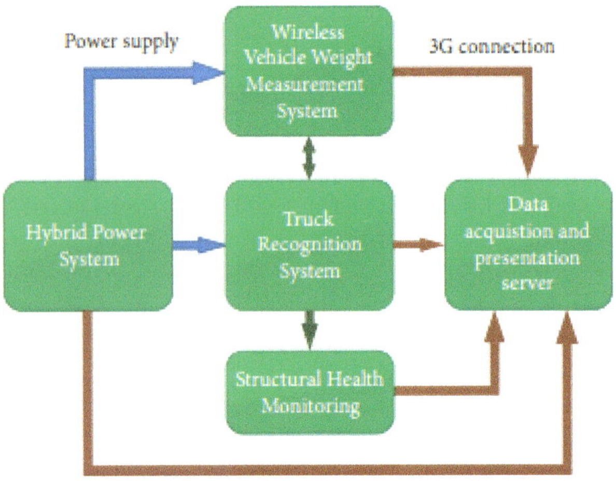

Figure 2: TULCOEMPA subsystems dependencies.

Truck Recognition System Overview. One of the main tasks in the TULCOEMPA project is a development of the Truck Recognition System. As described in the introduction, its aim is the detection of the trucks passing through the bridge. This event is sent to the Health Monitoring System in order to trigger high-frequency strain measurement. Vehicle recognition is done based on picture analysis, sound sensor, vibration sensor, and magnetic field sensor. Due to false positives detections, assessment mechanism for TRS is required.

A structure of the solution and its relation to SHM is presented in Figure 3. A management unit of the TRS will be an Open Multimedia Application Platform (OMAP). It is also responsible for image analysis, which has been described in [6]. Vibration, sound, and magnetic field detection is conducted by independent identification modules. A dedicated software will scan general input-output ports and wait for a detection signal from identification modules. These modules will work independently of each other and other parts of the system. A decision about triggering of the Structural Health Monitoring System will be made based on a voting system. The modules, after event detection, take a vote for starting a measurement. Not taking any action by the identification modules is interpreted as a vote against starting the measurement.

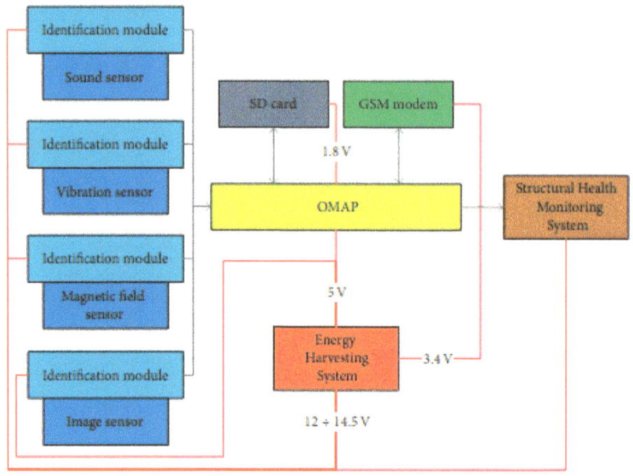

Figure 3: Block diagram of the Truck Recognition System with the Energy Harvesting System.

Accuracy and reliability (including stability) of the identification modules can be disturbed during a year. Changes of climatic and lighting conditions can affect the sensors reliability. In addition, an identification module with a damaged sensor should be excluded from the voting process. In order to ensure that the Structural Health Monitoring System is trigged only when a specified event occurs, influence of sensors inaccuracy and a number of false-positive and false-negative event detections has to be reduced. Based on the performed analysis and research work a universal method for the credibility evaluation of information from sensors was developed [6]. The Sensor Reliability Method based on Credibility Coefficients [7] purpose is to estimate a level of effectiveness of the identification modules and assign a proper weight to them during the voting process. In this case, the voting process takes a form of a weighted voting system [8].

In terms of a number of properly detected events, the performed simulations showed high effectiveness of this solution. The Sensor Reliability Evaluation Method has the advantage over simple majority voting by obtaining the results better by almost 20 percentage points and a reliability of sensors is estimated with an accuracy of 5%. Nevertheless, the only way to make a comprehensive test of the Sensor Reliability Evaluation Method is implementation in the real system. A reason for that is that some situations, which are hard to predict or simulate, are to be expected. Feedback from the vehicle weighting system will be used for the evaluation of this system.

WIRELESS VEHICLE WEIGHT MEASUREMENT SYSTEM

System Structure

Challenge of the architecture of vehicle weight measurement

system is to support multiple measurement points. A distance between road scales is expected to be hundreds of meters. In order to decrease energy consumption, single network gate is advisable. Additionally, this should decrease deployment cost of such system. On the other hand, communication in on-site Truck Recognition System should be with possibly short delays.

The authors propose wireless mesh networking solution. The system structure is presented in Figure 4. It includes measurement boards (weight pads) embedded in the road surface. Foundation is made of steel plates, on which electrical strain gauges are mounted. They provide both high precision measurements and reliability, while being relatively simple to use due to linear dependency between resistance and elongation. Weight measurement of passing vehicles is possible with a speed up to 160 km/h [9]. Therefore, this procedure will be transparent for users of the bridge, by not requiring any changes in the traffic organisation. Within the board (or two boards), the WVWMS module will be installed. Communication between the gate and measurement nodes is wireless with the use of IEEE 802.15.4 [10]. The network sink is required for data acquisition and forwarding through the cellular modem Internet connection. The usage of IEEE 802.15.4 transceivers for local module communication makes the presented system expandable. Multiple modules (and as a result weight measurement boards) can use single network gate. What is more, a range of multipoint WVWMS deployment can be significantly improved by mesh routing. The presented hardware solution is capable of handling this feature. In such mode of operation, it is possible to analyse multiple weight measurement points, provided that distance between outermost ones is smaller than a few hundred meters. The mentioned restriction results from the fact that modules work in the ISM (Industrial Scientific Medical) band, in which transmission power is formally limited.

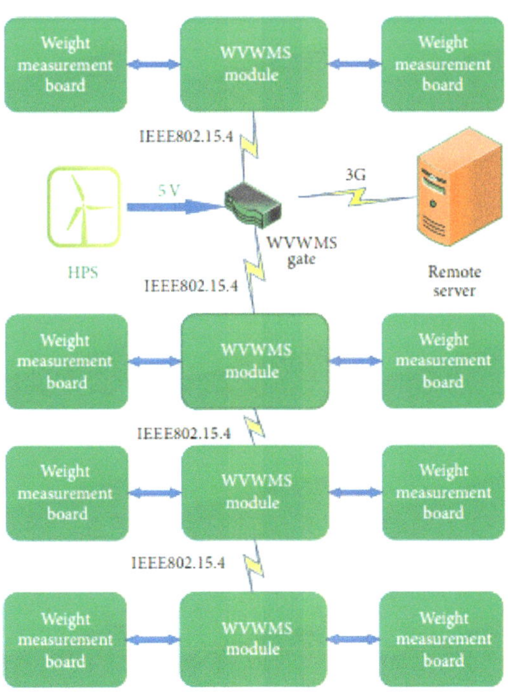

Figure 4: Architecture of Wireless Vehicle Weight Measurement System, single gate is used for multiple modules, which can participate in mesh routing.

Wireless Vehicle Weight Measurement Module Hardware

WVWMS requires several capabilities from the measurement module. Because of usage of electrical strain gauge bridge for weight measurement, a high resolution ADC with programmable gain and reference voltage excitation is advised. In order to achieve multipoint coverage, the module is expected to handle mesh networking capabilities. Selected microcontroller is advised to support open source Internet of Things operating system. Authors propose the module structure presented in Figure 6. Usage of two microcontrollers in single node comes with two benefits. First microcontroller handles communication and power management of the node. As a result, most of its time is dedicated to connectivity tasks, such as routing. This comes with decreased response time. Additional 32-bit ARM microcontroller enables local data processing without negative impact of networking activity. Compression of measurement data decreases network load, which improves its capacity and decreases delays. Moreover, in-node data analysis creates possibility of sending detected weight events directly to network actuator. This makes described system different from other available WSN solutions [11].

Texas Instruments MSP430 ultra-low power microcontroller was chosen for communication tasks. Its architecture combines five low-power modes, with 2.1 μA standby mode and wakeup time below 5 μs, making it an appropriate choice for battery-powered devices. Input voltage range is from 1.8 V to 3.6 V, enabling direct connection to 2 $\times$ 1.5 V batteries. The direct connection significantly decreases superfluous current drawn from a DC/DC converter, particularly when both microcontroller and transceiver are in sleep mode and consumes a few μA. A built-in real time clock can be used for periodical activation of the device for scheduled measurements. Due to the absence of signal processing tasks on this microcontroller, it can be managed by a lightweight operating system, Contiki [12]. As a result, MCU is focused on handling advanced communication mechanisms. The mesh routing or IPv6 (Internet Protocol Version 6) with UDP communication can be considered as an example. It simplifies development and integration process of software responsible for uploading and publishing measurement results. The Contiki OS advanced routing features, such as RIME stack, are used when WVWMS modules are forming mesh network. In that situation communication part of the module needs to be active not only for measurement purposes, but also for data forwarding.

The secondary microcontroller is responsible for measurement and local data processing. In addition to vehicle weight, it can also analyse environmental conditions with additional sensor, for instance, SHT-11 from Sensirion. The information about a humidity and temperature delivers feedback about impact of these parameters on the results from the weight pads. The data analysis performed by the microcontroller may cover the determination of vehicle weight based on dynamic axle pressure measurement. The weight-time characteristics from the WMB can be compressed for a further remote server upload. The compression of results can be more efficient than sending raw data through the radio interface [13]. The measurement MCU is also responsible for battery voltage control. The Atmel SAM4L ARM Cortex-M4 microcontroller was chosen for measurement tasks. It operates with a clock of 48 MHz and provides satisfying computational efficiency. Combined power control techniques reduce active current draw down to 90 μA per MHz. What is more, the WAIT and RETENTION modes consume only 3 μA and 1.5 μA, while sustaining 1.5 μs wake-up time.

The measurement module uses Analog Devices AD7195 24-bit sigma-delta ADC. It is a low noise analogue front-end for high precision measurements. A built-in support for AC excitation reduces DC-induced offset drift from bridge sensors, which increases achieved accuracy. Two differential inputs enable connection of two separate weight boards to a single device.

The weight measurement system communication is based on an integrated IEEE 802.15.4 2.4 GHz ISM band transceiver, the CC2520 from Texas Instruments. The main advantage of this device is its extensive hardware support for IEEE 802.15.4 frame handling. It is an essential feature in the situation, when modules are in sleep mode, awaiting for measurement trigger or packets to be forwarded. In this case, the only active device is CC2520, which analyses incoming packets. If a packet is related to the module, it awakes the communication MCU, which performs further packet analysis and actions.

The communication MCU and the transceiver are directly connected to the battery. Other digital ICs requiring higher voltage are powered by the DC/DC converter, which is controlled by the communication MCU. In order to connect both MCUs, a voltage level translator is applied. The second DC/DC converter is used for powering analogue circuit including strain gauge bridge, the weight pad.

The assembled measurement module is presented in Figure 5. Both microcontrollers are on single, two-sided PCB. Analogue digital converter has separate ground in order to avoid interference from DC/DC converters and digital part. The devices enable connection of separate antenna. An integrated SMD balloon is used. The detailed description of the module hardware structure is available [14].

Figure 5: WVWMS measurement module prototype on two-sided PCB.

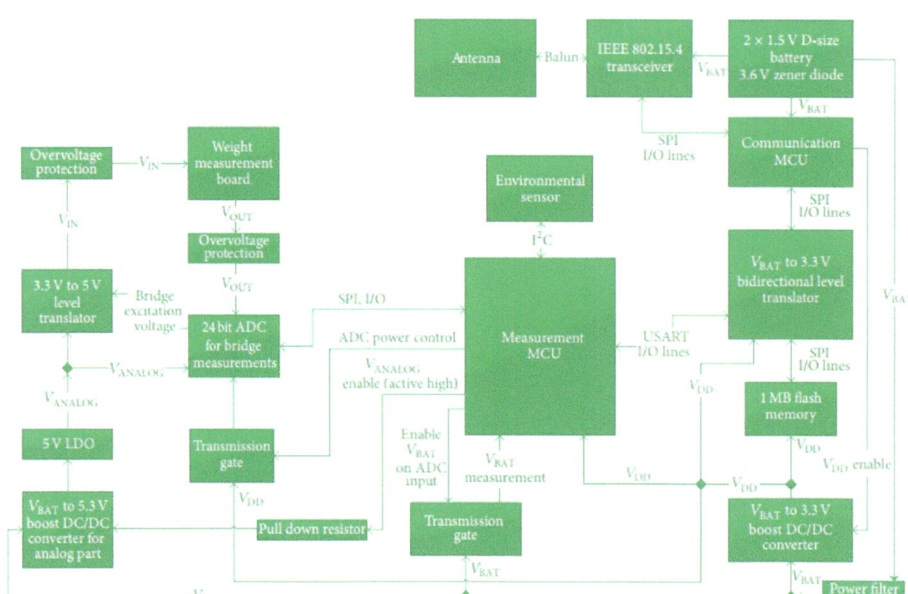

Figure 6: WVWMS node structure diagram with relation between two microcontrollers, radio transceiver, ADC, and flash memory.

WVWMS Network Gate

The network gate is responsible for acquisition of the measurement data and uploading them to the storage and server. Additionally, it should be able to receive remote configuration and calibration commands from the Internet. 3G modem is required for communication with the Internet. The authors propose usage of the ARM board, the BeagleBoard-xM from Texas Instruments. It is controlled by the custom built distribution of the GNU/Linux operating system. The IEEE 802.15.4 interface is connected to the ARM board through the USB and seen as a serial connection. It enables usage of Point-to-Point Protocol and packet forwarding with 6lbr. The Contiki slip-radio application is used. The IEEE 802.15.4 interface is based on the communication part of the measurement module and uses the same ICs and most of the PCB layout. Figure 7presents the hardware. USB board is detachable, enabling direct access to serial interface. There is a second interface for Contiki debugging and exposed JTAG interface.

Figure 7: The IEEE 802.15.4 interface similar to the communication part of measurement module, the USB interface is detachable.

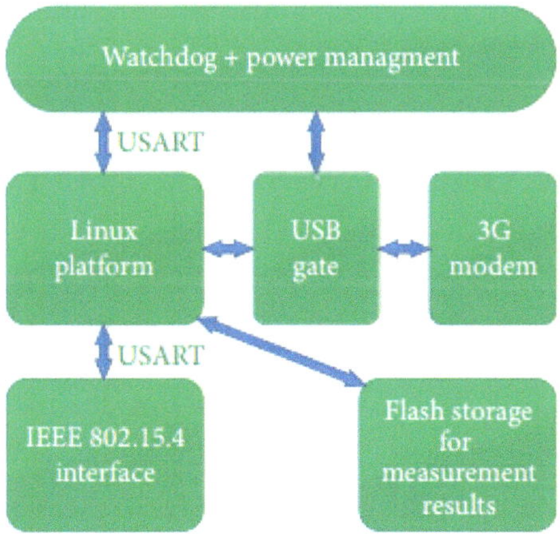

Figure 8: WVWMS gate structure, Linux platform connected to the WVWMS interface, USB modem, and Power Management Board.

The application running on the Linux board is responsible for receiving packets from the measurement network and uploading them to a remote server. The Internet connection is provided by the USB UMTS modem and controlled by the WvDial tool. The dedicated script controls connection status and resets the modem if necessary. What is more, when the Linux board is not responding, external hardware conducts hard reset of the board. The USB gate enables hard reset of the modem without rebooting the board. This solution improves reliability of the system and as a result, on-site maintenance time is decreased. If the Internet connection is unavailable, measurement data can be stored in internal SD flash storage. The structure of the WVWMS network gate is presented in Figure 8.

HYBRID POWER SYSTEM FOR WVWMS

Power supply subsystem, which is common for both the WVWMS gate and TRS, utilises energy harvesting due to a fact that the bridge is located in a long distance from the nearest power line and the systems need to operate uninterruptedly. Two various power sources are planned to be used in order to prevent energy shortage. A small wind generator (Hyacinth P-300 W, rated output power of 300 W at 12 V) will deliver power to cover deficiencies when weather conditions are not favourable for solar panels (four MH-30 panels, 30 W output power at 12 V) to generate sufficient amount of energy to maintain uninterrupted WVWMS and TRS operation (especially in winter and in nights). On the other hand, when wind is not strong enough for a wind generator to rotate, it is predicted that solar panels should produce necessary amount of energy. However, the system will utilise energy storage mechanism to solve the problem of short-term weather changes (several consecutive cloudy and windless days).

The power budget estimation was performed for both TRS with WVWMS gate before actual power system design. Power from the solar panels was calculated as their nominal power multiplied by a ratio of average insolation (for a given season in central Poland) to insolation value at which the nominal power can be achieved (1000 Wm^{-2}). Average insolation is measured for horizontal plane while the panel surface is inclined to this plane (with an angle of about 30 degrees). This makes direct sun light rays to form larger angle with the panel surface (maximum angle of incidence on horizontal plane varies from 15 to 62 degrees in the considered location). Hence, true irradiance of the panels will be higher.

Power from the wind generator was estimated using average seasonal wind speeds (central Poland, elevation of 5 meters above ground). The first step was calculation of probabilities for various wind speeds according to Weibull distribution with shape factor of 1.7. Next percentage power (relative to nominal power available at 12.5 ms^{-1}) was calculated for each wind speed assuming that power is proportional to a wind speed cubed. Minimum wind speed for the generator to rotate and maximum generated power level were taken into account. Finally, each estimated probability for a given wind speed and a corresponding percentage power level were multiplied and then all obtained values were summed to give an expected power level.

Predicted power consumption was calculated for all internal circuits taking into consideration their supply voltages and currents in active and sleep modes as well as average percent of time of staying in a given mode. An average number of trucks passing through the bridge were assumed to have a value of 5 per hour while a total number of cars were predicted as 50 per hour (according to on-site traffic observations). The first parameter was used in TRS power consumption calculations (activity of sensor communication and synchronization with

WSS). Moreover, one transmission per hour using 3G modem was assumed. The total number of cars per hour was used in WVWMS gate power consumption evaluation. Power processing efficiencies of necessary switching converters were also considered in the computation of consumed power. The results shown in Table 1 confirmed that a positive power margin can be expected in all conditions.

Table 1: Power budget estimation results.

Device	Conditions	Average power [W] Consumed	Produced
TRS	All	6.7	
WVWMS gate	All	4.5	
Solar panels	Summer		28.4
	Winter		3.3
Wind generator	Summer		3.5
	Winter		22.4
Power margin	Summer	+20.5	
	Winter	+14.7	

The Energy Harvesting Power system is based on Energy Harvesting Controller (EHC) and additional power processing circuits (DC-DC converters) for output voltage regulation to necessary levels. The EHC consists of DC-DC boost converters for the solar panels and wind generator, current summing circuit, and microcontroller measurement and control module with external communication interface. Block diagram of the EHC is presented in Figure 9.

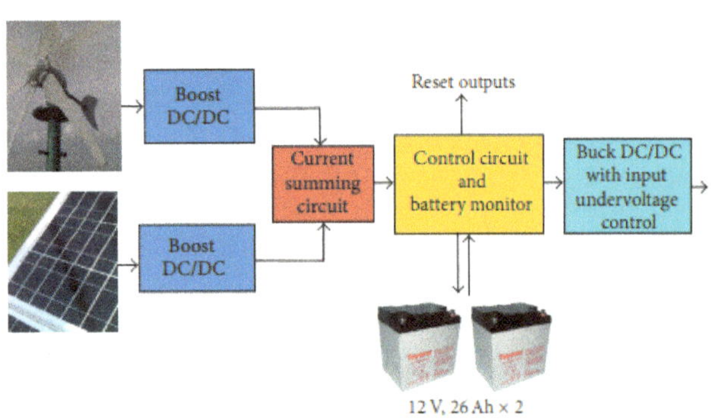

Figure 9: Block diagram of EHC with wind turbine, solar cell, and two batteries.

The power generated by both the wind generator and solar panels is stored in a set of two 12 V gel valve-regulated lead-acid batteries of 26 Ah capacity. The batteries are connected in series which makes it necessary to increase output voltage of both the solar panels and wind generator.

A DC-DC converter based on TPS43060 IC from Texas Instruments is used for boosting output voltage of the solar panels from an input V range to 29 V. The inputs of all solar panels are connected to the DC/DC converter via SPV1001 switches to avoid reverse current flow when some panels are shaded (not fully irradiated). The DC-DC converter for the wind generator is based on LT3757 IC from Linear Technology. It increases generated voltage to 29 V.

A MOSFET switch controlled by the microcontroller connects parallel resistor to wind generator output when wind is too strong. The current summing circuit is implemented as double parallel connections of STPS20SM60 power Schottky diodes and SPV1001 switches. The Schottky diodes can conduct current at startup. The MOSFET switches enable much lower forward voltage drop than the diodes.

The rest of power processing circuits is based on step-down switching converters with battery under-voltage control. One of them, based on LM43603 Simple Switcher IC from Texas Instruments, delivers 3.3 V supply for the microcontroller part.

Power monitoring is performed by the microcontroller measurement and control module. The microcontroller applied is ATmega644P from Atmel. It cooperates with ADE7953 energy measurement IC from Analog Devices and bq78412 battery monitor IC from Texas Instruments. The first IC is responsible for measurement of wind generator and solar panel output currents. The second IC measures battery voltage, current, and ambient temperature to calculate remaining charge and operational time. The microcontroller communicates with OMAP using USB interface (FT232RL IC). It can also reset 3G modem, TRS OMAP, and WVWMS gate when necessary.

PCB for the EHC was designed (according to EMC rules and with heat dissipation considered) as a two-layer board with 100 × 150 mm dimensions. Figure 10 presents the view of the EHC PCB with all its components.

Figure 10: Printed circuit board of EHC, dedicated to TULCOEMPA project.

DATA ACQUISITION, STORAGE, AND PRESENTATION

Data acquisition system is responsible for data transmission, storage, processing, and presentation.

It was designed to accommodate various measurement systems, one of them being the Wireless Vehicle Weight Measurement System.

Due to integration of data from various measurement systems main requirement for the data acquisition system is flexibility of implementation and support for different data transmission channels. Currently, the system has modules that support channels of Representational State Transfer (REST) [15, 16] and sockets [17], but virtually any communication channel can be added to the system. In order to provide scalability of the system, it is modularised into two main component types, broker and data processing and presentation. A diagram showing its structure is presented in Figure 11.

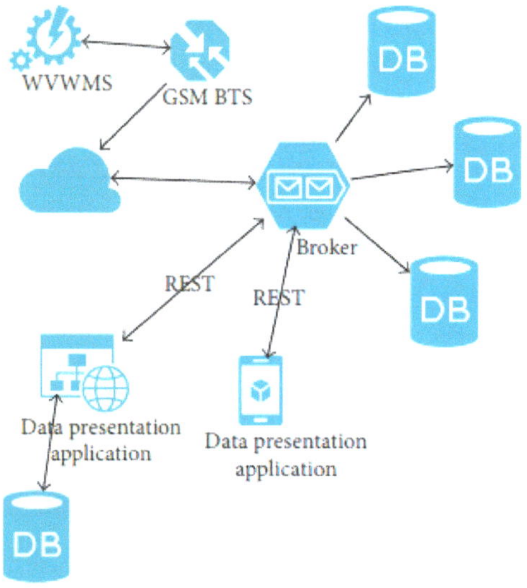

Figure 11: Data acquisition system diagram, WVWMS communicates with server through the Internet.

Each module has a different set of requirements imposed. The broker module is responsible for receiving measurement data from the remote system and providing data for data presentation software.

A module enabling connectivity for a specific measurement system is written as a separate library that can be easily included in the broker. Due to maintainability currently developed modules, including the one for WVWMS, are implemented as self-sufficient; that is, they do not require additional resources on the host server to work.

Java was chosen as the primary programming language of the project due to portability, good documentation, good community support, large repositories of free and open-source libraries and availability of tools, and last, but not least, an ease of programming.

While the broker component requires additional modules to be written in Java and integrated, data presentation application programming language is not limited to Java. Thanks to decoupling of components, it can be written in an arbitrary programming language.

Nevertheless, the current version of data presentation and processing application has been implemented as a web application using Java, HTML, JavaScript, and CSS.

The data presentation module is written using Model-View-View Model paradigm. This makes it possible to implement many different user interfaces, while using only one server backend, that is, desktop application and mobile application.

The server backend is responsible for retrieving data from the broker, storing it locally, processing using business logic, and providing processed data, ready for visualisation, to the user interface application.

The server backend is implemented as a Java Servlet [18] application. Only one resource is required, that is, a relational database. For the purpose of the project PostgreSQL implementation was chosen. WVWMS data transmission was designed and implemented to fulfil specific requirements: data integrity and support for security. The data integrity means assuring accuracy and consistency. This has also been considered with respect to data storage. In order to assure data integrity, the authors considered communication channel, data transmitter, and data receiver. GSM was chosen as the physical communication channel, because of distributed nature of the system. Due to the requirements for data integrity and communications reliability logical communications are conducted over HTTP [19] using REST. Because of reliability, a common approach of HTTP over TCP/IP [20] is used.

Security is implemented in two aspects: authorisation and encryption. The authorisation is implemented using HTTP Basic Authentication, which enables sending username and password in HTTP header. Because the Basic Authentication does not provide confidentiality of data, the channel is further secured using HTTP over SSL [21].

The selected REST methods with their responses are presented in Algorithm 1.

Algorithm 1:

POST /wvwms (application/json)

Body

[{“id”:1,“data”:[1.12343,..,..,..],..}]

Response 200 (application/json)

Body

{“message”:“Data saved”,“generatedIds”:[1,]}

Response 400 (application/json)

Body

[“message”:“..”,errors:[]}

Response 500 (application/json)

Body

{“message”:“..”}

GET /wvwms/read/between/{date}/{date}

GET /wvwms/read/from/{sampleId}

Response 200 (application/json)

Body

[{“id”:1,..},..]

GET /wvwms/{sampleId}

Response 200 (application/json)

Body

{“id”:1,..}

Response 400 (application/json)

Body

{“message”:“No sample”}

CONCLUSION

In conclusion, the paper has presented challenges and proposed solutions for architecture of the Wireless Vehicle Weight Measurement System, which is a part of the TULCOEMPA project. Authors objective was to design and develop a road scale system enabling multipoint vehicle weight with the single network sink. Proposed platform is capable of weight measurement based on electrical strain gauge weight pads and in-node data processing. Communication is wireless, with mesh networking. The block schema of designed WVWMS installation has been introduced and described in detail. Due to lack of mains, all installations are powered by Hybrid Power System. Energy Harvesting Controller cooperates with both solar cells and wind generators. The controller enables efficient energy harvesting from both power sources as well as monitoring of generated and consumed power levels. Additionally, the paper described design of vehicle weight data acquisition software architecture. The approach using REST API provides easy to understand and implement API, with readily available lightweight tools, that is, curl in Contiki, which enables easy client implementation.

CONFLICT OF INTERESTS

The authors declare that there is no conflict of interests regarding the publication of this paper.

ACKNOWLEDGMENTS

The authors would like to thank all members of the TULCOEMPA project, including Department of Concrete Structures, Swiss Federal Laboratories for Materials Science and Technology EMPA, and Department of Microelectronics and Computer Science.

Received January 27, 2015.

Revision received May 6, 2015.

Accepted May 27, 2015.

REFERENCES

1. K. Chintalapudi, T. Fu, J. Paek et al., "Monitoring civil structures with a wireless sensor network," IEEE Internet Computing, vol. 10, no. 2, pp. 26–34, 2006.
2. S. Kim, S. Pakzad, D. Culler et al., "Health monitoring of civil infrastructures using wireless sensor networks," in Proceedings of the 6th International Symposium on Information Processing in Sensor Networks (IPSN '07), pp. 254–263, April 2007.
3. S. N. Pakzad, "Development and deployment of large scale wireless sensor network on a long-span bridge," Smart Structures and Systems, vol. 6, no. 5-6, pp. 525–543, 2010.
4. M. Ceriotti, L. Mottola, G. P. Picco et al., "Monitoring heritage buildings with wireless sensor networks: the Torre Aquila deployment," in Proceedings of the International Conference on Information Processing in Sensor Networks (IPSN' 09), pp. 277– 288, April 2009.
5. TULCOEMPA Project, December 2014, http://www.tulcoempa .com.
6. K. Grabowski and J. Clapa, "A low-power embedded system for visual vehicle detection," in Proceedings of the 21st International Conference on Mixed Design of Integrated Circuits and Systems (MIXDES '14), pp. 471–476, June 2014.
7. P. Łęczycki, A. Andrzejczak, P. Pietrzak, B. Pękosławski, and A. Napieralski, "Extended sensor reliability evaluation method in multi-sensor control systems," Microelectronics Reliability, vol. 55, no. 3-4, pp. 671–678, 2015.
8. R. Podraza, M. Walkiewicz, and A. Dominik, "Credibility coefficients based on decision rules," in Proceedings of the International Multiconference on Computer Science and Information Technology, 22nd Autumn Meeting of Polish Information Processing Society, pp. 179–187, 2006.
9. TRAX elektronik Internet, 2014, http://www.traxelektronik.pl/.
10. J. A. Gutierrez, M. Naeve, E. Callaway, M. Bourgeois, V. Mitter, and B. Heile, "IEEE 802.15.4: a developing standard for lowpower low-cost wireless personal area networks," IEEE Network, vol. 15, no. 5, pp. 12–19, 2001.
11. A. Dunkels, B. Gronvall, and T. Voigt, "Contiki—a lightweight ¨ and flexible operating system for tiny networked sensors," in Proceedings of the 29th Annual IEEE International Conference on Local Computer Networks (LCN '04), pp. 455–462, November 2004.
12. J. Jung, B. Song, and S. Park, "The possibility of wireless sensor networks for commercial vehicle load monitoring,"ACM SIGBED Review, vol. 8, no. 4, pp. 30–34, 2011.
13. A. Willig, "Wireless sensor networks: concept, challenges and approaches," Elektrotechnik und Informationstechnik, vol. 123, no. 6, pp. 224–231, 2006.
14. A. Andrzejczak, P. Łecczycki, P. Pietrzak, B. Pe¸ckoslawski, and A. Napieralski, "Module hardware structure of wireless vehicle weight measurement system," in Proceedings of the 21st International Conference on Mixed Design of Integrated Circuits and Systems (MIXDES' 14), pp. 477–481, June 2014.
15. R. Fielding, Architectural styles and the design of network-based software architectures [Dissertation], 2014, https://www.ics.uci .edu/~fielding/pubs/dissertation/fielding dissertation.pdf.
16. W. Colitti, K. Steenhaut, N. De Caro, B. Buta, and V. Dobrota, "REST enabled wireless sensor networks for seamless integration with web applications," in Proceedings of the 8th IEEE International Conference on Mobile Ad-Hoc and Sensor Systems (MASS '11), pp. 867–872, October 2011.
17. The Definition of a Socket, 2014, https://tools.ietf.org/html/rfc1- 47.
18. Java Servlet Specification, December 2014, https://www.jcp.org/ en/jsr/detail?id=315.

19. Hypertext Transfer Protocol—HTTP/1.1, 2014, http://tools.ietf .org/html/rfc2616.
20. Transmission Control Protocol, 2014, https://www.ietf.org/rfc/ rfc793.txt.
21. A. Freier, P. Karlton, and P. Kocher, "The Secure Sockets Layer (SSL) protocol version 3.0," RFC 6101, Internet Engineering Task Force (IETF), 2011, https://tools.ietf.org/html/rfc6101.

CHAPTER 13

TLCD PARAMETRIC OPTIMIZATION FOR THE VIBRATION CONTROL OF BUILDING STRUCTURES BASED ON LINEAR MATRIX INEQUALITY

Linsheng Huo, Chunxu Qu, and Hongnan Li

INTRODUCTION

Civil structural buildings can be damaged under heavy excitation, such as earthquake. The obvious vibration induced by some slight excitations, such as slight earthquakes and strong winds, can make the people in the building feel uncomfortable. Passive control is an effective strategy to suppress the vibration [1–4]. However, a lot of passive control devices are needed to be instrumented additionally. If there is already some device as a part of the building, it will not cost so much to deal with the vibration control. Tuned liquid column damper (TLCD) is first proposed by Sakai et al. [5], which can change its dynamics by tuning the mass, stiffness, and damping. So TLCD can be designed to make its dynamics be suitable for the vibration control of a building. In the tall building, the liquid tank filled with fresh water has to be instrumented for the fire protection or daily supply. TLCD can be set in the building as a liquid tank and also used for vibration control.

Considering Building 1 in Dalian Xinghai Financial Business Building, this paper presents a building vibration control example with TLCD. The finite element model is built and compared with the lumped mass model. It is found that these two models are pretty close. The TLCD dynamics is introduced, and the building dynamics with TLCD is discussed. Motivated by the method proposed by Moreno and Thomson [6], the formulation is derived to transfer the parametric optimization problem of TLCD to design a feedback controller problem.

For the feedback controller problem, this paper focuses on how to design a static output feedback H_∞ controller. According to the bounded real lemma [7], designing the static H_∞ controller is an optimization problem with bilinear matrix inequality (BMI) constraints. However, BMI is not a convex problem and cannot be solved with the general off-the-shelf packages as linear matrix inequality (LMI). To solve the static output feedback H_∞ controller, Rubió-Massegú et al. [8] produced a simple expression of the controller, in which the corresponding symmetric variable matrix in the bounded real lemma is congruent with a block diagonal matrix. However, this variable matrix cannot represent all its possible results because there may exist some matrix that is not congruent with the block diagonal matrix. So the controller is conservative, and the desired controller may not be obtained. Zečević and Šiljak [9] present another method to design the static output feedback controller by making some transformations between the output matrix and controller matrix. Unfortunately, this method also supposed some relationship between output matrix and controller. So, the controller is also conservative. This paper employs the iterative linear matrix inequality (ILMI) method proposed by Cao et al. [10] and adds some value range restrictions into ILMI method. ILMI is solving a generalized eigenvalue minimization problem and does the iteration till satisfying the predetermined tolerance. After the desired controller is obtained, the optimal parameters of

Citation: Linsheng Huo, Chunxu Qu, and Hongnan Li, "TLCD Parametric Optimization for the Vibration Control of Building Structures Based on Linear Matrix Inequality," Journal of Applied Mathematics, vol. 2014, Article ID 527530, 10 pages, 2014. doi:10.1155/2014/527530.

TLCD are determined. The responses of the displacement and acceleration in frequency domain and time domain are compared to illustrate the performance of the optimized TLCD.

PRACTICAL MODEL

This paper considers the vibration control of a building subjected to the earthquake excitation, which has 46 stories and is 149.95 meters high, as shown in Figure 1.

Figure 1: The buildings in Dalian Xinghai Bay Financial Business District.

The building has a frame-shear-wall structure and can be modeled by the finite element software ETABS. The finite element model contains 5472 joints, 13691 frame elements, and 4472 cell elements as shown in Figure 2.

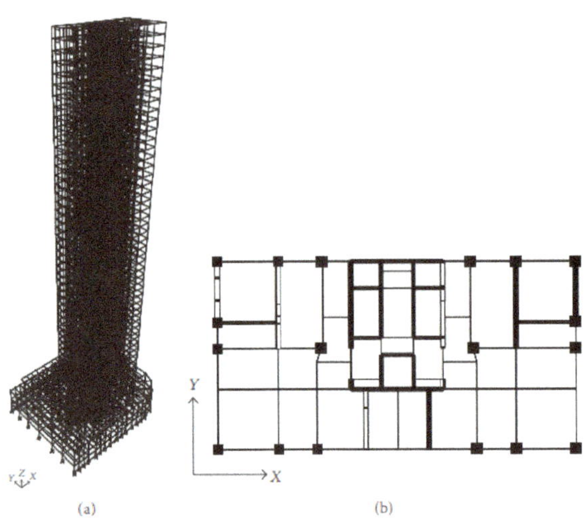

Figure 2: Structural diagram: (a) structural three-dimensional diagram; (b) structural top floor plan diagram.

Figure 2 displays the structural three-dimensional diagram as Figure 2(a) and structural top floor plan diagram as Figure 2(b). The first two vibration periods of the finite element model are 3.346829 s and 2.69537 s. The corresponding vibration modes are shown in Figures 3(a) and 3(b).

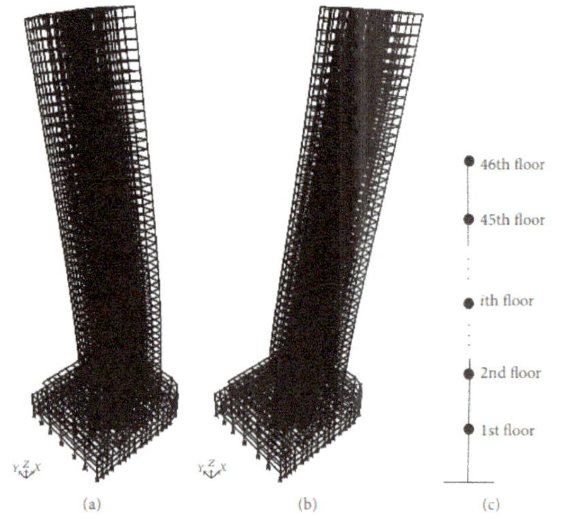

Figure 3: Structural diagram: (a) the first vibration mode; (b) the second vibration mode; (c) the lumped mass model.

However, the computation for the responses of the finite element model will cost a lot of time. Therefore, the lumped mass can be used to simplify the finite element model, which has 46 degrees of freedom (DOF) as shown in Figure 3(c). The structural masses are concentrated on each floor, and the interstory damping and stiffness between two lumped masses can be calculated from the finite element model. To describe the difference between the finite element model and the lumped mass model, the vibration periods have been compared in Table 1.

Table 1: Comparison with difference of the period between two models.

	First vibration period	Second vibration period
Finite element model	3.3468	2.6954
Lumped mass model	3.1312	2.6123
Difference	6.44%	3.08%

The difference in the 3rd row of Table 1 shows how different the two types of models are and can be calculated as the following expression:

$$\frac{|T_f - T_l|}{T_f} \times 100\%, \qquad (1)$$

where T_f and T_l are the vibration periods of the finite element model and lumped mass model, respectively. The above comparison illustrates that there is little difference between the finite element model and lumped mass model. So, the lumped mass model can be used to approximate the practical structure building. In Table 1, the first and second vibration periods are the first mode along Y direction and the first mode along X direction, respectively. It means that the structural stiffness along X direction is larger than the one along Y direction; that is to say, the structure along Y direction is weaker than that along X direction. So, the lumped mass model can be built only considering the stiffness and damping along Y direction. The structure is subjected to an earthquake, whose dynamics can be formulated as follows:

$$\mathbf{M}\ddot{\mathbf{x}} + \mathbf{C}\dot{\mathbf{x}} + \mathbf{K}\mathbf{x} = -\mathbf{M}\ddot{x}_g, \qquad (2)$$

Where

$$\mathbf{M} = \begin{bmatrix} m_1 & 0 & 0 \\ 0 & \ddots & 0 \\ 0 & 0 & m_n \end{bmatrix},$$

$$\mathbf{C} = \begin{bmatrix} c_1 + c_2 & -c_2 & 0 \\ -c_2 & \ddots & -c_n \\ 0 & -c_n & c_n \end{bmatrix},$$

$$\mathbf{K} = \begin{bmatrix} k_1 + k_2 & -k_2 & 0 \\ -k_2 & \ddots & -k_n \\ 0 & -k_n & k_n \end{bmatrix}, \quad (3)$$

where the elements mi, ci, and ki in matrices M, C, and K are the mass, damping, and stiffness of the ith floor ($i = 1, \ldots, n$, $n = 46$). It should be noted that matrices C and K are calculated along Y direction. $\ddot{x}\ g$ is the unidirectional earthquake acceleration. The relative displacement vector x ∈ $R^{n\times 1}$ is defined as

$$\mathbf{x} = [x_1\ x_2\ \cdots\ x_n]^{\mathrm{T}}. \quad (4)$$

For the tall building, the structural design proposed does not only make the building satisfy safety requirement during the earthquake excitation but also make the people in the building feel comfortable during the slight external excitation, such as wind excitation. The safe requirements mean that the interstory drifts of the vibrating building should be smaller. The comfortable requirement needs the acceleration of the building to be smaller.

A liquid tank with fresh water has to be set in a tall building, which would be used for the fire protection or daily supply. So, the structural vibration can be suppressed by the liquid tank with appropriate dynamics. The dynamics of the liquid tank can be designed by the tank shapes. This paper employs the tank with the shape of two columns that is called the tuned liquid column damper (TLCD).

TLCD DESIGN

The Dynamics of the Structure with TLCD

The structure of the tuned liquid column damper (TLCD) is shown as Figure 4(a). When it is located on the top floor of the building as shown in Figure 4(b), the TLCD dynamics can be described as the following equation, which is derived by Lagrange's method:

$$\rho A (2H + B)\ddot{h} + \frac{1}{2}\rho A\xi |\dot{h}| \dot{h} + 2\rho Agh = -\rho AB(\ddot{x}_{46} + \ddot{x}_g), \quad 5)$$

where ρ is the liquid density, h is the displacement of the liquid in the damper, H is the vertical length of the liquid column when the liquid is quiescent, A is the cross-sectional area of the tube, B is the horizontal length of the column, and ξ is the head-loss coefficient. $\ddot{x}_{\ g}$ is the unidirectional earthquake acceleration. $\ddot{x}_{\ 46}$ is the relative acceleration (with respect to the ground) for the top floor of the building. Equation (5) can be simplified as

$$m_T\ddot{h} + c_T\dot{h} + k_T h = -\alpha m_T(\ddot{x}_{46} + \ddot{x}_g), \quad (6)$$

where $m_T = {}_\rho AL$ means the mass of liquid in the tube, $c_T = 2m_T\omega_T\zeta_T$ is the equivalent damping of the liquid damper, ωT is the first resonant frequency of TLCD, ζ_T is the damping ratio of the liquid damper, $k_T = 2{}_\rho A_g$ expresses the stiffness of the liquid column, g is the gravitational constant, and $\alpha = B/L$ denotes the length ratio. According to (4) and (6), the dynamics of the structure instrumented with TLCD on the top floor can be formulated as

$$\mathbf{M}_s\ddot{\mathbf{q}} + \mathbf{C}_s\dot{\mathbf{q}} + \mathbf{K}_s\mathbf{q} = \mathbf{T}_w\ddot{x}_g, \quad (7)$$

where

$$\mathbf{M}_s = \begin{bmatrix} m_1 & 0 & 0 & 0 \\ 0 & \ddots & 0 & 0 \\ 0 & 0 & m_n + m_T & \alpha m_T \\ 0 & 0 & \alpha m_T & m_T \end{bmatrix},$$

$$\mathbf{C}_s = \begin{bmatrix} c_1 + c_2 & -c_2 & 0 & 0 \\ -c_2 & \ddots & -c_n & 0 \\ 0 & -c_n & c_n & 0 \\ 0 & 0 & 0 & c_T \end{bmatrix},$$

$$\mathbf{K}_s = \begin{bmatrix} k_1 + k_2 & -k_2 & 0 & 0 \\ -k_2 & \ddots & -k_n & 0 \\ 0 & -k_n & k_n & 0 \\ 0 & 0 & 0 & k_T \end{bmatrix},$$

$$\mathbf{T}_w = -\begin{bmatrix} m_1 & 0 & 0 & 0 \\ 0 & \ddots & 0 & 0 \\ 0 & 0 & m_n + m_T & 0 \\ 0 & 0 & 0 & \alpha m_T \end{bmatrix},$$

$$\mathbf{q} = [q_1\ \cdots\ q_n\ q_{n+1}]^{\mathrm{T}} = [x_1\ \cdots\ x_n\ h]^{\mathrm{T}} \in \Re^{(n+1)\times 1}. \quad (8)$$

From (7), it is found that the dynamics of the structure instrumented with TLCD can be influenced by the TLCD parameters, such as m_T, c_T, and k_T. When a TLCD is needed to be designed for a building structure, the value ranges for the mass ratio, frequency tuning ratio, and damping ratio are always given first, which can be determined uniquely by m_T, c_T, and k_T. In practical engineering, mass ratio is usually predetermined to satisfy the architectural requirements. Therefore, only frequency tuning ratio and damping ratio should be optimized during the TLCD designing. There are the following relationships with m_T, c_T, and k_T

$$\tau = \frac{\omega_T}{\omega_s} = \frac{\sqrt{k_T}}{\omega_s\sqrt{m_T}},$$

$$\zeta_T = \frac{c_T}{2m_T\omega_T} = \frac{c_T}{2\sqrt{m_T k_T}}, \quad (9)$$

where τ is the frequency tuning ratio, ωs is the first resonant frequency of the building, and ζ_T is the TLCD damping ratio as defined above. The TLCD designing objective is to optimize the parameters τ and ζ_T such that the structural vibration is reduced and satisfies the need for safety and comfort. From (9), c_T and k_T can replace the parameters τ and ζ_T to be optimized.

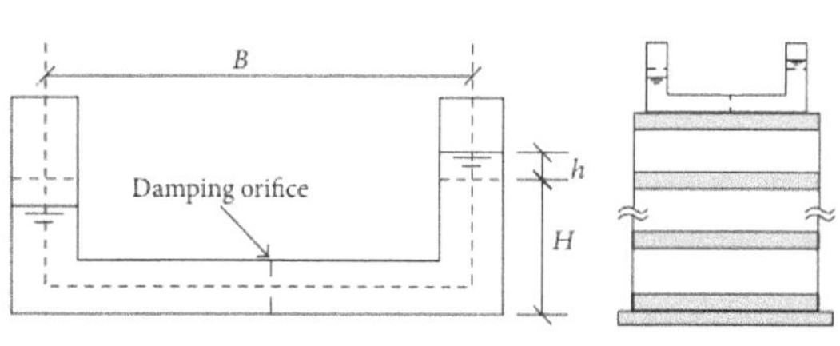

Figure 4: Tuned liquid column damper system: (a) tuned liq-

uid column damper; (b) the building with tuned liquid column damper.

Transfer Parametric Optimization to Controller Design

In order to optimize the parameters c_T and k_T in (8), they are separated from the matrices C_s and K_s. Equation (7) is equal to the following expression:

$$\mathbf{M}_s\ddot{\mathbf{q}} + \overline{\mathbf{C}}\dot{\mathbf{q}} + \overline{\mathbf{K}}\mathbf{q} = \mathbf{T}_w\ddot{x}_g + \mathbf{T}_u p(t), \tag{10}$$

Where

$$\overline{\mathbf{C}} = \begin{bmatrix} c_1 + c_2 & -c_2 & 0 & 0 \\ -c_2 & \ddots & -c_n & 0 \\ 0 & -c_n & c_n & 0 \\ 0 & 0 & 0 & 0 \end{bmatrix},$$

$$\overline{\mathbf{K}} = \begin{bmatrix} k_1 + k_2 & -k_2 & 0 & 0 \\ -k_2 & \ddots & -k_n & 0 \\ 0 & -k_n & k_n & 0 \\ 0 & 0 & 0 & 0 \end{bmatrix},$$

$$\mathbf{T}_u = \begin{bmatrix} \mathbf{0}_{1\times n} & -1 \end{bmatrix}^{\mathrm{T}}, \tag{11}$$

$$p(t) = \begin{bmatrix} k_T & c_T \end{bmatrix} \begin{bmatrix} h \\ \dot{h} \end{bmatrix}. \tag{12}$$

Equation (10) can be recognized as the structure with the mass matrix Ms, damping matrix $\overline{\mathbf{C}}$,, and stiffness matrix $\overline{\mathbf{K}}$ subjected to the earthquake excitation $\ddot{x}_g$ and the control force $p(t)$. An appropriate control force (t) can suppress the structural vibration. From (12), the value of the control force can be decided by the matrix $[k_T\ c_T]$ and would vary with the TLCD state $[h^{\cdot}\ h]^{\mathrm{T}}$. Equation (10) can be represented by the augmented system in state-space as

$$\begin{aligned} \dot{\mathbf{z}} &= \mathbf{A}\mathbf{z} + \mathbf{B}_1\ddot{x}_g + \mathbf{B}_2 p(t), \\ \mathbf{y}_z &= \mathbf{C}_1\mathbf{z} + \mathbf{D}_{11}\ddot{x}_g + \mathbf{D}_{12}p(t), \\ \mathbf{y}_m &= \mathbf{C}_2\mathbf{z} + \mathbf{D}_{21}\ddot{x}_g + \mathbf{D}_{22}p(t), \end{aligned} \tag{13}$$

Where

$$\mathbf{z} = \begin{bmatrix} \mathbf{q} & \dot{\mathbf{q}} \end{bmatrix}^{\mathrm{T}}; \qquad \mathbf{y}_m = \begin{bmatrix} h & \dot{h} \end{bmatrix}^{\mathrm{T}}. \tag{14}$$

Matrices C_1, D_{11}, and D_{12} can be determined according to objective output y_z, which is always the combination of the structural interstory drifts and the weighted control force. Other matrices in (13) are denoted as

$$\left[\begin{array}{c|c|c} \mathbf{A} & \mathbf{B}_1 & \mathbf{B}_2 \\ \hline \mathbf{C}_2 & \mathbf{D}_{21} & \mathbf{D}_{22} \end{array}\right] = \left[\begin{array}{cc|c|c} \mathbf{0}_{(n+1)\times(n+1)} & \mathbf{I}_{(n+1)\times(n+1)} & 0 & 0 \\ -\mathbf{M}_s^{-1}\overline{\mathbf{K}} & -\mathbf{M}_s^{-1}\overline{\mathbf{C}} & \mathbf{M}_s^{-1}\mathbf{T}_w & \mathbf{M}_s^{-1}\mathbf{T}_u \\ \hline \mathbf{0}_{1\times n}\ \ 1 & \mathbf{0}_{1\times n}\ \ 0 & 0 & 0 \\ \mathbf{0}_{1\times n}\ \ 0 & \mathbf{0}_{1\times n}\ \ 1 & 0 & 0 \end{array}\right]. \tag{15}$$

The control force (t) in (12) can be rewritten as

$$p(t) = \mathbf{G}\mathbf{y}_m = \begin{bmatrix} k_T & c_T \end{bmatrix} \begin{bmatrix} h \\ \dot{h} \end{bmatrix}. \tag{16}$$

So, the TLCD parametric optimization problem is transferred to the controller design problem, that is, to design a controller G to suppress the structural vibration efficiently

H_∞ Static Output Feedback Controller Design

Taking (16) into (13) to complete the feedback loop, the closed-loop system can be represented as

$$\begin{aligned} \dot{\mathbf{z}} &= \mathbf{A}_{\mathrm{CL}}\mathbf{z} + \mathbf{B}_{\mathrm{CL}}\ddot{x}_g, \\ \mathbf{y}_z &= \mathbf{C}_{\mathrm{CL}}\mathbf{z} + \mathbf{D}_{\mathrm{CL}}\ddot{x}_g, \end{aligned} \tag{17}$$

Where

$$\begin{aligned} \mathbf{A}_{\mathrm{CL}} &= \mathbf{A} + \mathbf{B}_2\mathbf{G}\mathbf{C}_2 \\ \mathbf{B}_{\mathrm{CL}} &= \mathbf{B}_1 + \mathbf{B}_2\mathbf{G}\mathbf{D}_{21} \\ \mathbf{C}_{\mathrm{CL}} &= \mathbf{C}_1 + \mathbf{D}_{12}\mathbf{G}\mathbf{C}_2 \\ \mathbf{D}_{\mathrm{CL}} &= \mathbf{D}_{11} + \mathbf{D}_{12}\mathbf{G}\mathbf{D}_{21}. \end{aligned} \tag{18}$$

This paper focuses on how to design a static output feedback H_∞ controller G in (16). Some concepts about H_∞ can refer to [11]. According to the bounded real lemma [7], an H_∞ controller can be designed to stabilize the closed-loop system in (17) and make the transfer function norm of the closedloop system smaller than a given scalar γ, if and only if there exists a symmetric positive definite matrix $X_{CL} > 0$ to satisfy the following matrix inequality:

$$\begin{bmatrix} \mathbf{A}_{\mathrm{CL}}^{\mathrm{T}}\mathbf{X}_{\mathrm{CL}} + \mathbf{X}_{\mathrm{CL}}\mathbf{A}_{\mathrm{CL}} & \mathbf{X}_{\mathrm{CL}}\mathbf{B}_{\mathrm{CL}} & \mathbf{C}_{\mathrm{CL}}^{\mathrm{T}} \\ \mathbf{B}_{\mathrm{CL}}^{\mathrm{T}}\mathbf{X}_{\mathrm{CL}} & -\gamma\mathbf{I} & \mathbf{D}_{\mathrm{CL}}^{\mathrm{T}} \\ \mathbf{C}_{\mathrm{CL}} & \mathbf{D}_{\mathrm{CL}} & -\gamma\mathbf{I} \end{bmatrix} < 0. \tag{19}$$

Substitute (18) into (19):

$$\begin{bmatrix} (\mathbf{A} + \mathbf{B}_2\mathbf{G}\mathbf{C}_2)^{\mathrm{T}}\mathbf{X}_{\mathrm{CL}} + \mathbf{X}_{\mathrm{CL}}(\mathbf{A} + \mathbf{B}_2\mathbf{G}\mathbf{C}_2) & \mathbf{X}_{\mathrm{CL}}(\mathbf{B}_1 + \mathbf{B}_2\mathbf{G}\mathbf{D}_{21}) & (\mathbf{C}_1 + \mathbf{D}_{12}\mathbf{G}\mathbf{C}_2)^{\mathrm{T}} \\ (\mathbf{B}_1 + \mathbf{B}_2\mathbf{G}\mathbf{D}_{21})^{\mathrm{T}}\mathbf{X}_{\mathrm{CL}} & -\gamma\mathbf{I} & (\mathbf{D}_{11} + \mathbf{D}_{12}\mathbf{G}\mathbf{D}_{21})^{\mathrm{T}} \\ \mathbf{C}_1 + \mathbf{D}_{12}\mathbf{G}\mathbf{C}_2 & \mathbf{D}_{11} + \mathbf{D}_{12}\mathbf{G}\mathbf{D}_{21} & -\gamma\mathbf{I} \end{bmatrix} < 0. \tag{20}$$

In practical engineering, there are always some value ranges for the frequency tuning ratio τ and damping ratio ζ_T. So, the value ranges for k_T and c_T in G can be calculated from (9) and formulated as

$$\begin{aligned} c_T &= \mathbf{L}_c\mathbf{G}\mathbf{R}_c, \quad c_T^l \le \mathbf{L}_c\mathbf{G}\mathbf{R}_c \le c_T^u, \\ k_T &= \mathbf{L}_k\mathbf{G}\mathbf{R}_k, \quad k_T^l \le \mathbf{L}_k\mathbf{G}\mathbf{R}_k \le k_T^u, \end{aligned} \tag{21}$$

Where

$$\begin{aligned} \mathbf{L}_c &= [1], \quad \mathbf{R}_c = \begin{bmatrix} 0 & 1 \end{bmatrix}^{\mathrm{T}}, \\ \mathbf{L}_k &= [1], \quad \mathbf{R}_k = \begin{bmatrix} 1 & 0 \end{bmatrix}^{\mathrm{T}}. \end{aligned} \tag{22}$$

In (21), c_T^u and c_T^l are denoted as the upper and lower boundary for c_T and k_T^u and k_T^l are the upper and lower boundary for k_T, respectively. So, the controller G should be searched with the constraints (20) and (21). However, it is a bilinear matrix inequality problem, which is nonconvex. Rubio-Masseg ´ u et al. [´ 8] and Zecevi ˇ c and ´ Siljak [ˇ 9] proposed new approaches to deal with the static output feedback problem, respectively. But the controllers obtained by these two methods are both conservative due to the predefined restrictions. This paper employs the iterative linear matrix inequality (ILMI) method proposed by Cao et al. [10] and adds the constraints (21) into this method to numerically calculate the static output feedback controller G. In order to use ILMI method, (20) should be re-written into another form:

$$\left(\overline{\mathbf{A}}+\overline{\mathbf{B}}\mathbf{G}\overline{\mathbf{C}}\right)^{\mathrm{T}}\mathbf{P}+\mathbf{P}\left(\overline{\mathbf{A}}+\overline{\mathbf{B}}\mathbf{G}\overline{\mathbf{C}}\right)<\mathbf{0}, \tag{23}$$

Where

$$\mathbf{P}=\begin{pmatrix}\mathbf{X}_{\mathrm{CL}} & \mathbf{0} & \mathbf{0}\\ \mathbf{0} & \mathbf{I} & \mathbf{0}\\ \mathbf{0} & \mathbf{0} & \mathbf{I}\end{pmatrix},\qquad \overline{\mathbf{A}}=\begin{pmatrix}\mathbf{A} & \mathbf{B}_1 & \mathbf{0}\\ \mathbf{0} & -\gamma\mathbf{I}/2 & \mathbf{0}\\ \mathbf{C}_1 & \mathbf{D}_{11} & -\gamma\mathbf{I}/2\end{pmatrix},$$
$$\overline{\mathbf{B}}=\begin{pmatrix}\mathbf{B}_2\\ \mathbf{0}\\ \mathbf{D}_{12}\end{pmatrix},\qquad \overline{\mathbf{C}}=\begin{pmatrix}\mathbf{C}_2 & \mathbf{D}_{21} & \mathbf{0}\end{pmatrix}. \tag{24}$$

Equation (23) is a Lyapunov inequality, which means that there is a static output feedback to stabilize the system $(\overline{\mathbf{A}},\overline{\mathbf{B}},\overline{\mathbf{C}})$. So, the ILMI method can be used for the system $(\overline{\mathbf{A}},\overline{\mathbf{B}},\overline{\mathbf{C}})$ and summarized as follows.

Step 1. According to the value range of the parameters τ and ζ_T, set the lower boundaries c_T^l and k_T^l and upper boundaries c_T^u and k_T^u for c_T and k_T, respectively. Select Q > 0 and solve P from the following algebraic Riccati equation:

$$\overline{\mathbf{A}}^{\mathrm{T}}\mathbf{P}+\mathbf{P}\overline{\mathbf{A}}-\mathbf{P}\overline{\mathbf{B}}\overline{\mathbf{B}}^{\mathrm{T}}\mathbf{P}+\mathbf{Q}=\mathbf{0}. \tag{25}$$

Set i=1 and X_1 = P.

Step 2. Solve the following convex problem for P_i, G, and α_i:

$$\begin{aligned}&\text{Minimize}\quad \alpha_i\\ &\text{s.t.}\quad \begin{pmatrix}\overline{\mathbf{A}}^{\mathrm{T}}\mathbf{P}_i+\mathbf{P}_i\overline{\mathbf{A}}-\mathbf{X}_i\overline{\mathbf{B}}\overline{\mathbf{B}}^{\mathrm{T}}\mathbf{P}_i-\mathbf{P}_i\overline{\mathbf{B}}\overline{\mathbf{B}}^{\mathrm{T}}\mathbf{X}_i+\mathbf{X}_i\overline{\mathbf{B}}\overline{\mathbf{B}}^{\mathrm{T}}\mathbf{X}_i-\alpha_i\mathbf{P}_i & \left(\overline{\mathbf{B}}^{\mathrm{T}}\mathbf{P}_i+\mathbf{G}\overline{\mathbf{C}}\right)^{\mathrm{T}}\\ \overline{\mathbf{B}}^{\mathrm{T}}\mathbf{P}_i+\mathbf{G}\overline{\mathbf{C}} & -\mathbf{I}\end{pmatrix}<\mathbf{0}\\ &\qquad \mathbf{P}_i=\mathbf{P}_i^{\mathrm{T}}>\mathbf{0}\\ &\qquad c_T^l\le \mathbf{L}_c\mathbf{G}\mathbf{R}_c\le c_T^u,\quad k_T^l\le \mathbf{L}_k\mathbf{G}\mathbf{R}_k\le k_T^u.\end{aligned} \tag{26}$$

Set α_i^* = the minimized αi.

Step 3. If $\alpha_i^*\le 0$, G is the desired static output feedback gain in (16). Iteration can be stopped.

Step 4. Solve the variable matrices P_i and G by minimizing trace (P_i) with the constraints in (26), where $\alpha_i=\alpha_i^*$. Set $\mathbf{P}_i^*$ = the solution P_i.

Step 5. Compare X_i and $\mathbf{P}_i^*$. If $\|\mathbf{X}_i-\mathbf{P}_i^*\|$ is smaller than a given tolerance, go to Step 6; else set i = i+1 and X_i = $\mathbf{P}_i^*$, and then go to Step 2.

Step 6. The system may not be stabilized through the output feedback gain in (20) with the constraints (21). Stop the iteration.

It should be noted that the matrix Q in (25) of Step 1 can influence the convergence of the ILMI method. If the iteration cannot be converged, another Q should be chosen.Q is always set to the identity matrix such that the iteration may obtain a good convergence. Another notice is that the convex problem in Step 2 is a generalized eigenvalue minimization problem. It can be solved through the PEMBMI solver in Yalmip [12].

CALCULATION RESULTS

For the building modeled in the first section, the TLCDs can be instrumented on the top floor and used for fire protection or daily supply, whose deployment is shown in Figure 5.

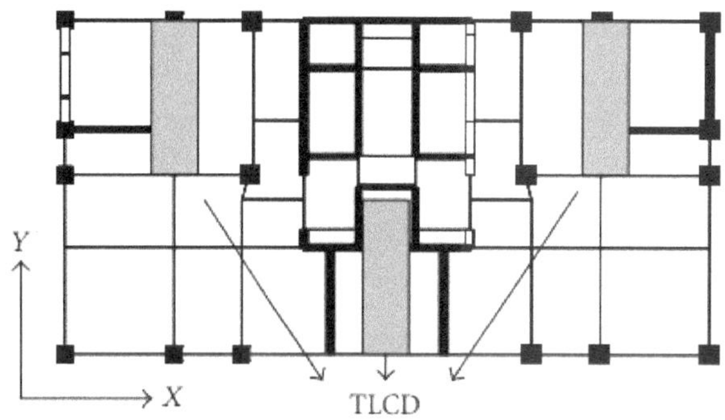

Figure 5: Deployment of three TLCDs on the top floor.

For the three TLCDs, they are first assumed to be one TLCD for the purpose of simplicity during optimization. After the assumed TLCD is optimized, it would be divided into three same TLCDs, whose total mass ratio, frequency tuning ratio, and damping ratio are, respectively, equal to the optimized parameters. To design the TLCD, its mass ratio is pregiven to be 0.015, which means that m_T is 1.5% of the structural total mass. And the value ranges for its frequency tuning ratio τ and damping ratio ζ_T are set to be 0~2.0 and 0~0.1. So, the ranges for k_T and c_T can be derived as $0=c_T^l<c_T<c_T^u=1.3373\times10^6$ and $0=k_T^l<k_T<k_T^u=4.1923\times10^7$ according to (9). The length ratio α in (6) is set to be 0.8. The parametric optimizing process described in the last section is used. After the iteration of the ILMI method, the static output feedback controller G in (16) is obtained:

$$\mathbf{G}=\begin{bmatrix}k_T & c_T\end{bmatrix}=\begin{bmatrix}7.83\times10^6 & 5.78\times10^5\end{bmatrix}. \tag{27}$$

Then, the parameters τ and ζ_T are determined as

$$\tau=0.86452;\qquad \zeta_T=0.09988. \tag{28}$$

To study the TLCD performance, the frequency response comparisons between the lumped model without TLCD and that with the optimized TLCD are shown in Figure 6.

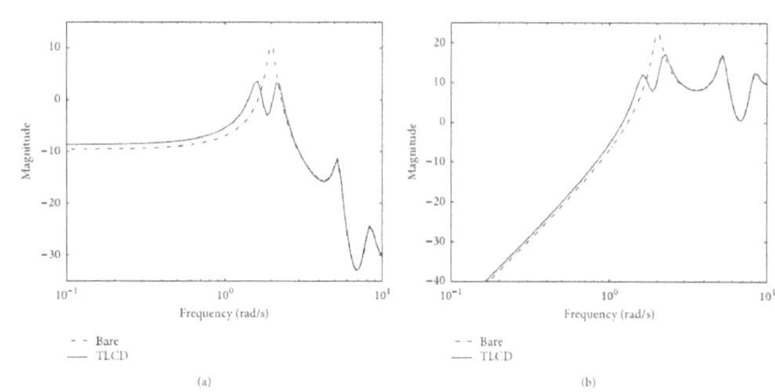

Figure 6: Frequency response curve comparisons: (a) frequency response for displacement; (b) frequency response for acceleration.

Figure 6(a) is the frequency response for the top floor displacement. The dashed line named Bare denotes the frequency response of the structural top floor displacement without TLCD, while the solid line named TLCD is for the response with TLCD. It is found that the frequency response peak is reduced by TLCD, which can reduce the vibration response in time domain indirectly. It means that the building instrumented with TLCD may satisfy the safety need more easily than that without TLCD. Figure 6(b) is the frequency response for the top floor acceleration. The dashed line and solid line have similar definition with that in Figure 6(a). It illustrates that the TLCD can reduce the vibration acceleration and make the building satisfy the comfortable requirement. The comparison in the frequency domain can reflect the inherent property of the structure with TLCD.

In practical engineering analysis, it is necessary to make some comparisons in the time domain. Some earthquake waves should be chosen to verify that the structural vibration can be reduced, subjected to any earthquake excitation. There are four earthquake waves to be chosen, which are Kobe wave, El Centro wave, North wave, and Hach wave. Their peak accelerations are scaled to 0.8337 m/s^2, 0.3495 m/s^2, 0.8428 m/s^2, and 0.2294 m/s^2, respectively. Due to the space limitation, only the top floor responses are compared as shown in Figures 7 and 8.

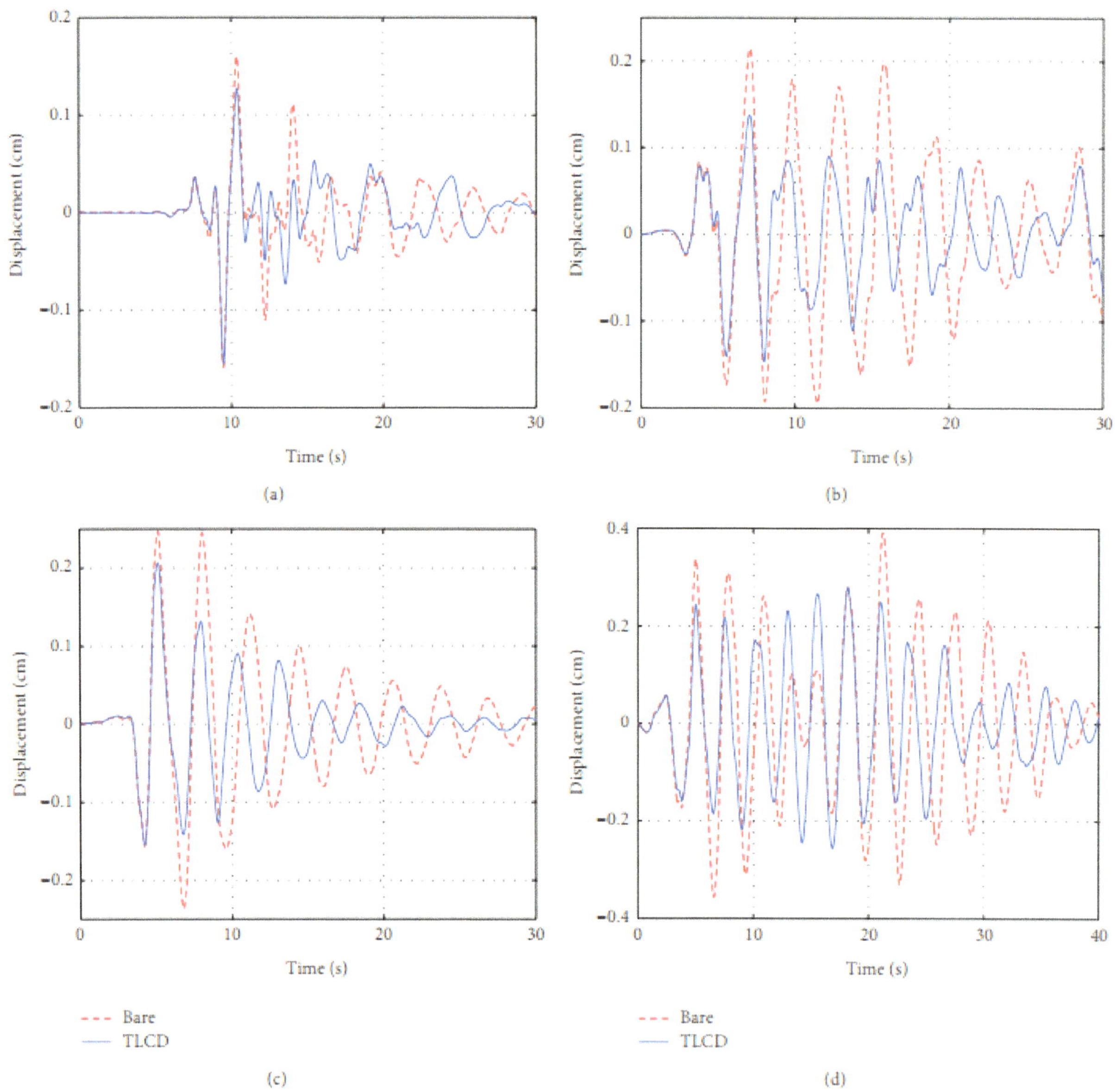

Figure 7: Time history comparison for the top floor displacement: (a) under Kobe wave; (b) under El Centro wave; (c) under North wave; (d) under Hach wave.

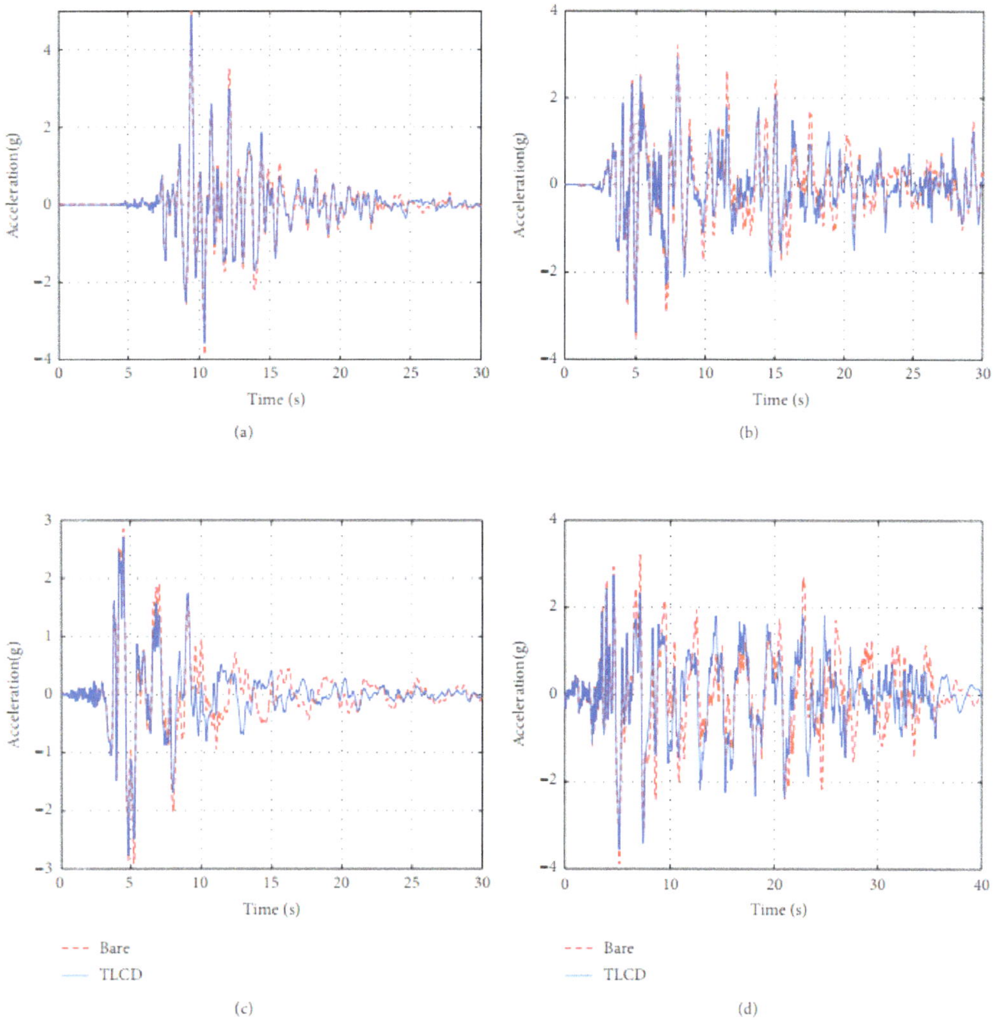

Figure 8: Time history comparison for the top floor acceleration: (a) under Kobe wave; (b) under El Centro wave; (c) under North wave; (d) under Hach wave.

In Figure 7, the dashed line named Bare means the structure without TLCD and the solid line named TLCD is the one with TLCD control. From this figure, all the solid line can reduce the corresponding dashed line a lot. In order to describe the reduction precisely, the reduction ratios for the top floor displacement under the different earthquake waves are calculated according to the following equation:

$$\text{reduction ratio} = \frac{(\text{Bare structural response}) - (\text{TLCD structural response})}{\text{Bare structural response}} \times 100\%. \tag{29}$$

The reduction ratios are 3.13%, 31.31%, 16.13, and 28.55% under the Kobe wave, El Centro wave, North wave, and Hach wave, respectively. It means that the TLCD can suppress the structural displacement vibration efficiently, which can decrease the component damage and make the structure safe. The acceleration comparisons are also made as shown in Figure 8.

In Figure 8, the dashed line and solid line have the same definition as that in Figure 7. In the four figures, all the solid line can reduce the peak value of the corresponding dashed line. The reduction ratios for the top floor acceleration under the four earthquake waves are also calculated as (29), which are 3.54%, 4.05%, 5.09%, and 8.84%, respectively. It means that TLCD can make the structure satisfy the requirement for comfort.

CONCLUSION

This paper optimizes the parameters of TLCD for Building 1 in Dalian Xinghai Financial Business Building with the consideration of the earthquake excitation. The building is built to the finite element model and the lumped mass model, respectively. After the comparison of these two models, they are pretty close. The lumped model is adopted due to the simple calculation process for optimization.

TLCD is used to suppress the structural vibration subjected to the earthquake excitation. The TLCD parameters are considered to be optimized. Through the equation derivation, the parametric optimization problem is transferred to the feedback controller design problem. The controller is the static output feedback controller with some value range constraints. This paper designs a static H_∞ controller. According to bounded real lemma, a bilinear matrix inequality (BMI) is constructed. To solve the BMI problem, the iterative linear matrix inequality (ILMI) is employed and added by the value range constraints. After the iteration, the feedback controller is calculated. Then the frequency tuning ratio and damping ratio of TLCD are determined. To illustrate the TLCD performance, the frequency responses of the top floor displacement and acceleration for the building without and with TLCD in frequency domain are compared. In time domain, the displacement and acceleration responses are compared, subjected to the four earthquake waves. It is found that TLCD can reduce both the displacements and accelerations of the building. It means that the TLCD can make the structure satisfy the need for safety and comfort.

CONFLICT OF INTERESTS

The authors declare that there is no conflict of interests regarding the publication of this paper.

ACKNOWLEDGMENTS

The authors would like to thank the support of the National Natural Science Foundation of China (Grant nos. 51121005 and 51261120375).

REFERENCES

1. G. W. Housner, L. A. Bergman, T. K. Caughey et al., "Structural control: past, present, and future,"Journal of Engineering Mechanics, vol. 123, no. 9, pp. 897–971, 1997.
2. C.-C. Lin, C.-L. Chen, and J.-F. Wang, "Vibration control of structures with initially accelerated passive tuned mass dampers under near-fault earthquake excitation," Computer-Aided Civil and Infrastructure Engineering, vol. 25, no. 1, pp. 69–75, 2010.
3. S. Pall and C. Marsh, "Response of friction damped braced frames," Journal of the Structural Division, vol. 108, no. 6, pp. 1313–1323, 1982.
4. J. T. P. Yao, "Concept of structural control," Journal of Structural Division, vol. 98, no. 7, pp. 1567–1574, 1972.
5. F. Sakai, S. Takaeda, and T. Tamaki, "Tuned liquid column damper-new type device for suppression of building vibrations," in Proceedings of the International Conference on Highrise Buildings, Nanjing, China, 1989.
6. C. P. Moreno and P. Thomson, "Design of an optimal tuned mass damper for a system with parametric uncertainty," Annals of Operations Research, vol. 181, pp. 783–793, 2010.
7. U. Mackenroth, Robust Control Systems: Theory and Case Studies, Springer, Berlin, Germany, 2004.
8. J. Rubió-Massegú, F. Palacios-Quiñonero, and J. M. Rossell, "Decentralized static output-feedback H∞controller design for buildings under seismic excitation," Earthquake Engineering and Structural Dynamics, vol. 41, no. 7, pp. 1199–1205, 2012.
9. I. Zečević and D. D. Šiljak, "Design of robust static output feedback for large-scale systems," IEEE Transactions on Automatic Control, vol. 49, no. 11, pp. 2040–2044, 2004.
10. Y.-Y. Cao, Y.-X. Sun, and W.-J. Mao, "Output feedback decentralized stabilization: ILMI approach,"Systems & Control Letters, vol. 35, no. 3, pp. 183–194, 1998.
11. Y. Wang, "Time-delayed dynamic output feedback H∞ controller design for civil structures: a decentralized approach through homotopic transformation," Structural Control and Health Monitoring, vol. 18, no. 2, pp. 121–139, 2011.
12. M. Kočvara and M. Stingl, "Pennon: a code for convex nonlinear and semidefinite programming,"Optimization Methods & Software, vol. 18, no. 3, pp. 317–333, 2003.

CHAPTER 14

NONINVASIVE SENSING TECHNIQUES AND GEOPHYSICAL METHODS FOR CULTURAL HERITAGE AND CIVIL INFRASTRUCTURES MONITORING

Francesco Soldovieri, Jean Dumoulin, Nicola Masini, and Erica Utsi

This special issue presents a good and interesting review of the state of art and novel methodologies, instrumentation, and data-processing approaches to perform non- or minimally destructive investigations with a specific focus to the applicative field of the cultural heritages and critical infrastructure monitoring. In addition, it is concerned with some interesting theoretical and "in real world" examples of how the integration between the classical geophysical techniques with new emerging surface and subsurface-sensing techniques enables a multidepth, multiresolution, and multiscale monitoring.

P. L. Cosentino et al. present the new concept of "microgeophysics" of relevant interest in both the fields of the civil engineering and cultural heritage studies. In particular, the paper is concerned with research efforts in the field of the instrumentation and methodologies so to gain the relevant applicative advantages of time reduction without any loss of information, by taking care to the adoption of multichannel systems.

L. Fragasso and N. Masini present a very interesting use of the infrared reflectography in cultural heritage; in particular, the survey regards the assessment of the polyptych panel attributed to Vivarini's workshop conserved in the "Sigismondo Castromediano" Museum in Lecce. An effort has been made in processing the IR images through Principal Component Analysis and spectral indices able to give information about different pictorial drafting and restoration works.

L. N. Eppelbaum gives an interesting review of the microgravity investigations with a focus to archaeological applications, which is a very challenging research field. In particular, an advanced methodology based on gravity anomalies analysis and 3D modelling is presented with the aim to delineate ancient targets. The performed model computations indicate that microgravity investigations might be successfully applied at least in 20–25% of archaeological sites in Israel.

Orlando et al's paper is concerned with the application of a multisensing approach based on geology, GPS, surveys of cracking, boreholes, seismic refraction and electrical resistivity tomography—to image the shallow stratigraphy and hypothesize the cause of instability of an historical centre. The synergy due to the combination of different geophysical surveys permits to depict underground features at small and medium scale.

The paper from G. Scardozzi et al. regards the contribution of the high-resolution multitemporal optical satellite imagery for change detection of a territory with a focus to the modifications that affected archaeological features and the landscape. In particular, the investigation was focused to five great Assyrian capital cities (Ashur, Kar-Tukulti-Ninurta, Kalhu, Dur-Sharrukin, and Nineveh, in northern Iraq) by using very high-resolution satellite images taken between 2001 and 2007 and panchromatic photographs taken by U.S. spy satellites operating between 1965 and 1969. The paper by G. Bitelli et al. is an example of integrated surveying and monitoring activities for the control of an ancient structure, the Casalecchio di Reno

Citation: Francesco Soldovieri, Jean Dumoulin, Nicola Masini, and Erica Utsi, "Noninvasive Sensing Techniques and Geophysical Methods for Cultural Heritage and Civil Infrastructures Monitoring," International Journal of Geophysics, vol. 2011, Article ID 487679, 2 pages, 2011. doi:10.1155/2011/487679.

sluice, located near Bologna, Italy, representing a case of "industrial archaeological site." Classical topography, high-precision spirit levelling, terrestrial laser scanning, digital close-range photogrammetry, and thermal imagery were used to monitor the stability and the movements of the structure. In addition, the creation of a 3D model of the structure was achieved and permitted to compare two situations, before and after the serious damages suffered by the sluice. Also, an analysis of the whole sluice, carried out at a regional scale, was done via the use of aerial photogrammetry.

The paper by R. Bernini et al. is concerned with the application of a new diagnostic technique for long-range strain and temperature monitoring, exploiting a a distributed fiber-optic sensor portable system based on Brillouin scattering. The experiment in laboratory has demonstrated the effectiveness of this novel technique as a structural assessment tool by measuring the strain on a 4 m long concrete beam and detecting the formation of a crack in the beam resulting from the external applied load.

Paper by A. Loperte et al. deals with an interesting case of use of the Ground Penetrating Radar (GPR) in monitoring of critical infrastructure by a high-resolution diagnostics. In particular, in this paper, the application of GPR has permitted the monitoring and diagnostics of one of the largest dams in the Basilicata region (Southern Italy) for safety management; in fact, the investigation had the aim to detect and localize underground sandstone banks that are potential ways of flow of water below the dam.

Paper by M. Marchetti et al. is concerned with the study of the performance of an infrared camera for the determination of ice road susceptibility, to build a new winter risk index, to improve the measurements rate, and to analyze its consistency with seasons and infrastructures environment. The comparison of this technique with the classical radiometer sensing technique shows a good effectiveness of the IR technique with promising perspectives.

Paper by S. Chambon et al. regards the use of automatic analysis of optical images for crack detection on the French national roads. The work gives three main contributions as: a state of the art of the image-processing tools applied to civil engineering, an approach based on multi-scale extraction and a Markovian segmentation for a fine-defect detection in pavement surface, the design and the validation of a protocol for evaluating the road pavement crack detection.

Paper by A. Palombo et al. regards an experiment at Montagnole site performed in the framework of FP7 ISTIMES project. Ground-based microwave radar interferometer and high-frequency InfraRed Thermography (IRT) were used at the same time to detect and characterize the dynamic displacement of a 16 m concrete beam affected by high-energy direct impacts of an iron ball weighting few tons. The results achieved by the two techniques were compared in terms of frequency analysis and show a very good correlation, which makes confident about the reliability of the diagnosis results.

The paper by G. Colangelo and A. Guariglia discusses a hazard assessment strategy and risk mitigation for rockfalls in a section of a national road, along the coast of Maratea (Southern Italy), using LIDAR technique and spatial modelling. The results represent a valid cognitive support to choose the most appropriate technical solution for topography strengthening and an example of good practice for the cooperation between innovative technologies and field emergency management.

The paper by S. Nordebo et al. represents a methodological contribution to the data fusion problem, which is of great interest in the integration of different techniques for critical infrastructure monitoring. In particular, it is presented a maximum likelihood-based approach to data fusion for electromagnetic (EM) and electrical resistive (ER) tomography. As a multiphysics problem formulation with applications in geophysics, the problem of tunnel detection based on EM and ER tomography is studied in this paper.

The paper by L. Pajewski et al. falls in the framework of the research activities for the development of physical-based data processing for GPR measurements. The effectiveness of the full-wave technique for the solution of the two-dimensional plane-wave scattering problem is shown for a set of perfectly conducting and dielectric cylinders buried in a dielectric half-space, or in a finite-thickness slab. In particular, metallic pipes and air cavities are simulated as buried utilities.

The last paper by A. C. D. Royal et al. presents the progress of a research project that aims to develop a multisensor geophysical platform that can improve the probability of detection of the infrastructure buried beneath the carriageway. The multisensor platform is being developed in conjunction with a knowledge-based system that aims to provide information on how the properties of the ground might affect the sensing technologies being deployed. The fusion of data sources (sensor data and utilities record data) is also being researched to maximize the probability of location. This paper describes the outcome of the initial phase of testing along with the development of the knowledge-based system and the fusion of data to produce utility maps.

Francesco Soldovieri

Jean Dumoulin

Nicola Masini

Erica Utsi

CHAPTER 15

ANALYTICAL SOLUTION FOR FREE VIBRATION ANALYSIS OF BEAM ON ELASTIC FOUNDATION WITH DIFFERENT SUPPORT CONDITIONS

Baki Ozturk and Safa Bozkurt Coskun

INTRODUCTION

The free vibration equation of an axially loaded beam on elastic foundation is a fourth-order partial differential equation. For this particular engineering problem, the analytical solutions will be implemented in this study. Although the governing equation seems to be a linear one, finding the eigenvalues for the free vibration analysis is still challenging. One may not simply obtain the eigenvalues sequentially and their corresponding eigen vectors even with a software.

Free vibration equation of the beam on partially elastic foundation including only bending moment effect was analytically solved [1] while the eigenvalues for free vibration of beam-column systems on elastic foundation were obtained using a numerical approach [2]. The separation of variables technique was used to obtain the free vibration circular frequencies of piles partially embedded in soils [3]. In addition, differential transform method (DTM) has been proposed to solve eigenvalue problems for free and transverse vibration problems of a rotating twisted Timoshenko beam under axial loading [4, 5]. Furthermore, the DTM was also used to find the nondimensional natural frequencies of tapered cantilever Bernoulli-Euler beam [6]. Free vibration equations for one end clamped and other end simply supported beam on elastic foundation were solved by using the DTM for various axial loads acting on the beam [7].

Meanwhile, both the variational iteration method (VIM) and homotopy perturbation method (HPM) were used to solve the free vibration equations of beam on elastic foundation for support conditions of one end clamped, and other end simply supported, both ends clamped and both ends simply supported considering various case studies [8–11]. The beam on elastic foundation was investigated for these three different support conditions considering various N_r values.

Recently, there have been also other studies which are helpful to better understand dynamic behavior of both infinite beams resting on elastic foundation [12] and tapered column with pinned ends embedded in Winkler-Pasternak elastic foundation [13]. In this study, the analytical solutions and analytical results for free vibration analysis of beam on elastic foundation are provided. The analytical results are in excellent agreement with the results of the particular problem solved using the VIM method and the HPM method available in the literature [8–10].

Problem Formulation

Beam on elastic foundation and internal forces and deformations of differential segment of the beam having a length of dx are depicted in Figures 1(a) and 1(b), respectively. The elastic foundation is idealized by Winkler model. Hence, the relationship between displacement function $y(x, t)$ of the beam on elastic foundation and the distributed force $q(x, t)$ applied on the elastic foundation beneath the beam can be

Citation: Baki Ozturk and Safa Bozkurt Coskun, "Analytical Solution for Free Vibration Analysis of Beam on Elastic Foundation with Different Support Conditions," Mathematical Problems in Engineering, vol. 2013, Article ID 470927, 7 pages, 2013. doi:10.1155/2013/470927.

written by $q(x, t) = C_s y(x, t)$. In this equation $C_s = C_o b$, where C_o is the modulus of subgrade reaction while b is the beam width.

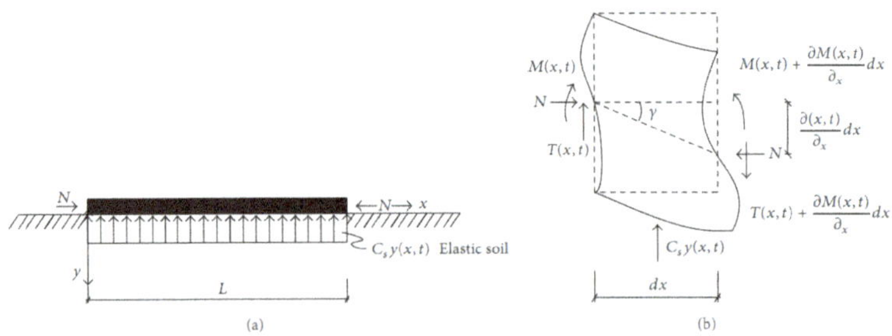

Figure 1: (a) beam on elastic foundation and (b) internal forces and deformations of the beam on elastic foundation.

The equilibrium equations of the internal forces acting on differential beam segment, dimensionless parameter z instead of position variable $x$with $0 \leq z \leq x/L$ and neglecting the second-order terms are used in order to define the motion equation of the beam on elastic foundation [3]:

$$\left\{\phi_1^{iv}(z) + \left[\pi^2 N_r + \frac{\left(\bar{m}\omega^2 - C_S\right)\bar{k}L^2}{AG}\right]\phi_1^{ii}(z) + \frac{\left(C_S - \bar{m}\omega^2\right)L^4}{EI}\phi_1(z)\right\}\sin(\omega t + \theta) = 0. \quad (1)$$

In (1), $\phi_1(z)$ is dimensionless displacement function of the beam considering axial and shear forces; t is the time variable; θ is the phase angle; $N_r = NL^2/(\pi^2 EI)$ is the ratio of axial load N acting on the beam to Euler buckling load; $\bar{m}$ is the distributed mass of the beam; ω is the beam circular frequency; $\bar{k}$ is the shape factor for the shape of the beam section considered; L is the beam length; A, G, E, and I are cross-section area, shear modulus, elastic modulus, and moment of inertia of the beam; and $\phi_1^{ii}(z) = d^2\phi_1(z)/dt^2$, $\phi_1^{iv}(z) = d^4\phi_1(z)/dt^4$, respectively. If the axial and shear force effects are neglected, the dimensionless equation of motion for the beam on elastic foundation becomes as follows [14]:

$$\left\{\phi_2^{iv}(z) + \frac{\left(C_S - \bar{m}\omega^2\right)L^4}{EI}\phi_2(z)\right\}\sin(\omega t + \theta) = 0. \quad (2)$$

In (2), $\phi_2(z)$ is the dimensionless displacement function of the beam neglecting axial and shear forces. Division of both sides of (1) and (2) by $\sin(\omega t+\theta)$ gives the following equations:

$$\left\{\phi_1^{iv}(z) + \left[\pi^2 N_r + \frac{\left(\bar{m}\omega^2 - C_S\right)\bar{k}L^2}{AG}\right]\phi_1^{ii}(z) + \frac{\left(C_S - \bar{m}\omega^2\right)L^4}{EI}\phi_1(z)\right\} = 0, \quad (3)$$

$$\left\{\phi_2^{iv}(z) + \frac{\left(C_S - \bar{m}\omega^2\right)L^4}{EI}\phi_2(z)\right\} = 0. \quad (4)$$

ANALYTICAL SOLUTIONS OF THE PROBLEM

The equation of motion in (3) can be rearranged as follows:

$$\phi_1^{iv}(z) + \left[\pi^2 N_r - \frac{EI\bar{k}}{AGL^2}\left(\lambda - \gamma^4\right)\right]\phi_1''(z) + \left(\lambda - \gamma^4\right)\phi_1(z) = 0, \quad (5)$$

where λ is relative stiffness, $\lambda = C_S L^4/EI$, and γ is frequency factor, $\gamma = \sqrt[4]{\bar{m}\omega^2 L^4/EI}$.

In order to simplify the symbolic representation of solution process, (5) is rewritten as

$$\phi_1^{iv}(z) + \xi_1\phi_1''(z) + \xi_2\phi_1(z) = 0, \quad (6)$$

Where

$$\xi_1 = \pi^2 N_r - \frac{EI\bar{k}}{AGL^2}\left(\lambda - \gamma^4\right)$$
$$\xi_2 = \lambda - \gamma^4. \quad (7)$$

Assuming a solution of the form of e^{rz}, characteristic equation for (6) becomes

$$r^4 + \xi_1 r^2 + \xi_2 = 0. \quad (8)$$

Equation (8) is a quadratic equation in terms of r^2 and its solution is

$$\left(r^2\right)_{1,2} = -\frac{\xi_1}{2} \pm \sqrt{\left(\frac{\xi_1}{2}\right)^2 - \xi_2}. \quad (9)$$

By taking the square root of (9), roots of (8) would be obtained as follows:

$$s_1 = \sqrt{\left(r^2\right)_1}$$
$$s_2 = -\sqrt{\left(r^2\right)_1}$$
$$s_3 = \sqrt{\left(r^2\right)_2}$$
$$s_4 = -\sqrt{\left(r^2\right)_2}. \quad (10)$$

Hence the solution finally takes the following form:

$$\phi_1(z) = C_1 e^{s_1 z} + C_2 e^{s_2 z} + C_3 e^{s_3 z} + C_4 e^{s_4 z}. \quad (11)$$

Coefficients C_1, C_2, C_3, and C_4 would be obtained by inserting the boundary conditions for the beam. In this study three different support configurations for the beam will be considered as shown in Figure 2. These are SS for Simply Supported, CC for Clamped-Clamped, and CS for ClampedSimply Supported beams. The boundary conditions related to these supports are given below

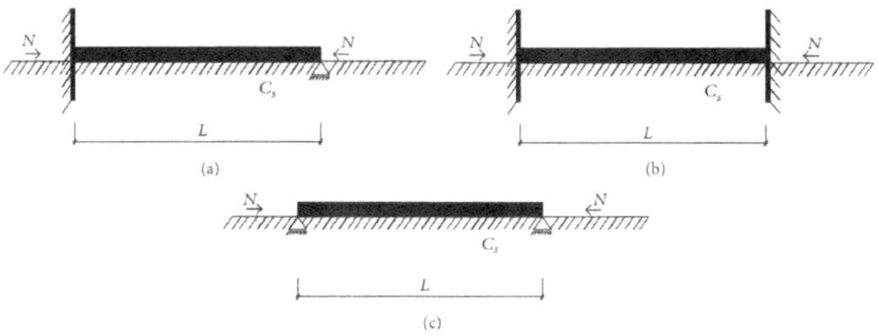

Figure 2: (a) CS beam, (b) CC beam, and (c) SS beam.

For a simple support:

$$\phi_1\big|_z = 0, \qquad \phi_1''\big|_z = 0. \tag{12}$$

For a clamped support:

$$\phi_1\big|_z = 0, \qquad \phi_1'\big|_z = 0. \tag{13}$$

Inserting the boundary conditions given in (12)-(13) into (11) for SS, CC, and CS beams, a nonlinear system of equations in the following form would be produced:

$$[K(\gamma)]\{C\} = \{0\}, \tag{14}$$

where $\{C\}^T = \{C_1\ C_2\ C_3\ C_4\}$ and includes the coefficients of (11). Coefficient matrix $[(\gamma)]$ is dependent on frequency factor γ and equating the determinant to zero gives the frequency factors for vibration modes of the corresponding beam:

$$\det[K(\gamma)] = 0. \tag{15}$$

Positive real roots of this equation are free vibration frequencies of the beams on elastic foundations shown in Figure 2.

NUMERICAL STUDY

Analytical solution for free vibration of a one end clamped, and one end pinned beam on elastic foundation was previously given by Çatal [7] with additional analyses by DTM. However, Çatal [7] gives analytical solutions for only CS beam, and in this study it is observed that these results fail to coincide with the exact results.

Previously, both the variational iteration method (VIM) and homotopy perturbation method (HPM) were used to solve the free vibration equations of beam on elastic foundation for support conditions of one end clamped, and other end simply supported, both ends clamped and both ends simply supported considering various case studies [8–10]. In this study, analytical solutions for three cases are provided which are shown in Figure 2. Each case was previously analyzed by the use of both VIM and HPM with the previously explained procedure [8–10]. Numerical values in this study are chosen as the same used in Çatal [7]. Hence, an IPB 500 steel profile resting on a Winkler foundation having a modulus of subgrade reaction of 50,000 $\mathrm{kN/m^2}$ is considered. Other numerical values are as follows:

$I = 107.2 * 10^{-5}\,\mathrm{m^4}$; $A = 2.39 * 10^{-2}\,\mathrm{m^2}$; $\overline{m} = 0.19\,\mathrm{kN\,s^2/m}$; $\overline{k} = 3.705$;

$E = 2.1 * 10^8\,\mathrm{kN/m^2}$; $G = 8.1 * 10^7\,\mathrm{kN/m^2}$.

Frequency factors $\gamma = \sqrt[4]{\overline{m}\omega^2 L^4/EI}$ are calculated taking bending moment, shear, and axial effects into consideration for $N_r = 0.25$, $N_r = 0.5$, and $N_r = 0.75$ due to circular frequencies of the beam.

Variation of frequency factor, γ is tabulated in Tables 1–3. Table 1 includes comparison of analytical results of this study for the CS beam with analytical results obtained by Çatal [7]. Tables 2 and 3 show the analytical results for both CC and SS beams which are not solved by Çatal [7].

Table 1: Variation of frequency factor, γ, with relative stiffness, λ, for a CS beam on elastic foundation. (The analytical results of C¸atal [7] are provided below in bold characters.)

	γ_1	γ_2	γ_3
$N_r = 0.25$			
$\lambda = 1$	2.81620011	3.97739569	4.82021485
	2.81574011	**3.97658348**	**4.81918573**
$\lambda = 10$	3.37458660	5.04748828	6.25843195
	3.37451696	**5.04723072**	**6.25809765**
$\lambda = 100$	4.05117292	6.09258233	7.83201825
	4.05107170	**6.09248874**	**7.83184147**
$\lambda = 1000$	5.87924217	7.32194493	9.34675005
	5.87935543	**7.32217598**	**9.34584181**
$N_r = 0.5$			
$\lambda = 1$	2.71457148	3.93377748	4.79443677
	2.71413994	**3.93298292**	**4.79341412**
$\lambda = 10$	3.26137662	4.99222355	6.22484874
	3.26133704	**4.99199057**	**6.22451830**
$\lambda = 100$	3.96066525	6.02920759	7.79040674
	3.96064973	**6.02918911**	**7.79025650**
$\lambda = 1000$	5.84614526	7.26848984	9.30219043
	5.84624672	**7.26853895**	**9.30226231**
$N_r = 0.75$			
$\lambda = 1$	2.59996785	3.88867219	4.76823873
	2.59956455	**3.88787082**	**4.76721819**
$\lambda = 10$	3.13469732	4.93509845	6.19072130
	3.13456678	**4.93488312**	**6.19038296**
$\lambda = 100$	3.86312198	5.96380562	7.74812917
	3.86301875	**5.96338463**	**7.74796867**
$\lambda = 1000$	5.81232280	7.21383882	9.25698856
	5.81241941	**7.21407604**	**9.25701332**

Table 2: Variation of frequency factor, γ, with relative stiffness, λ, for a CC beam on elastic foundation.

	γ_1	γ_2	γ_3
$N_r = 0.25$			
$\lambda = 1$	3.39742898	4.32815705	5.16088024
$\lambda = 10$	4.05721417	5.52373153	6.68428571
$\lambda = 100$	4.66987707	6.68497843	8.36140368
$\lambda = 1000$	6.15541347	7.87417668	9.94633264
$N_r = 0.5$			
$\lambda = 1$	3.33994125	4.29375653	5.14012431
$\lambda = 10$	3.99112373	5.48024811	6.65690121
$\lambda = 100$	4.60834095	6.63472888	8.32680496
$\lambda = 1000$	6.12479804	7.82851855	9.90820057
$N_r = 0.75$			
$\lambda = 1$	3.27929840	4.25850809	5.11911832
$\lambda = 10$	3.92148631	5.43569919	6.62918805
$\lambda = 100$	4.54408497	6.58329012	8.29178517
$\lambda = 1000$	6.09362560	7.78201148	9.86962997

Table 3: Variation of frequency factor, γ, with relative stiffness, λ, for an SS beam on elastic foundation.

	γ_1	γ_2	γ_1
$N_r = 0.25$			
$\lambda = 1$	2.24318643	3.57344623	4.48319716
$\lambda = 10$	2.71977348	4.53601512	5.82884027
$\lambda = 100$	3.56133061	5.49391862	7.29761546
$\lambda = 1000$	5.71760639	6.82177513	8.75859570
$N_r = 0.5$			
$\lambda = 1$	2.03688119	3.51272374	4.45090640
$\lambda = 10$	2.51190763	4.46032723	5.78712097
$\lambda = 100$	3.44330860	5.41052448	7.24679924
$\lambda = 1000$	5.68672681	6.75973740	8.70619682
$N_r = 0.75$			
$\lambda = 1$	1.73715809	3.44867815	4.41789714
$\lambda = 10$	2.23396379	4.38057743	5.74447900
$\lambda = 100$	3.31172675	5.32308548	7.19489096
$\lambda = 1000$	5.65533583	6.69594299	8.65283438

The graphical presentations of analytical solutions are provided in Figures 3, 4, and 5.

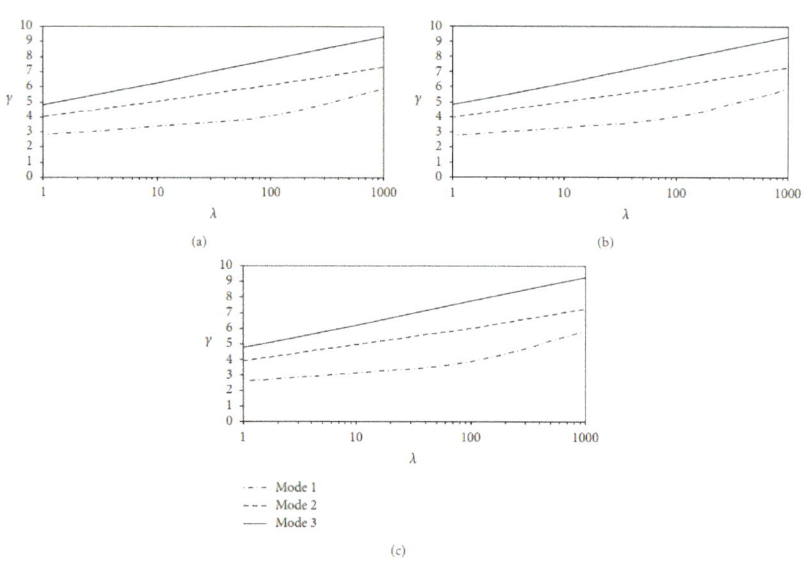

Figure 3: Variation of frequency factor, γ, with relative stiffness, λ, for a CS beam on elastic foundation: (a) for $N_r = 0.25$, (b) for $N_r = 0.5$, and (c) for $N_r = 0.75$.

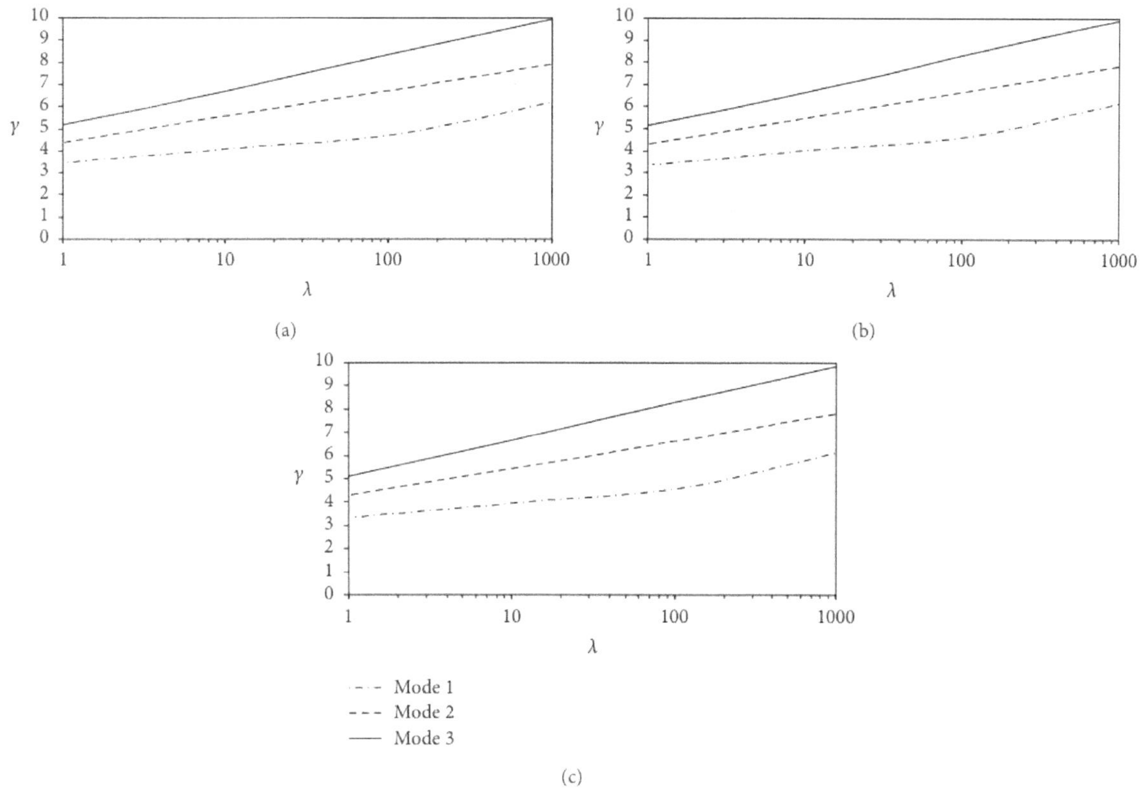

Figure 4: Variation of frequency factor, γ, with relative stiffness, λ, for a CC beam on elastic foundation: (a) for $N_r = 0.25$, (b) for $N_r = 0.5$, and (c) for $N_r = 0.75$.

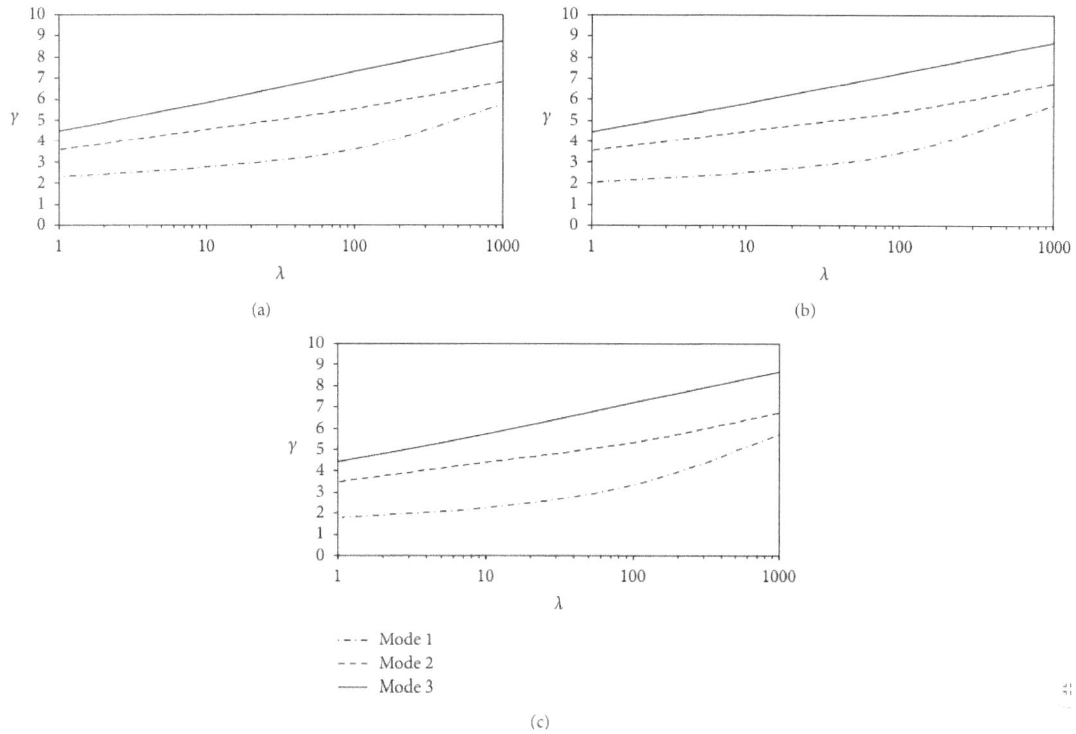

Figure 5: Variation of frequency factor, γ, with relative stiffness, λ, for an SS beam on elastic foundation: (a) for $N_r = 0.25$, (b) for $N_r = 0.5$, and (c) for $N_r = 0.75$.

CONCLUSION

In this study, the analytical results for free vibration analysis of beam on elastic foundation are provided for three different axially loaded cases which are namely one end clamped, the other end simply supported (CS beam); both ends clamped (CC beam), and both ends simply supported (SS beam) cases. In addition, the analytical solution and the frequency factors evaluated for different ratios of axial load N acting on the beam to Euler buckling load, N_r are obtained. In Figures 3, 4, and 5, it is observed that for the same values of λ, the largest values of γ are obtained for both ends clamped (CC) beam while the lowest values of γ are obtained for both ends pinned (SS) beam.

For the CS Beam, it is observed that analytical solutions given by C¸ atal [7] fail to coincide with the exact results provided in this study. Previous studies [8–10] employing both the variational iteration method (VIM) and homotopy perturbation method (HPM) to solve the free vibration equations of beam on elastic foundation in this study converge to the same results obtained in this study. The reason for that is the governing equation is a linear equation in nature; and as expected analytical approximate methods converge to the exact solution. In the study by C¸ atal [7], the DTM converges the analytical results provided in the same study. Presumably, the difference is due to the fact that C¸ atal [7] used the length of the beam, L, in three-digit accuracy while in this study the exact values of length, L, are used in all calculations conducted.

Accordingly, this study provides the analytical results for free vibration analysis of beam on elastic foundation for three different axially loaded cases which are namely one end clamped, the other end simply supported (CS beam); both ends clamped (CC beam) and both ends simply supported (SS beam) cases.

REFERENCES

1. P. F. Doyle and M. N. Pavlovic, "Vibration of beams on partial elastic foundations," Earthquake Engineering and Structural Dynamics, vol. 10, no. 5, pp. 663–674, 1982.
2. H. H. West and M. Mafi, "Eigenvalues for beam-columns on elastic supports," Journal of Structural Engineering, vol. 110, no. 6, pp. 1305–1320, 1984.
3. H. H. Çatal, "Free vibration of partially supported piles with the effects of bending moment, axial and shear force," Engineering Structures, vol. 24, no. 12, pp. 1615–1622, 2002.
4. C.-K. Chen and S.-H. Ho, "Application of differential transformation to eigenvalue problems," Applied Mathematics and Computation, vol. 79, no. 2-3, pp. 173–188, 1996.
5. C. K. Chen and S. H. Ho, "Transverse vibration of a rotating twisted Timoshenko beams under axial loading using differential transform," International Journal of Mechanical Sciences, vol. 41, no. 11, pp. 1339–1356, 1999.
6. O. Özdemir and M. O. Kaya, "Flapwise bending vibration analysis of a rotating tapered cantilever Bernoulli-Euler beam by differential transform method," Journal of Sound and Vibration, vol. 289, no. 1-2, pp. 413–420, 2006.
7. S. Çatal, "Solution of free vibration equations of beam on elastic soil by using differential transform method," Applied Mathematical Modelling, vol. 32, no. 9, pp. 1744–1757, 2008.
8. B. Öztürk, "Free vibration analysis of beam on elastic foundation by the variational iteration method,"International Journal of Nonlinear Sciences and Numerical Simulation, vol. 10, no. 10, pp. 1255–1262, 2009.
9. B. Öztürk and S. B. Coşkun, "The homotopy perturbation method for free vibration analysis of beam on elastic foundation Structural Engineering and Mechanics," An International Journal, vol. 37, no. 4, 2011.
10. B. Öztürk, S. B. Coşkun, M. Z. Koç, and M. T. Atay, "Homotopy perturbation method for free vibration analysis of beams on elastic foundations," IOP Conference Series, vol. 10, no. 1, 2010
11. S. B. Coşkun, M. T. Atay, and B. Öztürk, "Transverse vibration analysis of euler-bernoulli beams using analytical approximate techniques," in Advances in Vibration Analysis Research, chapter 1, pp. 1–22, InTech, Vienna, Austria, 2011.
12. G. Raftoyiannis, T. P. Avraam, and G. T. Michaltsos, "Dynamic behavior of infinite beams resting on elastic foundation under the action of moving loads. Structural Engineering and Mechanics," An International Journal, vol. 35, no. 3, 2010.
13. Ö. Civalek and B. Öztürk, "Free vibration analysis of tapered beam-column with pinned ends embedded in Winkler-Pasternak elastic foundation," Geomechanics and Engineering, vol. 2, no. 1, pp. 45–56, 2010.
14. J. Tuma and F. Cheng, Theory and Problems of Dynamic Structural Analysis, Schaum's Outline Series, McGraw-Hill, New York, NY, USA, 1983.

CHAPTER 16

ENGINEERING SOLUTIONS TO ENHANCE TRAFFIC SAFETY PERFORMANCE ON TWO-LANE HIGHWAYS

Lina Wu, Jiangwei Chu, Yusheng Ci, Shumin Feng, and Xingwang Liu

INTRODUCTION

Improving two-lane highway traffic safety conditions is of practical importance and has attracted significant research attention within the last decade. Many cost-effective and proactive solutions such as low-cost treatments and roadway safety monitoring programs have been developed to enhance traffic safety performance under prevailing conditions. A two-lane Highway Safety Enhancement Project (HSEP) was implemented on the major highway network in Beijing from 2004 to 2006 by the Ministry of Transport of the People's Republic of China. A series of engineering solutions was proposed and integrated to improve traffic safety performance network-wide. More than 170,000 hazardous sections of national or provincial highways were improved over 50,000 kilometers, which included improper curves, heavy gradients, insufficient sight distances, and vague clear zones. Cost-effective countermeasures against severe injuries and fatal crashes also included adding or replacing traffic signs, painting markings on pavement, installing different barriers on roadsides, and channelizing intersections. Substantial work has been conducted on engineering-based countermeasure development for improvement of safety performance. Bagdade et al. [1] introduced specific traffic engineering improvements targeted at improving safety for seniors, which were developed to popularize the use of a number of engineering countermeasures throughout Michigan. Labi [2] investigated the efficacy of roadway improvements in terms of crash reduction at various subclasses of rural two-lane highways using the empirical analysis method of the negative binomial modeling technique. However, most of these previous studies focused on only a single improvement that is, widening shoulders, adding passing lanes, or installing center left-turn lanes. Schneider et al. [3] and Bauer and Harwood [4] presented safety prediction models and conducted evaluations of safety effects of horizontal curves on rural two-lane highways. Their findings showed that the radius and length of each horizontal curve significantly influenced the frequency of crashes. Gross et al. [5, 6] and Schrock et al. [7] explored the use of observational data to estimate safety effectiveness for changes in lane and shoulder width and found that adding passing lanes was an economical and effective safety improvement for rural two-lane highways. These findings can be used to help determine cost-benefit ratios to compare competing alternatives. Park et al. [8] examined the safety effect of wider edge lines by analyzing crash frequency data for road segments with and without wider edge lines based on different statistical approaches. Their results showed positive safety effects of wider edge lines installed on rural, two-lane highways. Lyon et al. [9, 10] and Moreno et al. [11] incorporated the Empirical Bayes method to determine the safety effectiveness of installing left-turn lanes and obtained operational effectiveness of passing zones from passing frequency. Their results indicated that these were cost-effective treatments for two-lane rural highways. Yuan et al. [12], Yuan and Lu [13], and Montella and Mauriello [14]

Citation: Lina Wu, Jiangwei Chu, Yusheng Ci, Shumin Feng, and Xingwang Liu, "Engineering Solutions to Enhance Traffic Safety Performance on Two-Lane Highways," Mathematical Problems in Engineering, vol. 2015, Article ID 762379, 7 pages, 2015. doi:10.1155/2015/762379.

used the Empirical Bayesian method and Safety Indexes to evaluate safety benefits and suggest relevant improvements for highway intersections. Cruzado and Donnell [15] evaluated the effectiveness of dynamic speed display signs in transition zones of two-lane, rural highways, and the results indicated that signs were effective in reducing free-flow passenger car operating speeds. There are many methods of evaluating the effectiveness of special improvements. The Bayes methods are widely used and effective. Li and Washburn [16] presented an improved methodology for two-lane highway facility analysis and a new version of CORSIM with the capability of modeling two-lane highways. Zhu et al. [17] discussed the relationship among drivers' heart rates, variability of the vehicle in the running process, design speed consistency, operating speed coordination, speed reduction coefficients, and speed gradients in two-lane highways. Persaud et al. [18], Mujalli and De Oña [19], and Vlahogianni and Golias [20] used the Empirical Bayes and Bayesian Network method to estimate the safety of specific sites and analyze the injury severity of traffic accidents. Wu et al. [21] developed two Mixed Logit models based on crash data collected in New Mexico to analyze driver injury severities in SV and MV crashes on rural two-lane highways.

In this study, the HSEP research findings are presented to evaluate the engineering solutions and their effectiveness in reducing traffic crash severities on Highway G109 in Beijing, China. The potential causal factors were identified based on the proposed evaluation criteria, and primary countermeasures were developed against inferior driving conditions. Six cost-effective engineering solutions were specifically implemented to improve two-lane highway safety conditions: (1) traffic sign replacement, (2) repainting of pavement markings, (3) installation of roadside barriers, (4) intersection channelization, (5) drainage optimization, and (6) sight distance improvement. An Empirical Bayes (EB) model-based before-after study was conducted. The results indicate that the proposed engineering solutions effectively improved traffic safety performance by significantly reducing crash occurrence risks and crash severities.

DATA COLLECTION

G109 highway is a major corridor connecting the city of Beijing and the province of Hebei with a length of 119.2 kilometers. The two-lane highway system is comprised mostly of Class 2, Class 3, and Class 4 highway segments. The terrain conditions are mainly mountainous with some plains and hills along the highway. All traffic engineering solutions for G109 were designed and finished in 2004.

Detailed crash data for G109 were collected from the traffic accident database of the Beijing Traffic Management Bureau (2000–2008), and part of the minor accident data was gathered from the Beijing Municipal Roadway Administration Bureau. Traffic volume data and geometric data were obtained from the Annual National Highways Traffic Volume Statistical Handbook (2000–2008) and the Design Documents of G109 managed by the Transport Planning and Research Institute of the Ministry of Transport, China, respectively. Accident and traffic volume data for similar segments were obtained from the Traffic Police Corps of Guangxi Department of Public Security (2004-2005).

ENGINEERING TREATMENTS IMPLEMENTED

Criteria for Development of Crash Hotspot Identification

HSEP mainly uses traffic engineering treatments to improve or eliminate potential crash hotspots to enhance traffic safety on highways. This is different from highway reconstruction and extension projects and aims to rehabilitate and improve traffic facilities. The objectives of HSEP are to decrease traffic accident fatality rates, especially in serious accidents and to supply a better highway traffic and safety environment with safer facilities. HSEP targets to maximize safety benefits with the lowest cost, least amount of engineering, and least time.

Assessments were first conducted due to limited construction funds. Five aspects of geometric features and roadway characteristics were chosen, including the horizontal curve radius and longitudinal gradient. Detailed assessment criteria information is shown in Table 4.

Engineering Treatments Implemented

Based on the design documents of G109, all latent dangerous entities were identified following the above judgment criteria and accident data. A total of 612 traffic signs were installed or replaced; 27,000 m^2 of pavement markings were painted; 15,137 m of barriers were built or replaced; 8 intersections were channelized. The average cost to implement these changes was only 100,713 RMB per kilometer.

Traffic Signs

The problems with the traffic sign system before treatment were discontinuous signs, lack of warning signs, mistakes on locations and content of signs, occlusion problems, and too many characters on a tourism sign.

Based on the existing problems and characteristics of the highway curvatures and gradients, new traffic signs were designed and installed systematically to avoid redundant information. Furthermore, locations of different signs were established according to the operation speed of traffic flow and sight distance.

In total, 612 new traffic signs were installed, including 19 solar energy signs. Several oversized warning signs and indication signs with high visibility were fixed as reminders to reduce speed and drive carefully before segments of curves and slopes. Relevant solar horizontal induction signs and flat top tubular delineators and chevron signs were set up at borders of the roadside along curves with the combination of median delineators. In addition, more easily recognized guide signs and tourist spot signs replaced older signs along the entire highway. Finally, stop and yield signs were installed on minor road approaches.

Pavement Markings

More than 27,000 m^2 of pavement markings were painted, including 1,400 m^2 of special pavement markings.

Night performance pavement markings such as yellow lane-lines and white edge-lines were painted at general segments. Channelizing lines, word markings, crosswalks, and symbol markings were introduced at the intersections. Thin red lines were painted in the vertical direction of sharp curves or accident prone locations. Optical illusion deceleration markings were painted at lane borders to remind drivers of special geometric conditions and at lane mediums for pedestrians.

Roadside Barriers

Almost 20% of accidents were Run-off-road (ROR) crashes along the highway. Thus, 15,137 m of barriers were built or replaced according to the degree of roadside hazard, accident frequency, operation speed, traffic volume and composition, and landscape demand in several special segments along the highway. The forgiveness design principle was one of the most critical methods we followed. However, because of terrain conditions, barrier equipment was essential in certain circumstances.

A total of 2,850 m of high strength concrete barriers were cemented beside cliff edges at segments of small, continuous horizontal curves and areas with a high frequency of side collision accidents. Serrated concrete barriers with a light appearance for guidance were built at sharp horizontal curve segments with fewer crashes.

A total of 5,158 m of corrugated beam barriers or steel guardrails were installed at cliff segments, winrows, and high retaining walls without enough earth pressure outside of the shoulder. A-level corrugated beam barriers were chosen for downgraded sections with higher operation speeds and serious consequences of rollover accidents. B-level barriers were prepared for upgraded sections with a lower possibility of crashes.

A total of 7,129 m of cable barriers were placed at sections of landscape with enough clearance and earth pressure outside of the shoulder. Yellow reflectors were pasted on barriers to enhance their visibility. A-level protection barriers were implemented at sections with rollover accident records or higher risk coefficients, and B-level cable barriers were used to protect areas without crash records or roadside shallows.

Intersection Channelization

There were many 3-legged and 4-legged intersections along the highway. There were previously no channelization pavement markings, and vehicles operated without rules. Many of these intersections had poor visibility, so the frequency of frontal and side collisions was much higher in these areas than in others (Figure2).

Therefore, (1) channelization pavement markings were painted to split different direction flows and standardize traffic order; (2) stop signs and/or yield signs were placed at minor-legged roads according to highway classes and traffic volumes for clearer priority and fewer conflicts; (3) a left-turn pocket lane was created within the confines of an existing roadway; (4) several solar yellow flashing lights were added to enhance night safety for vehicles, pedestrians and cyclists. As a result, 8 main intersections were channelized and the others were modified with stop or yield signs.

Drainage Optimization

The highway side ditch in G109 was in the shape of a rectangle with a wide cross section and deep depth of 40 to 50 centimeters. This type of side ditch was a latent conflict and had the potential to aggravate the severity of ROR crashes.

According to the real terrain conditions and roadside qualifications, the side ditch was transformed into a flat butterfly side ditch at several special segments to reduce the severity of ROR crashes.

Sight Distance Improvement

Poor visibility problems were serious on G109, especially at segments of intersections, switch-backs, and sharp curves. Slope flattening is an effective way to improve road visibility. Therefore, buildings, structures, mounds, trees, or wire poles were removed at sight triangles of intersections or at the inner sides of horizontal curves in some circumstances. This increased the visibility of convex mirrors, warning signs, and pavement markings (Figure 1).

Figure 1: Warning signs and pavement markings.

Figure 2: Intersection channelization.

Other Implementations

In addition to the treatments described, there were other significant implementations. For instance, roadside improvement, scenic view platform treatment, and tunnel management were extensively implemented along G109.

SAFETY EFFECTS OF TREATMENTS

Empirical Bayes Analysis

Few resources have been devoted to evaluation in the literature. In this study, the safety effects of the treatments were evaluated by the Empirical Bayes (EB) before-after analysis based on the severity of traffic accidents. In the EB estimate, the joint use of two clues is implemented by a weighted average, that is,

$$m = w \cdot x + (1 - w) \cdot P,$$
$$\sigma(m) = \sqrt{(1 - w) \cdot m}, \qquad (1)$$

where m is an estimate of expected accidents; x is the number of accidents expected on similar segments; P is the number of accidents on G109; w is the weight and $\sigma(m)$ is the standard deviation of m. Consider

$$x = \mu l,$$
$$w = \frac{1}{1 + (\mu \cdot Y)/\varphi}, \qquad (2)$$

where μ is the safety performance function, the average number of accidents/(km-year); l is the length of the segment, km; Y is the number of years during which the accident count materialized; φ is the over dispersion parameter

Here, the Safety Performance Function (SPF) was an equation of the Average Daily Traffic (ADT) and length of the segment. Consider

$$\mu = a \times \text{ADT}^b, \qquad (3)$$

where a, b are regression parameters.

After regression analysis with general and major accident data and the ADT data in Tables 1 and 3, two regression parameters were obtained, $a = 0.0073$ and $b = 0.5146$. Then, $\mu = 0.436$ for 2004 to 2005 and $\mu = 0.481$ for 2007 to 2008 were computed. The resulting weights were 0.82 and 0.81, meaning that 52.13 and 57.46 accidents were expected on similar segments, respectively. Finally, the average estimates of expected accidents were 45.27 ± 2.85 and 49.58 ± 3.07 accidents. There were only 14 and 16 accidents that occurred on G109 in Beijing, showing a significant improvement in safety after implementation of the engineering treatments.

Table 1: Statistics of accidents and traffic volumes of G109, Beijing

Durations	Total accidents	Minor accidents	General accidents	Major accidents	Deaths	Injuries	Traffic volumes
2000.5–2001.4	103	33	63	7	8	61	2,708
2001.5–2002.4	160	109	42	9	10	70	2,358
2002.5–2003.4	173	126	32	15	15	61	2,090
2003.5–2004.4	289	239	46	4	4	110	2,222
2004.5–2005.4	431	417	13	1	1	27	2,830
2007.5–2008.4	519	503	10	6	6	175	3,420

Accident Severity Analysis after Treatment Implementation

The number of minor accidents increased quickly. Possible reasons for this may be increased traffic volume, a better traffic environment causing relaxed driver vigilance, and a gradually improved accident record system. Figure 5 shows the variations of accident count versus traffic volume before and after the engineering treatments.

Accident severity analysis was evaluated in detail to find the safety effect evidence of engineering treatment effectiveness after implementation. Detailed information is listed in Table 5.

Six fatal accidents and 8 major accidents were caused by driver factors, totaling 100 percent of such accidents. All fatal accidents and one half of the major accidents occurred during the daytime. All fatal accidents and 5 major accidents occurred during clear weather conditions. This suggests that highway factors, such as sight distance, pavement condition, and geometric features were not the main causes of these types of accident. This indirectly proves the effectiveness of the treatments (Table 2).

Table 2: Geometric features of G109, Beijing.

Design indexes	Units	Class 2	Class 3
Lane	—	2	2
Section	—	Origin to K28 + 560	K28 + 560 to destination
Design speed	km/h	60	30
Paving width	m	12.0	8.5
Shoulder width	m	0.5	0.5
Length	km	119.15	
The minimum horizontal curve radius	m	10 (<60 m, 209 segments)	
The maximum longitudinal gradient	%	9.49 (>5%, 87 segments)	

Table 3: Statistics of accident data and ADT in similar segments.

Highway segments	Length/km	ADT		General and major accidents	
		2004	2005	2004	2005
G207	331.171	3,906	4,618	164	246
G209	520.258	3,058	3,236	245	303
G210	250.377	4,356	4,593	126	175
G321	375.709	3,308	3,190	122	141
G322	554.348	5,716	6,330	354	452
G323	712.332	5,804	6,779	289	310
G324	714.561	10,116	11,930	705	788
G325	244.731	2,928	3,217	90	109

Table 4: Assessment criteria for potential crash hotspots on G109, Beijing.

Categories	Indexes	Units	Class 2	Class 3
Sharp curve	The horizontal curve radius	m	$R \leq 125$	$R \leq 60$
Heavy gradient	The longitudinal gradient	%	$I \geq 6.0$	$I \geq 7.0$
Continuous downgrade	The longitudinal gradient	%	$I \geq 4.5$ (≥3 km)	$I \geq 5$ (≥3 km)
Bad sight distance	Decision sight distance	m	$L \leq 150$	$L \leq 80$
Bad clear zone	The height of slope The distance to shoulder border	m	$H \geq 4.0$ $W \leq 3.0$ (Lakes, rivers, swamps, ditches)	

Table 5: Variables, values, and actual classification by severity (2007-2008).

Variables	Categories	Coding/values	Total	Fatality	Major injury	Minor injury	Property damage
			519 (100%)	6 (1.16%)	8 (1.54%)	98 (18.88%)	407 (78.42%)
1	Driver's gender	Man = 1	498 (95.95%)	6 (1.20%)	7 (1.41%)	95 (19.08%)	390 (78.31%)
		Woman = 2	21 (4.05%)	0 (0.00%)	1 (4.76%)	3 (14.29%)	17 (80.95%)
2	Driver's age	<25 = 1	38 (7.32%)	1 (2.63%)	1 (2.63%)	7 (18.42%)	29 (76.32%)
		25~34 = 2	154 (29.67%)	3 (1.95%)	3 (1.95%)	38 (24.68%)	110 (71.42%)
		35~44 = 3	125 (24.08%)	1 (0.80%)	2 (1.60%)	32 (25.60%)	90 (72.00%)
		45~54 = 4	61 (11.76%)	1 (1.64%)	1 (1.64%)	17 (27.87%)	42 (68.85%)
		>54 = 5	141 (27.17%)	0 (0.00%)	1 (0.71%)	4 (2.84%)	136 (96.45%)
3	Type of vehicle	Passenger car = 1	220 (42.38%)	1 (0.45%)	2 (0.91%)	39 (17.73%)	178 (80.91%)
		Bus = 2	4 (0.77%)	0 (0.00%)	0 (0.00%)	0 (0.00%)	4 (100.00%)
		Light truck = 3	18 (3.47%)	0 (0.00%)	0 (0.00%)	2 (11.11%)	16 (88.89%)
		Truck = 4	231 (44.52%)	4 (1.73%)	1 (0.43%)	45 (19.48%)	181 (78.36%)
		Tractor trailer = 5	6 (1.16%)	0 (0.00%)	2 (33.33%)	2 (33.33%)	2 (33.33%)
		Tractor = 6	36 (6.94%)	1 (2.78%)	2 (5.56%)	10 (27.78%)	23 (63.88%)
		Cycles = 7	4 (0.77%)	0 (0.00%)	1 (25.00%)	0 (0.00%)	3 (75.00%)
4	Vehicle involved	1 = 1	162 (31.21%)	2 (1.23%)	1 (0.62%)	26 (16.05%)	133 (82.10%)
		2 = 2	340 (65.51%)	3 (0.88%)	6 (1.76%)	69 (20.29%)	262 (77.07%)
		>2 = 3	17 (3.28%)	1 (5.88%)	1 (5.88%)	3 (17.65%)	12 (70.59%)
5	Weather condition	Clear = 1	377 (72.63%)	6 (1.59%)	5 (1.33%)	73 (19.36%)	293 (77.72%)
		Cloudy = 2	78 (15.03%)	0 (0.00%)	1 (1.28%)	13 (16.67%)	64 (82.05%)
		Rainy = 3	58 (11.18%)	0 (0.00%)	2 (3.45%)	11 (18.97%)	45 (77.58%)
		Snowy = 4	6 (1.16%)	0 (0.00%)	0 (0.00%)	1 (16.67%)	5 (83.33%)
6	Month	Spring = 1	63 (12.14%)	0 (0.00%)	0 (0.00%)	0 (0.00%)	63 (100.00%)
		Summer = 2	144 (27.75%)	1 (0.69%)	3 (2.08%)	30 (20.83%)	110 (76.40%)
		Autumn = 3	154 (29.67%)	2 (1.30%)	2 (1.30%)	30 (19.48%)	120 (77.92%)
		Winter = 4	158 (30.44%)	3 (1.90%)	3 (1.90%)	38 (24.05%)	114 (72.15%)
7	Crash time	0~5 = 1	51 (9.83%)	0 (0.00%)	2 (3.92%)	9 (17.65%)	40 (78.43%)
		6~11 = 2	189 (36.42%)	3 (1.59%)	1 (0.53%)	37 (19.58%)	148 (78.30%)
		12~17 = 3	187 (36.02%)	3 (1.60%)	3 (1.60%)	38 (20.33%)	143 (76.47%)
		18~23 = 4	92 (17.73%)	0 (0.00%)	2 (2.17%)	14 (15.22%)	76 (82.61%)
8	Crash type	Frontal collision = 1	139 (26.78%)	3 (2.16%)	2 (1.44%)	38 (27.34%)	96 (69.06%)
		Side collision = 2	37 (7.13%)	0 (0.00%)	4 (10.81%)	4 (10.81%)	29 (78.38%)
		Rear-end collision = 3	74 (14.26%)	1 (1.35%)	0 (0.00%)	9 (12.16%)	64 (86.49%)
		Scape = 4	102 (19.65%)	1 (0.98%)	1 (0.98%)	21 (20.59%)	79 (77.45%)
		Run-off-road = 5	167 (32.18%)	1 (0.60%)	1 (0.60%)	26 (15.57%)	139 (83.23%)
9	Cause	Driver factors = 1	480 (92.48%)	6 (1.25%)	8 (1.67%)	89 (18.54%)	377 (78.54%)
		Vehicle factors = 2	10 (1.93%)	0 (0.00%)	0 (0.00%)	1 (10.00%)	9 (90.00%)
		Environment factors = 3	29 (5.59%)	0 (0.00%)	0 (0.00%)	8 (27.59%)	21 (72.41%)

ROR accidents occupied 32.18% of the total number of accidents, with 1 fatal and 1 major accident occurring. This indicates that most vehicles involved in ROR accidents were protected by treatments such as drainage optimization, shoulder improvements, and clear zone improvements (Figures 3 and 4).

Figure 3: Drainage optimization.

Figure 4: Roadside shoulder improvements.

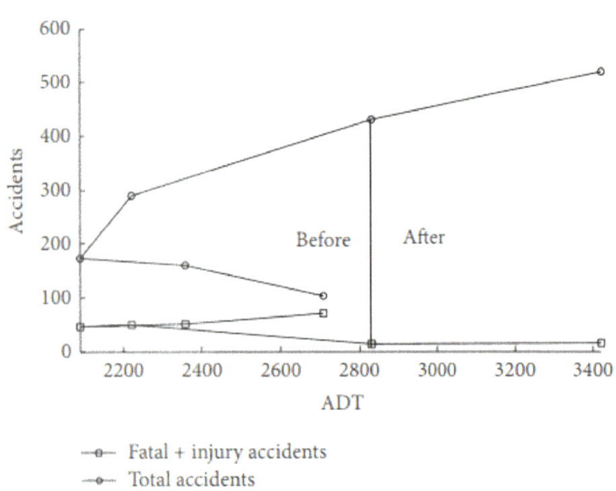

Figure 5: Traffic accidents versus traffic volume of G109, Beijing.

Truck and tractors had higher crash incidence rates. The majority of accidents included one or two vehicles and were frontal or side collisions. No fatal or major accidents occurred at intersection areas. Drivers over the age of 54 and female drivers were safer. Driver violations, carelessness, and false estimation of motor performance were the main causes of driver error. These results suggest that the engineering treatments played a role in protecting lives from serious crashes and reduced crash severity.

CONCLUSIONS

HSEP is a cost-effective traffic safety improvement project initialized by the Ministry of Transport, China. Various engineering treatments were implemented. Safety effects were evaluated using EB-based before-after studies, and crash severity was analyzed using the data collected from the G109 highway in Beijing. The major contributions of this study to the state of the practice include the following. (1)Traffic crash data and traffic dynamic data were collected to support EB-based before-after studies and crash severity analyses.(2)All potential crash hotspots were identified according to the proposed assessment criteria, which included sharp curves, heavy gradients, continuous downgrades, insufficient sight distance, and vague clear zones.(3)Seven categories of engineering improvements for enhanced safety were implemented: traffic signs, pavement markings, roadside barriers, intersection channelization, drainage optimization, sight distance improvement, and other treatments.(4)An evaluation of the safety effects of completed treatments was conducted. The EB-based before-after method and the crash severity analysis indicated that the proposed engineering treatments had significant effects on enhancing safety.

CONFLICT OF INTERESTS

The authors declare that there is no conflict of interests regarding the publication of this paper.

ACKNOWLEDGMENTS

This work was financially supported by the Grant from the National High Technology Research and Development Program of China (863 Program, no. 2014AA110304) and the Scientific Research Fund of Heilongjiang Provincial Education Department of China (no. 12541650).

REFERENCES

1. J. S. Bagdade, K. Lariviere, and D. Morena, "Michigan's senior driver showcase corridor: implementation of low-cost safety improvements for senior drivers," Institute of Transportation Engineering, vol. 76, no. 1, pp. 37–42, 2006.
2. S. Labi, "Efficacies of roadway safety improvements across functional subclasses of rural two-lane highways," Journal of Safety Research, vol. 42, no. 4, pp. 231–239, 2011.
3. W. H. Schneider, P. T. Savolainen, and D. N. Moore, "Effects of horizontal curvature on single-vehicle motorcycle crashes along rural two-lane highways," Transportation Research Record, no. 2194, pp. 91–98, 2010.
4. K. M. Bauer and D. W. Harwood, "Safety effects of horizontal curve and grade combinations on rural two-lane roads," in Proceedings of the 92nd Transportation Research Board Annual Meeting, pp. 1–22, Washington, DC, USA, 2013.

5. F. Gross and P. P. Jovanis, “Estimation of safety effectiveness of changes in shoulder width using case-control and cohort methods,” in Proceedings of the 86th Transportation Research Board Annual Meeting, pp. 1–18, Washington, DC, USA, 2007.

6. F. Gross, P. P. Jovanis, and K. Eccles, “Safety effectiveness of lane and shoulder width combinations on rural, two-lane, undivided roads,” Transportation Research Record, vol. 2103, pp. 42–49, 2009.

7. S. Schrock, R. Parsons, and H. Zeng, “Estimation of safety effectiveness of widening shoulders and adding passing lanes on rural two-lane roads,” Transportation Research Record, no. 2203, pp. 57–63, 2011.

8. E. S. Park, P. J. Carlson, R. J. Porter, and C. K. Andersen, “Safety effects of wider edge lines on rural, two-lane highways,” Accident Analysis and Prevention, vol. 48, pp. 317–325, 2012.

9. C. Lyon, B. Persaud, N. Lefler, D. Carter, and K. A. Eccles, “Safety evaluation of installing center two-way left-turn lanes on two-lane roads,” Transportation Research Record, no. 2075, pp. 34–41, 2008.

10. B. Persaud, C. Lyon, F. Gross, and K. Eccles, “Safety evaluation of offset improvements for left-turn lanes,”Transportation Research Record, no. 2171, pp. 44–51, 2010.

11. T. Moreno, C. Llorca, A. Garcia, and A. M. Perez-Zuriaga, “Operational effectiveness of passing zones depending on their length and traffic volume,” in Proceedings of the 92nd Transportation Research Board Annual Meeting, pp. 1–16, Washington, DC, USA, 2013.

12. F. Yuan, J. N. Ivan, X. Qin, N. W. Garrick, and C. F. Davis, “Safety benefits of intersection approach realignment on rural two-lane highways,” Transportation Research Record, no. 1758, pp. 21–29, 2001.

13. L. Yuan and J. Lu, “Safety evaluation and improvements for highway intersections,” Transportation Research Record, no. 2060, pp. 46–52, 2008.

14. Montella and F. Mauriello, “Procedure for ranking unsignalized rural intersections for safety improvement,” Transportation Research Record, no. 2318, pp. 75–82, 2012.

15. Cruzado and E. T. Donnell, “Evaluating effectiveness of dynamic speed display signs in transition zones of two-lane, rural highways in Pennsylvania,” Transportation Research Record, no. 2122, pp. 1–8, 2009.

16. Li and S. S. Washburn, “Improved operational performance assessment for two-Lane highway facilities,” Journal of Transportation Engineering-ASCE, vol. 140, no. 6, pp. 1–9, 2014.

17. Z.-D. Zhu, Y.-C. Lu, C.-H. Fu, and T. Xu, “Research on the safety audit methods for two-lane highway based on HRV,” Mathematical Problems in Engineering, vol. 2014, Article ID 308028, 6 pages, 2014.

18. B. Persaud, C. Lyon, and T. Nguyen, “Empirical Bayes procedure for ranking sites for safety investigation by potential for safety improvement,” Transportation Research Record, no. 1665, pp. 7–12, 1999.

19. R. O. Mujalli and J. De Oña, “A method for simplifying the analysis of traffic accidents injury severity on two-lane highways using Bayesian networks,” Journal of Safety Research, vol. 42, no. 5, pp. 317–326, 2011.

20. E. I. Vlahogianni and J. C. Golias, “Bayesian modeling of the microscopic traffic characteristics of overtaking in two-lane highways,” Transportation Research Part F: Traffic Psychology and Behaviour, vol. 15, no. 3, pp. 348–357, 2012.

21. Q. Wu, F. Chen, G. H. Zhang, et al., “Mixed logit model-based driver injury severity investigations in single- and multi-vehicle crashes on rural two-lane highways,” Accident Analysis and Prevention, vol. 72, pp. 105–115, 2014.

CHAPTER 17 COMPARISON OF FUZZY AHP AND FUZZY TOPSIS FOR ROAD PAVEMENT MAINTENANCE PRIORITIZATION: METHODOLOGICAL EXPOSITION AND CASE STUDY

Yashon O. Ouma, J. Opudo, and S. Nyambenya

INTRODUCTION

Road pavement condition survey and assessment are an important component in the decision making procedure in pavement management systems (PMS). The two main reasons for the assessment of pavement condition are to recognize and prioritize maintenance requirements in specific road sections and to identify the pavement network conditions and, if any, the rehabilitation requirements [1]. In urban road maintenance and rehabilitation (M&R) projects, the evaluation of road pavement conditions is the initial phase that guarantees suitable and sustainable maintenance programs [2].

For pavement M&R projects, specific pavement maintenance management systems (PMMS) should be proposed, as suitable systems for the improvement of the efficiency of decision making. Furthermore, such systems should be able to provide feedbacks on the impacts of the decisions in terms of short and long term maintenance strategies and with respect to cost-benefit analysis [3]. A typical structure that integrates PMS, PMMS, and M&R is depicted in Figure 1.

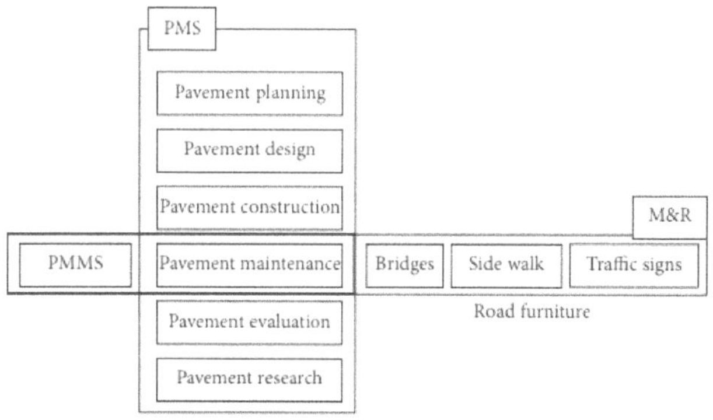

Figure 1: Schematic structure of a typical pavement management system (PMS) and pavement maintenance and management system (PMMS).

From Figure 1, it is seen that a PMS is supposed to be a coordinated and systematic process for carrying out all activities related to providing and maintaining pavements. On the other hand a PMMS should be able to predict the pavement condition and the cost associated with its M&R over a given time frame and also to aid in the planning and programming of the associated M&R works. In implementing PMMS, multicriteria priority ranking and analysis can be used to rank and select pavement sections due for maintenance in the order of urgency. PMMS is considered to be a timely, cost-beneficial, and preventive M&R approach in the maintenance as compared to the traditional reactive M&R approach.

Pavement Maintenance Prioritization Using Multicriteria Decision Approach. The main multicriteria decision making approach is the Analytic Hierarchy Process (AHP) which was

Citation: Yashon O. Ouma, J. Opudo, and S. Nyambenya, "Comparison of Fuzzy AHP and Fuzzy TOPSIS for Road Pavement Maintenance Prioritization: Methodological Exposition and Case Study," Advances in Civil Engineering, vol. 2015, Article ID 140189, 17 pages, 2015. doi:10.1155/2015/140189.

introduced by Saaty in the 1970s [4, 5]. Saaty developed AHP to primarily meet the complexities and challenges of decision situations that are brought out by multiple and conflicting criteria as a multicriteria decision making (MCDM) tool. AHP aids decision makers in determining the best criteria out of multiple sets of criteria, based on crisp or exact numerical values.

On road pavement maintenance priority ranking and analysis, AHP has been widely used (e.g., [6, 7]). However, due to the fact that in most practical situations the human preference is marred by uncertainty, decision makers would find it difficult to explicitly assign exact numerical values to the comparison judgments [8]. Therefore the decision maker would theoretically find it very difficult to express the strength of his preferences and confidence in terms of the AHP in pairwise comparison judgments. As such AHP has been argued to be ineffective when employed to ambiguous or vaguely represented real-life problems that consist of uncertainty and subjectivity [9].

In decision making processes, uncertainty, imprecision, and subjectivity can be handled by means of probability theory and or fuzzy set theory (FST) [10]. While probability focuses on the stochasticity of the decision making process, FST tends to formalize the subjective and imprecise nature of human behavior by representing and quantifying vague information through a membership grade function [10]. As such, probabilistic theory is not adequately suited for subjectivity and imprecision in the human decision making process [11–14].

The advantage of fuzzy based pairwise comparison is that it allows decision makers to be more flexible in their judgments. This is attained through the different degrees of fuzzification. Furthermore if there is need for consideration of the attitude toward risk, FTS has the possibility of overlapping preferences of criteria in case the expert is not confident about the degree of importance within a set of decisions and there is the provision for interval decision as expressed by a fuzzy membership function [15]. FST can also be considered useful in decision problems where the degrees of uncertainty are expected to increase or change in time, for example, in pavement failure and deterioration, where different conditions are expected to change with time, but there is no predictive information on the future state.

In order to accommodate uncertainty in experts' heuristic and experiential judgments, Shah et al. [3] used the ranking of road maintenance factors based on subjective rating. And to take care of subjectivity in judgment, fuzzy pairwise comparison deduction technique can be integrated with AHP [3, 16, 17]. Used independently, fuzzy AHP results are directly ranked according to the normalized weights. However, in real life decision, decision makers' degrees of confidence and attitudes to risks should be taken into consideration.

Fuzzy Technique for Order Preference by Ideal Situation (TOPSIS) technique operates on the basis that the most preferred decision alternative should not only have the shortest distance from a positive ideal solution (PIS), but also have the farthest distance from the negative ideal solution (NIS), hence its capability and efficiency in handling uncertainty [18, 19]. However the use of Euclidean Distance as determined from PIS and NIS in TOPSIS does not consider the correlation of attributes and that makes it difficult to derive the weights of the decision attributes while keeping the judgment consistency [20]. Thus the integration fuzzy TOPSIS with AHP is considered in this study, in addition to taking into account the degree of confidence in decision making.

In this study, fuzzy approach, allowing experts to use linguistic variables, is considered for prioritization by applying and comparing the results from fuzzy AHP and fuzzy TOPSIS. The classical AHP is utilized to determine the weight vector of road maintenance factors with respect to the set objectives. In principle, therefore, the two approaches differ primarily in the way the fuzzy ratings are aggregated and normalized but do impact on the data collection procedure.

First, a methodological overview and approach for fuzzy AHP and fuzzy TOPSIS for the formulation and automation of cost-effective PMMS for consistent and reliable prioritization of M&R of road pavements in urban areas are presented. In the implementation, a case study in Eldoret Town in Kenya is used. For thematic display of the road network, a graphical user interface was designed in GIS and SQL database management systems. In summary, Figure 2 presents the schematic framework for the methodological approach in this study.

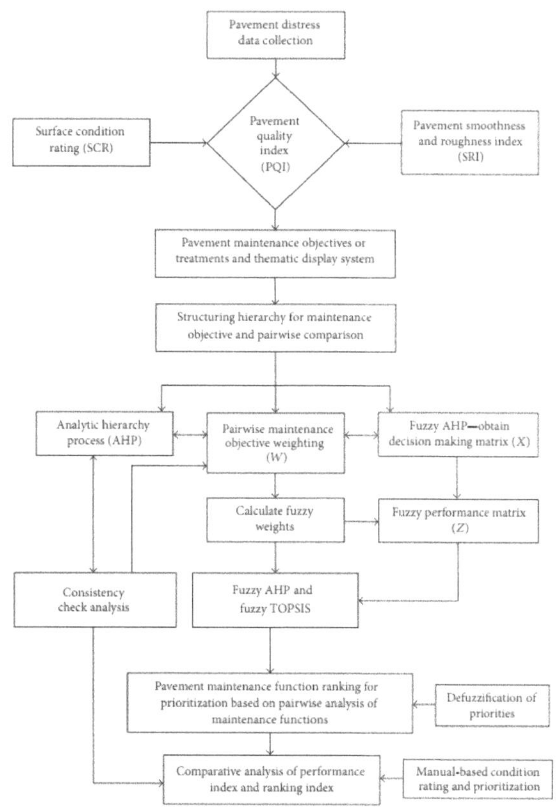

Figure 2: Schematic framework for road pavement prioritization using fuzzy AHP and fuzzy TOPSIS.

The rest of the paper is organized as follows. Section 2 is on the case study pavement distress data collection, cataloging, and display. Section 3 presents the methodological overview and approach based on fuzzy AHP and fuzzy TOPSIS. Section 4 presents the results of the case study, and Section 5 presents the analysis and discussions of the results. The study conclusions and recommendations are presented in Section 6.

CASE STUDY PAVEMENT DISTRESS CATALOGUE AND DISPLAY

There are two main approaches for the collection of pavement distress data: manual (visual inspection) distress data collection and automated distress data collection [21]. For cost-effective and simplified pavement condition surveys, this study adopted empirically manual-based inspection approach to identify the pavement failures or distresses. GPS was used to obtain the chainages of the observed pavement failures and to locate the associated pavement distresses. A summary of the structure for the pavement distress catalogue and display system is presented in Figure 3. The result of this stage of mapping is presented in Figure 4 for the case study of Nandi road in Eldoret Town (Kenya).

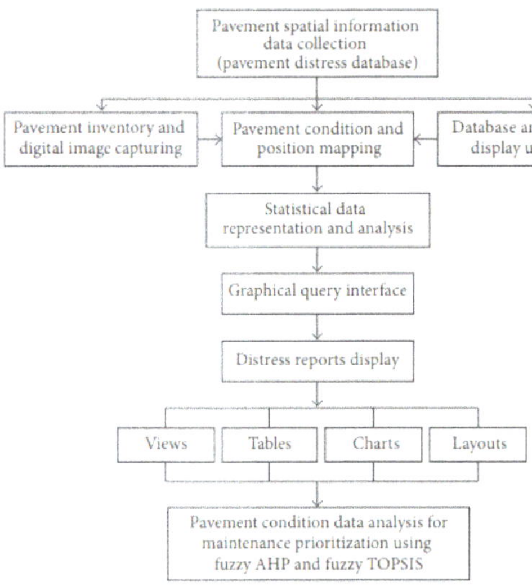

Figure 3: Proposed framework for pavement inventory and condition survey thematic display system.

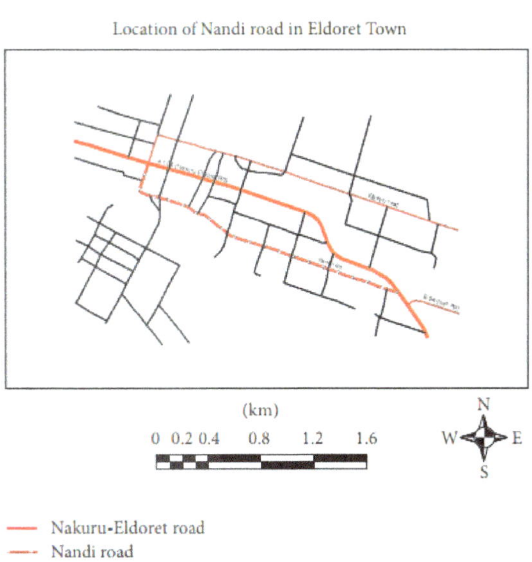

Figure 4: Study road and the adjoining roads digitized and georeferenced from Google Maps image and GPS field observations.

Pavement Distress Inventorying and Failure Index

Since the collection of detailed pavement condition data is very costly and time-consuming, innovative approaches for rapid data collection are in increasing demand among highway agencies with limited PMS budgets [22]. For the case study, visual based empirical inspection and rating (ECR), for distresses data collection, was adopted for the 3 km Nandi road, with observations made at 300 m segments. For each inspected segment of the road, a digital image of the observed distresses was also captured for thematic display.

In order to determine the degree of distress failure two approaches are used. One is a surface condition rating (SCR) to define the failure criteria for cracking and pothole based distresses. According to the proposed SCR, a rating scale of 0–3 was adopted, with a new road surface (no pothole, no cracking) having an SCR of 3. As the amount and severity of the defect increase, the SCR also drops. That implies that low cracking has an SCR of 3, while medium to severe cracking ranges from 2 to 0. In terms of the defect extent, defined by the depth, area, and/or length, an SCR of 3 implies negligible defects, while SCR index values of 2–0 imply wide or long defects that require repairs.

The second component of pavement failure is the pavement surface ride quality. This was characterized by patching and raveling distresses indicators and expressed in terms of the degree of smoothness or roughness index (SRI). The SRI is also ranked between 3 and 0, such that a new road pavement would have an SRI of 3. SRI indices of between 2 and 1 imply that the road has deteriorated to the extent where motorists feel that it is uncomfortable driving on and therefore needs rehabilitation. An SRI of 0 then means that there is complete surface failure requiring major rehabilitation, as is characterized by continuous cracking, potholes, patching, and raveling distresses. The pavement distresses that make up the general pavement condition are determined by the expert judges comprising of pavement engineers and road inspectors from the road management department.

From the experts' judgments and analyses, it was deduced that the overall road pavement quality index (PQI) indicator is defined by the geometric mean of SCR and SRI as expressed in (1). In (1), PQI values range from 3 to 0 with a PQI value of between 2 and 0 implying that the pavement is subject to maintenance and/or rehabilitation:

$$\mathrm{PQI} = \sqrt{\mathrm{SCR} * \mathrm{SRI}}. \tag{1}$$

In the implementation, it is recommended that PQI indices should not be rounded off; that is, a PQI of $\sqrt{1 * 3} = \sqrt{3} = 1.732$ is not equivalent to 2. This precaution is taken in order to avoid minor repair maintenances at any given time in the pavement life-span for economic reasons, at the expense of more immediate and larger M&R cases. From the PQI, the road pavement maintenance objectives and functions can then be derived.

A summary of the observed distress results for the 3 km road is presented in Table 1. The results in Table 1show that the distresses were mainly characterized by cracking, potholes, patching, and raveling. The results in Table 1 will be subjected to pairwise comparison and prioritization by the pavement management experts in order to determine the priority areas for M&R activities. Figure 5 characteristically shows some of the functional and structural failures that are observed on the case study road, as summarized in Table 1.

Table 1: Road condition survey data showing pavement segment chain age and the observed distresses, as derived from the PQI calculations.

Segment or chainage (m)	Observed distresses
0–300	(i) Potholes (ii) Longitudinal cracks (iii) Transverse cracks (iv) Alligator cracks
300–600	(i) Longitudinal cracks (ii) Transverse cracks (iii) Block cracking
600–900	(i) Potholes (ii) Longitudinal cracks (iii) Transverse cracks (iv) Alligator cracks (v) Patching
900–1 200	(i) Potholes (ii) Longitudinal cracks (iii) Alligator cracks
1 200–1 500	(i) Potholes (ii) Longitudinal cracks (iii) Edge cracking (iv) Raveling
1 500–1 800	(i) Potholes (ii) Longitudinal cracks (iii) Transverse cracks (iv) Block cracking
1 800–2 100	(i) Potholes (ii) Longitudinal cracks (iii) Alligator cracks
2 100–2 400	(i) Potholes (ii) Longitudinal cracks (iii) Transverse cracks (iv) Block cracking
2 400–2 700	(i) Potholes (ii) Longitudinal cracks (iii) Alligator cracks
2 700–3 000	(i) Potholes (ii) Longitudinal cracks (iii) Transverse cracks (iv) Alligator cracks (v) Block cracking (vi) Raveling

Figure 5: Photographs of typically observed road pavement distresses.

METHODOLOGY

AHP for Decision Prioritization

AHP is often used to guide decision makers coping with multicriteria decisions involving multiattribute situations [5, 23–25]. Decision judgments as articulated by pairwise comparisons are the fundamental inputs for facilitating the AHP procedure. Every pairwise comparison results in a numerical value a_{ij} representing the ratio between the weights of the two decision criteria, i and j. The Saaty preference scale is used to assign numerical values to different levels of preferences. As a standard, the Saaty scale used for AHP ranges from 1 to 9 and reflects the importance of one factor over another as represented in Table 2.

Table 2: The nine-point Saaty scale of relative importance [4].

Intensity of importance	Definition	Explanation
1	Equal importance	Two activities contribute equally to the objective.
3	Weak importance of one over the other	Experience and judgment slightly favour one over the other.
5	Essential or strong importance	An activity is strongly favoured and its dominance is demonstrated in practice.
7	Demonstrated importance	The evidence favouring over another is of highest possible order of affirmation.
9	Absolute importance	When compromise is needed.
2, 4, 6, 8	Intermediate values	

Supposing that $C = \{C_j | j= 1, 2. . . n\}$ is the set of criteria, then the evaluation matrix can be obtained, in which every element a_{ij} $(i, j = 1, 2, . . . , n)$ represents the relative weights of the criteria illustrated:

$$A = \begin{bmatrix} a_{11} & a_{12} & \cdots & a_{1n} \\ a_{21} & a_{22} & \cdots & a_{2n} \\ \vdots & \vdots & \vdots & \vdots \\ a_{n1} & a_{n2} & \cdots & a_{nn} \end{bmatrix}. \quad (2)$$

In (2), n is the number of criteria in a decision process, such that A is the judgment which contains the pairwise comparison value a_{ij} for all $i, j \in \{1, 2. . . n\}$.

For multiple decision makers or experts as is often the case, if h is the number of decision makers and a_{ij}^k is the pairwise comparison value of criteria i and j given by decision maker k, where $k = 1, 2, . . . , h$, then by using geometric mean of the a_{ij}^k conducted by each decision maker, a new compound judgment matrix is derived, with the corresponding elements determined as follows:

$$a_{ij} = \left(a_{ij}^1 * a_{ij}^2 * \cdots a_{ij}^k * \cdots a_{ij}^h\right)^{1/h} = \left(\prod_{k=1}^{h} a_{ij}^k\right)^{1/h}. \quad (3)$$

The basic procedure for AHP approach as represented by the mean of normalized values method can be summarized in the following steps according to Saaty [26]: (i) normalization of each column to get a new judgment matrix (A'), (ii) summing up of each row of normalized judgment matrix (A') to get weight vector (V), and (iii) finally defining the final normalization weight vector (W). Detailed and stepwise implementation of the AHP techniques can be found in Ouma and Tateishi [25].

According to the relative importance in Table 2, a decision maker who, for example, assigns a value of 3 to a pairwise comparison implies that the judgment slightly favours one alternative over the other. However, because of the subjective nature in linguistic decisions (e.g., very high, high, moderate, low, and very low), such an assessment cannot be accurate. Furthermore, when the decision maker's judgments are uncertain, determining such precise Saaty's static nine-point scale values may be very difficult. This means that the static crisp values may lack the ability to capture the decision makers' ambiguous and/or vague preferences. The logical way to overcome this limitation is to define the comparison ratios in terms of fuzzy numbers.

Fuzzy Sets Theory and Fuzzy AHP

In contrast to classical set theory, FST allows for the assessment of the membership of elements with respect to a set represented by μ= [0, 1] [10]. The fuzzy AHP is an extension of AHP technique for multicriteria decision making in pairwise based comparisons that deals with qualitative and imprecise real-world decisions problems. In essence fuzzy AHP is a synthetic extension of the classical AHP method when the fuzziness of the decision maker's is considered. Fuzzy AHP is comprised of the following steps in implementation: (i) structuring of the levels of hierarchy for decision making, (ii) prioritization based on fuzzy pairwise comparisons, (iii) checking for consistency of the preference judgments by the experts, (iv) synthesis of pairwise priorities, and (v) defuzzification of the determined priorities according to Chen and Hsieh [27]. Defuzzification is necessary because the fuzzy arithmetic means are not crisp values and hence cannot directly be ranked. Defuzzification thus refers to the fuzzy ranking method that is employed so as to obtain the desired no fuzzy or crisp values [28].

A fuzzy AHP decision problem consists of a number of alternatives $[M_i\,(i = 1, 2. . . m)]$, a set of evaluation criteria $[C_j\,(j = 1, 2. . . n)]$, a linguistic judgment (r_{ij}) representing the relative importance of each pair of criteria, and a weighting vector $[W = (w_1, w_2,...)]$. Like the classical AHP, fuzzy AHP also has a judgment matrix. However the extent fuzzy approach uses the triangle fuzzy numbers (TFNs) instead of a constant pairwise comparison value [29]. The extent analysis based fuzzy AHP depends on the degree of possibilities of each criterion used in the decision process. TFNs for the linguistic variables from the judgments are placed according to the responses and for a particular level on the decision hierarchy, from which the pairwise comparison matrix is constructed [30, 31].

According to Kaufmann and Gupta [32], a fuzzy number $\tilde{F}$ is a triangular fuzzy number if its membership function $\mu_{\tilde{F}}(x) : \Re \rightarrow [0, 1]$, is defined according to (4), and is depicted in Figure 6. As in (4), the comparison ratios between the success factors i and j are characterized with TFN which describes the judgment about a_{ij} and are denoted by $\tilde{a}_{ij}$. Hence it is possible to describe some degree of vague or ambiguous human perception and decision in pairwise comparisons. In Figure 6(a), a TFN is denoted by either $(m_1/m_2, m_2/m_3)$ or (m_1, m_2, m_3). For a fuzzy event, the parameters m_1, m_2, and m_3 in Figure 6(a), respectively, denote the smallest possible value, the most promising value, and the largest possible value. Each TFN has linear representations on its left and right sides such that its membership function can generally be defined as follows:

$$\mu_{\tilde{A}}(x) = \begin{cases} \left(x - \dfrac{l}{m} - l\right), & l \leq x \leq m, \\ \left(u - \dfrac{x}{u} - m\right), & m \leq x \leq u, \\ 0, & \text{otherwise}, \end{cases} \tag{4}$$

Where l $(= m_1)$ and u $(= m_3)$, respectively, represent the lower and upper bounds of the fuzzy number $\tilde{A}$, while m $(= m_2)$ is the median value as shown in Figure 6(a). Thus a TFN is generally denoted by $\tilde{A} = (l, m, u)$.

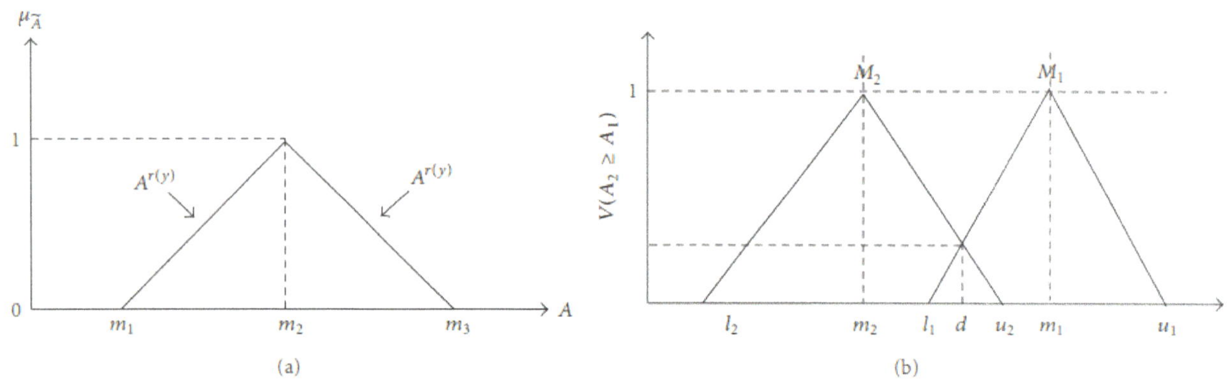

Figure 6: A triangular fuzzy number, M, and (b) the intersection between M_1 and M_2 in extent fuzzy analysis.

As shown in Figure 6(a), a fuzzy number can be given by its corresponding left and right representation of each degree of membership, defined by

$$\begin{aligned} \tilde{A} &= \left(A^{l(y)}, A^{r(y)}\right) \\ &= \left(m_1 + (m_2 - m_1)\,y, m_3 + (m_2 - m_3)\,y\right), \\ &\qquad \forall y \in [0, 1], \end{aligned} \tag{5}$$

where $l(y)$ and $r(y)$ denote the left side representation and the right side representation of a fuzzy number, respectively. Matrix $\tilde{A}$ represents an $(n \times n)$ judgment matrix containing TFNs $(\tilde{a}_{ij})$ for all $i, j \in \{1, 2... n\}$, as expressed in the following:

$$\tilde{A} = (\tilde{a}_{ij}) = \begin{bmatrix} (1,1,1) & \tilde{a}_{12} & \cdots & \tilde{a}_{1n} \\ \tilde{a}_{21} & (1,1,1) & \cdots & \tilde{a}_{2n} \\ \vdots & \vdots & \vdots & \vdots \\ \tilde{a}_{n1} & \tilde{a}_{n2} & \cdots & (1,1,1) \end{bmatrix}. \tag{6}$$

In general, $\tilde{a}_{ij} = (l_{ij}, m_{ij}, u_{ij})$, with l_{ij} being the lower limit, u_{ij} the upper limit, and mij the most likely value, such that $m_{ij} = \sqrt{l_{ij} * u_{ij}}$. Assume that $\tilde{A}_1$ and $\tilde{A}_2$ are two triangular fuzzy numbers with $\tilde{A}_1 = (l_1, m_1, u_1)$ and $\tilde{A}_2 = (l_2, m_2, u_2)$. Some of the fuzzy functional operators (additive, multiplicative, and inverse) are represented in the following:

$$\tilde{A}_1 \oplus \tilde{A}_2 = (l_1 + l_2, m_1 + m_2, u_1 + u_2), \tag{7}$$

$$\tilde{A}_1 \otimes \tilde{A}_2 = (l_1 l_2, m_1 m_2, u_1 u_2), \tag{8}$$

$$\tilde{A}_1^{-1} = \left(\frac{1}{u_1}, \frac{1}{m_1}, \frac{1}{l_1}\right). \tag{9}$$

The outcome of each set of pairwise comparisons is expressed as a fuzzy positive reciprocal (9) matrix: $\tilde{A} = \{\tilde{a}_{ij}\}$, such that if $\tilde{a}_{ij} = 1$ then $\tilde{a}_{ij} = 1/\tilde{a}_{ji}$ for all $i, j \leq n$, where n denotes the number of alternatives being compared within one set of pairwise comparisons, $\tilde{a}_{ij}$ denotes the importance of alternative i over alternative j, $\tilde{a}_{ij}$ denotes the importance of alternative j over alternative i, and $\tilde{a}_{ij} = (l_{ij}, m_{ij}, u_{ij})$.

The definitions of TFN values in Table 3 are used to facilitate the use of pairwise comparisons in fuzzy based decision processes. TFN $(\tilde{a})$ in Table 3 means that the decision maker thinks the importance ratio of two alternatives is about a. In Table 3(a), particular linguistic assessment terms called fuzzy linguistic variables are introduced to represent the underlying fuzzy numbers employed for factor evaluations. For this study, five fuzzy linguistic variables are used to capture the subjective judgments about the relative importance of a factor versus another. The linguistic variables are summarized in Table 3(a) as equally important, moderately important, more important, strongly important, and extremely important. In Tables 3(a) and 3(b), a summary of the fuzzy linguistic variable set with lower, most likely, and upper values of underlying TFN and their definitions are given.

Table 3: (a) Fuzzy linguistic variables and corresponding fuzzy numbers. (b) Fuzzy number memberships of pairwise comparisons [8].

(a)

Linguistic variable	Fuzzy number	Membership function	Definition
Equally important	$\tilde{1}$	(1, 1, 2)	Practical knowledge and experience imply that factor i is equally important when compared to factor j.
Moderately important	$\tilde{3}$	(2, 3, 4)	Practical knowledge and experience imply that factor i is moderately more important when compared to factor j.
More important	$\tilde{5}$	(4, 5, 6)	Practical knowledge and experience imply that factor i is more important when compared to factor j.
Strongly important	$\tilde{7}$	(6, 7, 8)	Practical knowledge and experience imply that factor i is strongly important when compared to factor j.
Extremely important	$\tilde{9}$	(8, 9, 9)	Practical knowledge and experience imply that factor i is extremely important when compared to factor j and totally outweighs it.

(b)

a	Fuzzy number	$\bar{a}$	Fuzzy number
1	(1, 1, 2)	0.50	(0.33, 0.5, 1)
2	(1, 2, 3)	0.33	(0.25, 0.33, 0.5)
3	(2, 3, 4)	0.25	(0.2, 0.25, 0.33)
4	(3, 4, 5)	0.20	(0.17, 0.2, 0.25)
5	(4, 5, 6)	0.17	(0.14, 0.17, 0.2)
6	(5, 6, 7)	0.14	(0.13, 0.14, 0.17)
7	(6, 7, 8)	0.13	(0.11, 0.13, 0.14)
8	(7, 8, 9)	0.11	(0.11, 0.11, 0.13)
9	(8, 9, 9)	Diagonal elements	(1, 1, 1)

After the decision criteria have been determined as depicted in Figure 6(b), a judgment approach for determining the importance levels of these criteria is established. To evaluate the decision judgments, the decision makers select the related linguistic variables; then for calculations they are converted into the TFN and generalized for such analysis as presented in Table 3. The fundamental step in fuzzy AHP methodology is the prioritization procedure, whereby the prioritization problem is defined as deriving the unknown priority vector from the judgment set $\tilde{A}$. The basic procedure of the fuzzy AHP can be summarized in the following four main steps [33].

Step 1 (developing and structuring of the decision hierarchy). This first step comprises the restructuring the complex decision making problem into a hierarchical structure. The derived structural framework is significant in understanding the interactions amongst the elements involved in each decision level and aids the decision makers in exploring the impacts of the different decision components on the evaluation system.

Step 2 (building of the fuzzy pairwise comparison matrices). First, the relative importance of the criteria is determined through pairwise comparison of the decision criteria. After expert evaluations, the relative importance is transformed into triangular fuzzy numbers [25]. By considering a prioritization problem at a decision level with n elements, each set of comparisons for a given level requires (n − 1)/2 judgments, which are used to construct a positive fuzzy reciprocal comparison matrix $\tilde{A} = \{\tilde{a}_{ij}\}$ as given in (6).

Step 3 (consistency check and fuzzy weights priorities derivation). This step checks for decision consistency and determines the priorities from the pairwise comparison matrices [25]. A fuzzy comparison matrix $\tilde{A} = \{\tilde{a}_{ij}\}$ is consistent if $\tilde{a}_{ik} \otimes \tilde{a}_{kj} \approx \tilde{a}_{ij}$, where $i, j, k = 1, 2, \ldots, n$ [34]. After the consistency check, the fuzzy priorities $\widetilde{w}_i$ are calculated, from which the priority vectors $(w_1, w_2, \ldots, w_n)^T$ are obtained from the comparison matrix by applying a prioritization ranking approach [31].

Step 4 (defuzzification—conversion of the fuzzy weights to crisp weights). Defuzzification can be solved using the graded mean integration representation (GMIR) method by Chen and Hsieh [27] and Hsieh et al. [35]:

$$\widetilde{r}_i = \left(\widetilde{a}_{i1} \otimes \cdots \otimes \widetilde{a}_{ij} \otimes \cdots \otimes \widetilde{a}_{in}\right)^{1/n},$$

$$\widetilde{w}_i = \widetilde{r}_i \otimes \left[\widetilde{r}_1 \oplus \cdots \oplus \widetilde{r}_i \oplus \cdots \oplus \widetilde{r}_n\right]^{-1}, \qquad (10)$$

where $\tilde{a}_{ij}$ is fuzzy comparison value of dimension i to criterion j, $\widetilde{r}_j$ is a geometric mean of fuzzy comparison value of criterion i to each criterion, and $\widetilde{w}_i$ is the fuzzy weight of the ith criterion, which can be indicted by a triangular fuzzy number $\widetilde{w}_i = (lw_i, mw_i, uw_i)$ [35].

Step 5 (calculation and normalization of the weight vector and alternatives ranking). The crisp values are normalized in this stage. This final step aggregates the local priorities obtained at the different levels of the decision structure hierarchy into the composite priorities for the alternatives according to the weighted sum $A_i = \sum_{j=1}^{n} w_j a_{ij}$, where w_j is the weight of criterion j. The higher is the value of A_i, the more preferred is the alternative.

Fuzzy TOPSIS

The fundamental principle of the TOPSIS method is that the chosen alternative should have the shortest distance from the positive ideal solution (PIS) and the farthest distance from the negative ideal solution (NIS). If in a MCDM problem decision process there are n criteria (C_1, C_2,...,C_n) amongst m alternatives (A_1, A_2,...,A_m), then according TOPSIS principle, the performance of the alternative i at the criterion j, is such that the best alternative should have the shortest distance from the PIS and the farthest distance from the NIS. The TOPSIS considers the PIS and NIS as the reference points and does not take into consideration the relative importance of the distances from these points [36].

Fuzzy TOPSIS, where TOPSIS refers to Technique for Order Preference by Similarity to Ideal Situation, is a multicriteria decision evaluation method for selected criteria. The principle of TOPSIS approach is that an alternative that is nearest to the Fuzzy Positive Ideal Solution (FPIS) and farthest from the Fuzzy Negative Ideal Solution (FNIS) is chosen as optimal [37, 38].

The following steps give a summary of the fuzzy TOPSIS multicriteria decision making method [36, 39].

Step 1 (assignment of ratings to the criteria and the alternatives). If there are J possible candidates denoted by $A = \{A_1, A_2,...,A_j\}$, to be evaluated against n criteria $C = \{C_1, C_2,...,C_j\}$, and the criteria weights are denoted by wi $\{i = 1, 2, \ldots, m\}$, then the performance ratings of each decision maker D_k $\{k = 1, 2, \ldots, K\}$ for each alternative A_j $\{j = 1, 2, \ldots, n\}$ with respect to criteria C_i $\{i = 1, 2, \ldots, m\}$ are denoted by $\tilde{R}_k = \tilde{x}_{ijk}$ $(i = 1, 2, \ldots, m; j = 1, 2, \ldots, n; k = 1, 2, \ldots, K)$ with the membership function $\mu_{\tilde{R}_k}(x)$.

Step 2 (compute and aggregate the fuzzy ratings for the criteria and the alternatives). If the fuzzy ratings of all decision makers are described as triangular fuzzy number $\tilde{R}_k = (ak, bk, ck)$, then the aggregated fuzzy rating is given by $\tilde{R} = (a, b, c)$, where $k = 1, 2... K$, where

$$a = \min_k \{a_k\},$$

$$b = \frac{1}{K}\sum_{k=1}^{K} b_k,$$

$$c = \max_k \{c_k\}. \tag{11}$$

If the fuzzy rating and importance weight of the kth decision maker are $\tilde{x}_{ijk} = (a_{ijk}, b_{ijk}, c_{ijk})$ and the corresponding weights are $\tilde{w}_{ijk} = (w_{i1k}, w_{i2k}, w_{i3k})$, then the aggregated fuzzy ratings xij of the alternatives with respect to each criterion are given by $\tilde{x}_{ij} = (a_{ij}, b_{ij}, c_{ij})$, where

$$a_{ij} = \min_k \{a_{ijk}\},$$

$$b_{ij} = \frac{1}{K}\sum_{k=1}^{K} b_{ijk},$$

$$c_{ij} = \max_k \{c_{ijk}\}. \tag{12}$$

The aggregated fuzzy weights $\tilde{w}_{ij}$ of each criterion are then calculated by $\tilde{w}_j = (w_{j1}, w_{j2}, w_{j3})$, where

$$w_{j1} = \min_k \{w_{jk1}\},$$

$$w_{j2} = \frac{1}{K}\sum_{k=1}^{K} w_{jk2},$$

$$w_{j3} = \max_k \{w_{jk3}\}. \tag{13}$$

Step 3 (compute the fuzzy decision matrix). The fuzzy performance/decision matrix for the alternatives $(\tilde{D})$ and the criteria $(\tilde{W})$ is constructed as follows. This is followed by the choice of the appropriate linguistic variables for the alternatives with respect to criteria:

$$\tilde{D} = \begin{array}{c} \\ A_1 \\ A_2 \\ \vdots \\ A_m \end{array} \begin{array}{c} \begin{array}{cccc} C_1 & C_2 & \cdots & C_n \end{array} \\ \begin{bmatrix} \tilde{x}_{11} & \tilde{x}_{12} & \cdots & \tilde{x}_{1n} \\ \tilde{x}_{21} & \tilde{x}_{22} & \cdots & \tilde{x}_{21} \\ \cdots & \cdots & \cdots & \cdots \\ \tilde{x}_{m1} & \tilde{x}_{m1} & \cdots & \tilde{x}_{mn} \end{bmatrix} \end{array}, \tag{14}$$

where $\tilde{x}_{ij} = (1/K)(\tilde{x}_{ij}^1 \oplus \cdots \oplus \tilde{x}_{ij}^k \oplus \cdots \oplus \tilde{x}_{ij}^K)$ with $\tilde{x}_{ij}^k$ being the performance rating of alternative Ai with respect to C_i as evaluated by the kth decision maker, $\tilde{x}_{ij}^k = (l_{ij}^k, m_{ij}^k, u_{ij}^k)$ and $\tilde{W} = (\tilde{w}_1, \tilde{w}_2, \ldots, \tilde{w}_n)$ [40].

Step 4 (normalize the fuzzy decision matrix). The raw data are normalized using linear scale transformation in order to bring the various criteria scales into a comparable scale. The normalized fuzzy decision matrix $\tilde{R}$ is given by

$$\tilde{R} = [\tilde{r}_{ij}]_{m\times n}, \quad i = 1, 2, \ldots, m; \; j = 1, 2, \ldots, n, \tag{15}$$

Where $\tilde{r}_{ij} = (a_{ij}/c_j^*, b_{ij}/c_j^*, c_{ij}/c_j^*)$ and $c_j^* = \max_i c_{ij}$ are called the benefit criteria and $\tilde{r}_{ij} = (a_j^-/c_j^*, a_j^-/b_j^*, a_j^-/a_j^*)$ and $a_j^- = \min_i a_{ij}$ are called the cost criteria.

Step 5 (compute the weighted normalized matrix). The weighted normalized matrix $(\tilde{V})$ for criteria is computed by multiplying the weights $(\tilde{w}_j)$ of evaluation criteria with the normalized fuzzy decision matrix r_{ij}:

$$\tilde{V} = [\tilde{v}_{ij}]_{m\times n} = [\tilde{r}_{ij} (\cdot) w_{ij}],$$

$$i = 1, 2, \ldots, m; \; j = 1, 2, \ldots, n. \tag{16}$$

Step 6 (compute the fuzzy positive ideal solution (FPIS) and fuzzy negative ideal solution (FNIS)). The FPIS and FNIS of the alternatives are computed as follows. From the weighted normalized fuzzy-decision matrix, the elements $\tilde{v}_{ij}$ are normalized positive TFNs within the closed interval [0, 1]. Then, FPIS (A^+) and FNIS (A^-) are calculated [41]:

$$A^+ = (\tilde{v}_1^+, \tilde{v}_2^+, \ldots, \tilde{v}_n^+),$$

$$\text{where } v_j^+ = \max_i \{v_{ij3}\},\ i = 1,2,\ldots,m;\ j = 1,2,\ldots,n,$$

$$A^- = (\tilde{v}_1^-, \tilde{v}_2^-, \ldots, \tilde{v}_n^-),$$

$$\text{where } v_j^- = \max_i \{v_{ij1}\},\ i = 1,2,\ldots,m;\ j = 1,2,\ldots,n. \qquad (17)$$

Step 7 (compute the distance of each alternative from FPIS and FNIS). The distance (d_i^+, d_i^-) of each weighted alternative i = 1, 2... m from the FPIS and the FNIS is computed as follows, using the area compensation method:

$$d_i^+ = \sum_{j=1}^{n} d_v(\tilde{v}_{ij}, \tilde{v}_j^+) = \left[\frac{1}{3}\sum_{j=1}^{n}(\tilde{v}_{ij} - \tilde{v}_j^+)^2\right]^{1/2}, \quad i = 1,2,\ldots,m,$$

$$d_i^- = \sum_{j=1}^{n} d_v(\tilde{v}_{ij}, \tilde{v}_j^-) = \left[\frac{1}{3}\sum_{j=1}^{n}(\tilde{v}_{ij} - \tilde{v}_j^-)^2\right]^{1/2}, \quad i = 1,2,\ldots,m, \qquad (18)$$

where $d_v(\tilde{a}, \tilde{b})$ is the distance measurement between two fuzzy numbers $\tilde{a}$ and $\tilde{b}$.

Step 8 (compute the closeness coefficient (CC^i) of each alternative). The closeness coefficient (CC^i) or the degree of relative gaps that represents the distances to the fuzzy positive ideal solution (A^+) and the fuzzy negative ideal solution (A^-) is computed. The closeness coefficient of each alternative is calculated as

$$CC_i = \frac{d_i^-}{d_i^- + d_i^+} = \left(1 - \frac{d_i^+}{d_i^- + d_i^+}\right), \quad \text{for } i = 1,2,\ldots,m, \qquad (19)$$

where $d_i^-/(d_i^- + d_i^+)$ is the fuzzy satisfaction degree in the ith alternative and $d_i^+/(d_i^- + d_i^+)$ is the fuzzy gap degree in the ith alternative.

Step 9 (ranking the alternatives). In the last step, the different alternatives are ranked according to the closeness coefficient (CC_i) in decreasing order. The best alternative is closest to the FPIS and farthest from the FNIS.

CASE STUDY RESULTS AND ANALYSIS

GIS-SQL Database for Thematic Displays of Pavement Distress Data

Figure 7 shows the typical output of the results for the GIS-SQL based graphical thematic display of the road condition survey for the Nandi road distress between segments 2 100 m and 2 400 m. The segment query results display the following information regarding the segment: the chain age of the road segment surveyed, the distresses observed within the segment, and the distress images. Such pavement condition data representation can easily be leveraged into a web access platform, so as to be easily accessible to the different levels of urban road maintenance and management.

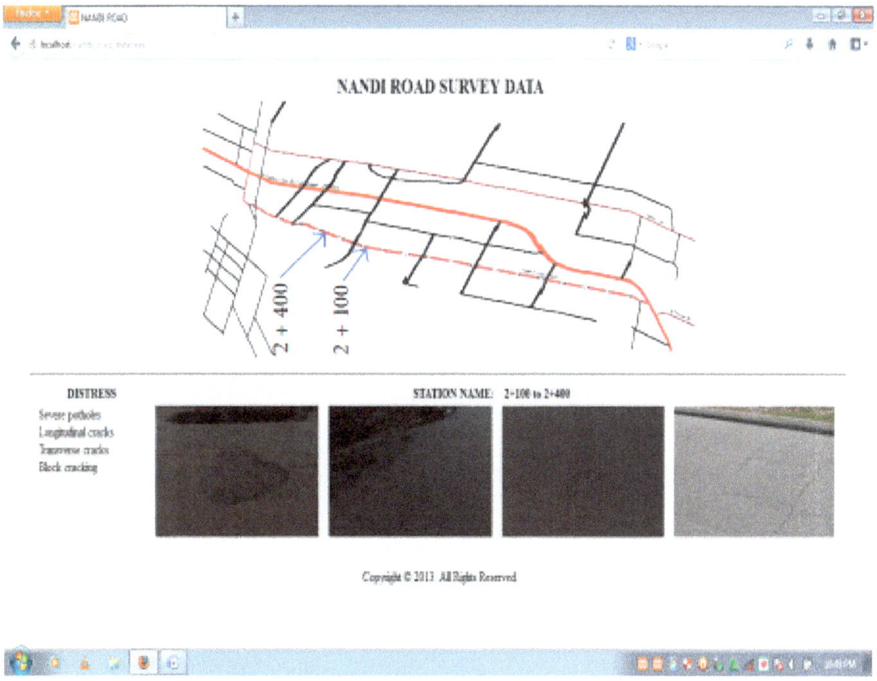

Figure 7: An SQL based thematic graphical display of road pavement distress at specified segments.

A summary statistic on the frequency of occurrence of distress types is also generated for the entire surveyed road pavement as presented in Figure 8. Results in Figure 8 show the most and least frequent distresses on the road as longitudinal cracks and patching, respectively. A plot of the degree of distress that is low, medium, or high can also be derived from the empirically observed and categorized severity of distresses, though such linguistic information is captured in the distress survey by the condition assessment experts.

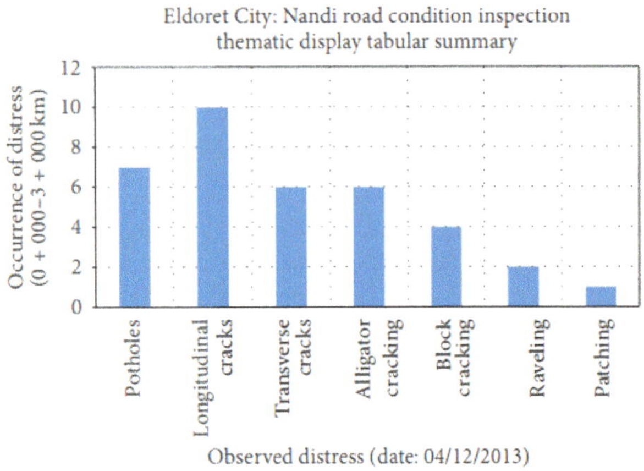

Figure 8: Summary of the distresses report of for the case study.

Case Study Road Maintenance Treatments

From the observed distresses presented in Table 1 and in Figure 5, the possible maintenance treatments and functions are determined, with the results presented in Table 4. From the condition survey results, it was observed that there were seven main pavement distresses or maintenance functions along the 3 km road. From the results in Table 4 it was noted that if the distresses identified in the pavement condition survey are related to structural deficiencies, then the road section is most likely not to be a candidate for a preventive basic maintenance treatment but rather for rehabilitation.

Table 4: Case study maintenance functions for the observed distresses.

Distress treatment (maintenance function) for observed distress	Fog seal	Micro surfacing	Slurry seal	Cape seal	Chip seal	Thin HMA overlay	Resurfacing
Rutting		×				×	
Alligator and block cracking	×	×		×	×	×	×
Longitudinal and transverse cracking		×		×	×	×	
Pothole		×				×	×
Raveling	×	×	×	×	×		
Resurfacing		×				×	×

The results in Table 4 are summarized as follows: (i) fog seal: it is an application of diluted emulsion (normally 1 to 1) to enrich the pavement surface and hinder raveling and oxidation and is considered as a temporary application; (ii) crack sealing: this treatment is used to prevent water and debris from entering cracks in the pavement; the treatment might include routing to clean the entire crack and to create a reservoir to hold the sealant; (iii) chip seal: this treatment is used to waterproof the surface, seal small cracks, and improve friction; although it is typically used on low volume roads and streets, it can also be used on high volume roads and highways; (iv) thin cold mix seals: these treatments include slurry seals, cape seals, and micro surfacing which are used on all types of facilities to fill cracks, improve friction, and improve ride quality; (v) Thin Hot Mix Asphalt (HMA) overlays: these include dense-, open-, and gap-graded mixes (as well as surface recycling) that are used to improve ride quality, provide surface drainage and friction, and correct surface irregularities; Thin HMA overlays are generally 37 mm in thickness; and (vi) resurfacing: resurfacing of a road pavement in totality or section thereof is normally carried out in order to add strength to the asphalt layer and to prolong its life and or to correct a specific section or the entire surface profile and in order to improve on the riding quality and surface water drainage.

Maintenance Objectives Determination, Weighting, and AHP Hierarchical Structure

To determine the maintenance objectives, from the maintenance functions, ten experts comprising pavement engineers, inspectors, and transport planners from Kenya Urban Roads Authority (KURA) gave their assessments and judgments on the maintenance objectives weighting for routine urban road maintenance. This was carried out by first selecting the maintenance objectives, where the experts were asked to nominate a number of maintenance objectives based on the observed distresses (Table 4). From the expert evaluation results, the top four maintenance objectives are adopted as the case study maintenance criteria.

The selected road maintenance objectives were determined, respectively, as road safety (RS); pavement surface preservation (PSR); road operational status and standards (ROS) or quality; and road aesthetics (RA) (Table 5). After the maintenance objectives are obtained, the weights of the selected maintenance objectives are then determined.

Table 5: Round 1 of decision based on importance of allocation or weighting.

Road pavement maintenance objectives	Pavement expert (engineer/inspector/planner)										Maintenance relative weights
	1	2	3	4	5	6	7	8	9	10	
Road safety	35	45	45	30	40	45	35	50	30	35	0.39
Pavement surface preservation	40	30	20	40	30	25	30	20	30	40	0.31
Road operational status and standards	15	15	20	20	15	20	25	20	20	15	0.19
Road aesthetics	10	10	15	10	15	10	10	10	10	10	0.11

The same group of experts gave their estimates on the four maintenance objectives weighting. Each of the experts assigned weights to the maintenance objectives in three rounds, each round using a different approach comprising the following elements. The rationale of using the three rounds is to vary and objectively, without biases, determine the weights based on different ranking scales, from percentages, incremental decimals linear scale (1.0–5.0), and whole number linear scale (1–5).(i)Round 1: in this round weighting is carried out in terms of percentages and applied to each maintenance objective.(ii)Round 2: by using linear incremental scale with steps of 0.25, the objective with the lowest importance is set at 1.0 and the one with the greatest importance is recorded at 5.0, depending on the function.(iii)Round 3: on a linear scale of crisp values of 1–5 (increasing at intervals of 1), the importance of the objectives is ranked with 1 representing the least important objective and 5 representing the most important. Because of human objectivity, reasoning, and subjectivity, at the end of three rounds, the results will vary based on the individual decision maker's perception of the criterion and in accordance with ranking scale used. At the end, the results are aggregated to minimize on subjectivity and to objectively determine the overall weights. Results for the three ranking steps are presented in Tables 5–7 for the four maintenance objectives. In the pavement expert columns, 1 to 10 represent the ten decision experts.

Table 6: Round 2 of judgment based on incremental linear ranking with 1.0 as the lowest importance.

Road pavement maintenance objectives	Pavement expert (engineer/inspector/planner)										Maintenance relative weights based on linear scale
	1	2	3	4	5	6	7	8	9	10	
Road safety	3	4	5	5	4	2	2	2.5	4	4	0.36
Pavement surface preservation	2.5	3	3	4	4	3	2	4	3	2	0.30
Road operational status and standards	2	3	3	4	3	1.5	2	3	3	1	0.25
Road aesthetics	1	1	1	1	1	2	1	1	1	1	0.12

Table 7: Round 3 of decision according to crisp scale based ranking (with 1 being the least important and 5 the most important).

Road pavement maintenance objectives	Pavement expert (engineer/inspector/planner)										Maintenance relative weights based on 1-5 crisp scale
	1	2	3	4	5	6	7	8	9	10	
Road safety	5	5	4	5	5	4	5	5	5	5	0.39
Pavement surface preservation	4	4	3	4	5	4	4	5	4	4	0.32
Road operational status and standards	3	2	3	3	4	3	2	3	2	3	0.22
Road aesthetics	1	1	1	1	1	2	1	1	1	1	0.07

From the averaged AHP and fuzzy weights based evaluation matrices represented, respectively, in Tables 8 and9, the final aggregated weighting vector was determined as s $W = [0.37, 0.31, 0.22, 0.10]^T$ for the four maintenance objectives and is presented in Table 10. The results in Table 10 show that, according to the experts, road safety and surface condition accounted for 68% of the maintenance objectives preferences, while quality and aesthetics together contributed to 32% of the maintenance preferences.

Table 8: Averaged AHP based evaluation matrix for the maintenance objectives.

Road maintenance objectives	Road safety	Pavement surface preservation	Operational status and standards	Road aesthetics
Road safety	1	2	1	2
Pavement surface preservation	0.5	1	0.5	0.75
Operational status and standards	1	2	1	1
Road aesthetics	0.5	1.33	1	1

Table 9: Equivalent fuzzy based evaluation matrix for maintenance objectives.

Road maintenance objectives	Road safety	Pavement surface preservation	Operational status and standards	Road aesthetics
Road safety	(1, 1, 1)	(1.33, 2, 4)	(0.8, 1, 1.33)	(1, 2, 3)
Pavement surface preservation	(0.33, 0.5, 0.75)	(1, 1, 1)	(0.25, 0.5, 1)	(0.5, 0.75, 1)
Operational status and standards	(0.75, 1, 1.25)	(1, 2, 3)	(1, 1, 1)	(0.5, 1, 1.25)
Road aesthetics	(0.33, 0.5, 1)	(1, 1.33, 2)	(0.8, 1, 1.33)	(1, 1, 1)

Table 10: Average maintenance objectives weighting vector (W) for the case study.

Road maintenance objectives	Road safety	Pavement surface preservation	Operational status and standards	Road aesthetics
Mean relative maintenance objective weights	0.37	0.31	0.22	0.10

From the above results the final AHP hierarchical structure of the maintenance prioritization levels is derived for the case study as presented in Figure 9. As already stated, hierarchical structuring of a pavement maintenance prioritization problem is an efficient way of dealing with the decision complexity and is useful in identifying the major components of the PMMS. From Figure 9, it is seen that the structure for the case study comprises three levels, with the final level being the overall goal, and the other two levels: the first, lower hierarchy, is the maintenance functions, and the second, higher hierarchy, is the maintenance objectives. For realistic cost-benefit evaluation, every M&R project should have its own hierarchical structure.

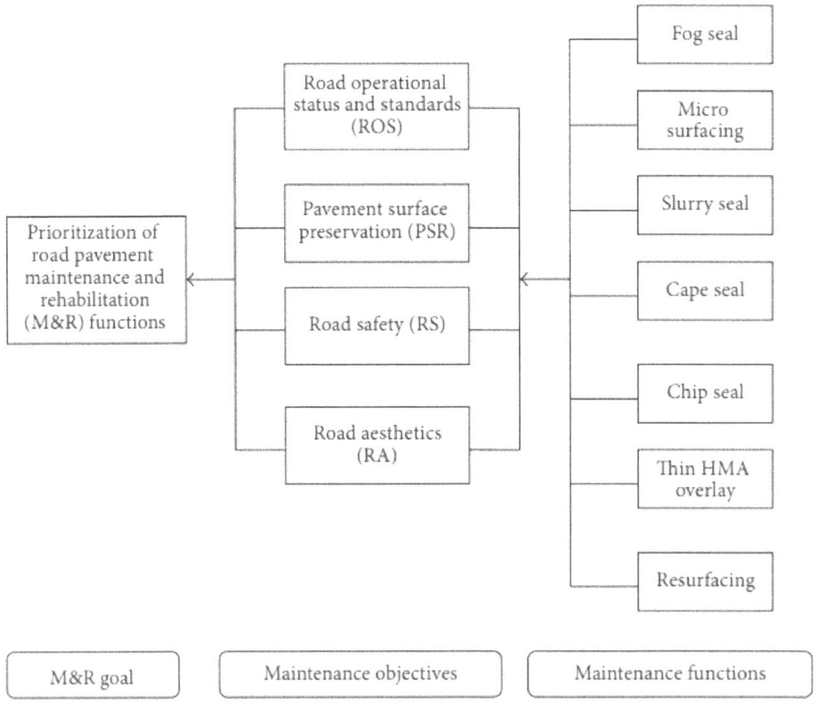

Figure 9: Three-level hierarchical structure of maintenance function prioritization process for the case study road.

Multiattribute Analysis for Pairwise Comparisons of Maintenance Objectives

For the pairwise comparative evaluation, ten pavement engineers from KURA gave their judgments using pairwise comparisons of the four maintenance objectives. All the ten engineers gave pairwise comparisons to the seven maintenance functions with respect to the predetermined four maintenance objectives. The pairwise comparisons are based on nine-point Saaty scale of relative importance [26] as presented in Table 2. Using the fuzzification and defuzzification rules as presented in Table 3, substitutes for the crisp numbers and the fuzzy numbers are obtained.

The optimal level of confidence is the degree of accuracy with which one is able to execute a given decision making task. That is, due to the subjective nature of the exercise, an optimal level from experienced researchers is empirically considered to be neither too high nor too low, and therefore a value above mean threshold of 60% is chosen. On the other hand, the attitude towards risk is the effect of executing the particular task within an error margin. For this case study, the error margin for the decision makers is set a maximum of 50% chance, and the errors are attributed to mistakes (gross), systematic and random occurrences. This means that an optimal performer should be able to execute the task with an accuracy of at least 60% and is liable to maximum error level of 50%.

By choosing an optimal level of confidence of (α = 60%) and assuming a moderate attitude to risk of ($\lambda = 0.5$), the fuzzy performance matrix (Z) is defuzzified, under the optimal boundary conditions of [$\alpha = 0.6, \lambda = 0.5$]. The matrix is normalized to satisfy the condition that the summation of all the seven maintenance functions under each criterion is 1. Therefore the crisp normalized performance matrix Z_α^λ is obtained as below for the four pavement maintenance objectives and as determined by the ten experts:

The final maintenance functions ranking is then derived and the results are presented in Tables 11 and 12 for fuzzy AHP and fuzzy TOPSIS. The final performance index for each maintenance function is obtained by averaging the respective priorities as determined by each of the decision makers. It is observed that the results in Tables 11 and 12 are nearly the same except for the rakings 5 and 6. Compared with the manual-based evaluations and ranking, which are based on the set out guiding principles such distress type, the dimensions (area, length, and depth), and significance of the road, for the aggregation of individual ranking of each distress and scoring, the fuzzy TOPSIS results in Table 12 were found to be consistent with the prioritization experiences in practice. The top three maintenance functions (Thin HMA overlay, resurfacing, and slurry seal) are all pavement related functions and are related to the pavement deficiencies and have more direct impacts on the road pavement functions.

Table 11: Maintenance function ranking and performance ranking using fuzzy AHP.

Function number	Maintenance function	Performance index	Ranking index
1	Fog seal	0.755	7
2	Micro surfacing	0.798	6
3	Slurry seal	0.870	3
4	Cape seal	0.816	5
5	Chip seal	0.852	4
6	Thin HMA overlay	0.883	1
7	Resurfacing	0.879	2

Table 12: Maintenance function ranking and performance index ranking using fuzzy TOPSIS.

Function number	Maintenance function	$D_{\alpha}^{\lambda+}$	$D_{\alpha}^{\lambda-}$	Separation score CC_i	Ranking
1	Fog seal	0.0580	0.0220	0.2750	7
2	Micro surfacing	0.0497	0.0485	0.4255	5
3	Slurry seal	0.0310	0.0270	0.4655	3
4	Cape seal	0.0459	0.0340	0.3921	6
5	Chip seal	0.0378	0.0291	0.4350	4
6	Thin HMA overlay	0.0347	0.0420	0.5480	1
7	Resurfacing	0.0432	0.0477	0.5248	2

COMPARISON OF FUZZY AHP AND FUZZY TOPSIS RESULTS

To compare variance in the decisions, the fuzzy AHP and fuzzy TOPSIS were plotted on the same scale, by rescaling the performance indices of all the seven maintenance functions to the same scale such that the normalized sum equals 1. Figure 10 presents the results of the comparative ranking indices. The results in Figure 10 show that the fuzzy AHP tends to overestimate the ranking process as compared to the fuzzy TOPSIS and the manual-based decision validation procedure. The results in Figure 10 show that the fuzzy TOPSIS results are quite close to manual-based approach, which implies that the TOPSIS approach is much more able to capture the user degree of confidence. Nonetheless, different case studies can be carried out in order to generalize the effectiveness of two fuzzy based prioritization techniques, especially in different geographical regions.

Figure 10: Comparison of maintenance prioritization results from fuzzy AHP and fuzzy TOPSIS.

There are however similarities and differences in the rankings as obtained from the two prioritization techniques as seen in Figure 10. For example, the top three maintenance functions are the same and only in the bottom three is there a slight difference. The results also show that there is consistency in the fuzzy TOPSIS ranking process, as compared to the variability in ranking values in fuzzy AHP.

Although fuzzy AHP and fuzzy TOPSIS have the same objective of prioritizing the maintenance objectives and functions through MCDM principles, they have fundamental differences. According to fuzzy AHP, pairwise comparisons by decision makers are made for the criteria and alternatives under each criterion. The resulting comparisons are integrated and decision makers' pairwise comparison values are transformed into triangular fuzzy numbers. The priority weights of the criteria and alternatives are then derived. According to the combination of the priority weights of criteria and alternatives, the best alternative is determined.

In fuzzy TOPSIS, the decision makers use the linguistic variables to evaluate each alternative with respect to each criterion in order to assess the importance of the criteria. The linguistic variables are converted into triangular fuzzy numbers, from which the fuzzy decision matrix is determined. The normalized fuzzy decision matrix and the weighted normalized fuzzy decision matrix are then formed. After determining the FPIS and FNIS, the distance (d) of each alternative to FPIS and FNIS is also determined. Therefrom the closeness coefficient of each alternative is individually computed. From the seven alternatives, the closeness coefficient revealed the results in Table 12 for the fuzzy TOPSIS and in Table 11 for the fuzzy AHP.

The fundamental differences between these two methods could be based on the fact that TOPSIS considers the linguistic variables at the onset of the pairwise comparison and therefore the results are not affected by the AHP drawbacks in fuzzy AHP. Nonetheless, according to the results, the top three results were similar for both methods. This in part shows that, apart from the subjectivity that is not critically captured by AHP in the pairwise comparison process, both TOPSIS and AHP as multicriteria decision making techniques are completive in performance. But, as presented in Figure 10, the subjectivity in AHP is noticed when the results are compared on standardized scale.

In overall, the appropriate preventive maintenance strategy will largely be influenced by the type, severity, and extent of the pavement surface distresses and the structural and functional condition of the pavement. In this regard, the influences of the categories of pavement maintenance and the performance of preventive maintenance treatments, as recommended by Johanns and Craig [42], should be taken into consideration in balancing between preventive, corrective, and emergency maintenances schedules.

CONCLUSIONS

The current case study presents a methodological overview on the use of multiattribute decision making using fuzzy AHP and fuzzy TOPSIS for the prioritization of pavement maintenance alternatives. By using different sets of fuzzy membership functions, the current approach enables decision makers to capture the often unpredictable and complex pavement maintenance prioritization procedure, in a more objective way.

For the case study, it is observed that maintenance measures relate to the following pavement failures: cracking, potholes, raveling, and patching. From the prioritization results, and with manual-based opinion, it is concluded that the top three maintenance objective functions in the case study urban roads are characterized by Thin HMA overlay, resurfacing, and slurry seals, for the following pavement maintenance objectives: road safety conditions, pavement surface preservation, road operational status and standards, and the road aesthetics. The results show that road safety is closely related to the actual pavement preservation status, and both contribute in nearly the same ratio to more than 60% of the maintenance objectives.

The ranking procedures yielded nearly similar results, with fuzzy TOPSIS performing slightly better than fuzzy AHP, when the results were compared against actual empirical-manual-based prioritization. Further, the results show that, at optimal degrees of confidence and risk, the fuzzy AHP tended to overestimate the prioritization ranking process, as compared to the fuzzy TOPSIS approach.

To improve on the current study, it is recommended that automated distress surveys and measurements be considered. This is important in improving on the speeds of condition surveys, accuracy and reliability, and cost-effectiveness. Developments on pavement maintenance prioritization approach that can enable decision makers to quantify and minimize the total risk due to canceling or deferring maintenance functions during budget cuts should also be considered in the prioritization process. Additionally, studies on the spatial-temporal modelling of the rate of pavement deterioration are recommended in order to integrate the maintenance prioritization approach with the asset's whole-life economic analysis. And, in addition, detailed studies on the variations and relationships between the decision makers' level of confidence and attitude to risk in multicriteria decision making process can also be a detailed subject of investigation.

CONFLICT OF INTERESTS

The authors declare that there is no conflict of interests regarding the publication of this paper.

ACKNOWLEDGMENTS

The authors would like to acknowledge all those who were involved in the field surveys on the road condition inspection project. Further acknowledgements go to the contributions in terms of comments by the anonymous reviewers who took their time to pinpoint key areas to improve on in this paper.

REFERENCES

1. L. Sun and W. Gu, "Pavement condition assessment using fuzzy logic theory and analytic hierarchy process," Journal of Transportation Engineering, vol. 137, no. 9, pp. 648–655, 2011.
2. S. Cafiso, A. Graziano, and S. Battiato, "Evaluation of pavement surface distress using digital image collection and analysis," in Proceedings of the 7th International Congress on Advances in Civil Engineering, p. 10, Yildiz Technical University, Istanbul, Turkey, October 2006.
3. Y. U. Shah, S. S. Jain, and M. Parida, "Evaluation of prioritization methods for effective pavement maintenance of urban roads," International Journal of Pavement Engineering, vol. 15, no. 3, pp. 238–250, 2014.
4. T. L. Saaty, "A scaling method for priorities in hierarchical structures," Journal of Mathematical Psychology, vol. 15, no. 3, pp. 234–281, 1977.
5. T. L. Saaty, Multi-Criteria Decision Making: The Analytic Hierarchy Process, McGraw-Hill, New York, NY, USA, 1980.
6. P. K. Agarwal, A. Das, and P. Chakroborty, "A rational approach of prioritization of highway sections for maintenance," in Proceedings of the 6th International Conference on Managing Pavements, p. 12, Brisbane, Australia, October 2004.
7. J. Farhan and T. F. Fwa, "Pavement maintenance prioritization using analytic hierarchy process,"Transportation Research Record, vol. 2093, pp. 12–24, 2009.
8. W. Liu and Z. Zhang, "Prioritization of highway maintenance functions using multi-attribute decision making with fuzzy pairwise comparison," Tech. Rep. SWUTC/11/161128–1, Center for Transportation Research, University of Texas, Austin, Tex, USA, 2012.
9. H. Deng, "Multi-criteria analysis with fuzzy pairwise Comparison," International Journal of Approximate Reasoning, vol. 21, no. 3, pp. 215–231, 1999.
10. L. A. Zadeh, "Fuzzy sets," Information and Computation, vol. 8, pp. 338–353, 1965.
11. J. Efstathiou, "Practical multi-attribute decision-making and fuzzy set theory," in TIMS/Studies in Management Science, vol. 20, pp. 307–320, Elsevier, 1984.
12. D. Dubois and H. Prade, "Recent models of uncertainty and imprecision as a basis for decision theory toward less normative frameworks," in Intelligent Decision Support in Process Environments, E. Hollnagel, G. Mancini, and D. Woods, Eds., pp. 3–24, Springer, Berlin, Germany, 1985.
13. S. J. Chen and C. L. Hwang, Fuzzy Multiple Attribute Decision Making: Methods and Applications, Springer, New York, NY, USA, 1992.
14. H.-J. Zimmermann, Fuzzy Set Theory and Its Applications, Kluwer Academic, Boston, Mass, USA, 1996.
15. M. Kordi and S. A. Brandt, "Effects of increasing fuzziness on analytic hierarchy process for spatial multicriteria decision analysis," Computers, Environment and Urban Systems, vol. 36, no. 1, pp. 43–53, 2012.
16. L. Mikhailov, "Deriving priorities from fuzzy pairwise comparison judgements," Fuzzy Sets and Systems, vol. 134, no. 3, pp. 365–385, 2003.
17. D. Moazami and R. Muniandy, "Fuzzy inference and multi-criteria decision making applications in pavement rehabilitation prioritization," Australian Journal of Basic and Applied Sciences, vol. 4, no. 10, pp. 4740–4748, 2010.
18. C. L. Hwang and K. Yoon, Multiple Attribute Decision Making: Methods and Applications. A State-of-the-Art Survey, Springer, New York, NY, USA, 1981.
19. C. L. Hwang and K. S. Yoon, Multiple Attribute Decision Making: Methods and Applications, Springer, Berlin, Germany, 1981.
20. Vega, J. Aguarón, J. García-Alcaraz, and J. M. Moreno-Jiménez, "Notes on dependent attributes in TOPSIS," Procedia Computer Science, vol. 31, pp. 308–317, 2014.
21. K. C. P. Wang, C. Nunn, C. Mackey et al., "Network level crack survey with the automated real-time distress analyzer," in Proceedings of the 83rd Annual Meeting of Transportation Research Board (TRB '03), CD-ROM, p. 22, 2003.
22. N. Bandara and M. Gunaratne, "Current and future pavement maintenance prioritization based on rapid

visual condition evaluation," Journal of Transportation Engineering, vol. 127, no. 2, pp. 116–123, 2001.

23. F. Zahedi, "The analytic hierarchy process: a survey of the method and its applications," Interfaces, vol. 16, no. 4, pp. 96–108, 1996.
24. O. S. Vaidya and S. Kumar, "Analytic hierarchy process: an overview of applications," European Journal of Operational Research, vol. 169, no. 1, pp. 1–29, 2006.
25. Y. Ouma and R. Tateishi, "Urban flood vulnerability and risk mapping using integrated multi-parametric AHP and GIS: methodological overview and case study assessment," Water, vol. 6, no. 6, pp. 1515–1545, 2014.
26. T. L. Saaty, "Decision making with analytic hierarchy process," International Journal of Services Sciences, vol. 1, no. 1, pp. 83–98, 2008.
27. S. H. Chen and C. H. Hsieh, "Representation, ranking, distance, and similarity of L-R type fuzzy number and application," Australian Journal of Intelligent Information Processing Systems, vol. 6, no. 4, pp. 217–229, 2000.
28. M. T. Tang, G. H. Tzeng, and S. W. Wang, "A hierarchy fuzzy MCDM method for studying electronic marketing strategies in the information service industry," Journal of International Information Management, vol. 8, no. 1, pp. 1–22, 1999.
29. D.-Y. Chang, "Applications of the extent analysis method on fuzzy AHP," European Journal of Operational Research, vol. 95, no. 3, pp. 649–655, 1996.
30. R. Aggarwal and S. Singh, "AHP and extent fuzzy AHP approach for prioritization of performance measurement attributes," International Journal of Social, Education, Economics and Management Engineering, vol. 7, no. 1, pp. 43–48, 2013.
31. Y. O. Ouma, C. Yabann, M. Kirichu, and R. Tateishi, "Optimization of urban highway bypass horizontal alignment: a methodological overview of intelligent spatial MCDA approach using fuzzy AHP and GIS,"Advances in Civil Engineering, vol. 2014, Article ID 182568, 26 pages, 2014.
32. Kaufmann and M. M. Gupta, Introduction to Fuzzy Arithmetic Theory and Applications, Van Nostrand Reinhold, New York, NY, USA, 1991.
33. M. B. Javanbarg, C. Scawthorn, J. Kiyono, and B. Shahbodaghkhan, "Fuzzy AHP-based multicriteria decision making systems using particle swarm optimization," Expert Systems with Applications, vol. 39, no. 1, pp. 960–966, 2012.
34. J. J. Buckley, "Fuzzy hierarchical analysis," Fuzzy Sets and Systems, vol. 17, no. 3, pp. 233–247, 1985.
35. T.-Y. Hsieh, S.-T. Lu, and G.-H. Tzeng, "Fuzzy MCDM approach for planning and design tenders selection in public office buildings," International Journal of Project Management, vol. 22, no. 7, pp. 573–584, 2004.
36. G. R. Jahanshahloo, F. H. Lotfi, and M. Izadikhah, "An algorithmic method to extend TOPSIS for decision-making problems with interval data," Applied Mathematics and Computation, vol. 175, no. 2, pp. 1375–1384, 2006.
37. T.-C. Wang and T.-H. Chang, "Application of TOPSIS in evaluating initial training aircraft under a fuzzy environment," Expert Systems with Applications, vol. 33, no. 4, pp. 870–880, 2007.
38. Ertugrul and N. Karakasoglu, "Performance evaluation of Turkish cement firms with fuzzy analytic hierarchy process and TOPSIS methods," Expert Systems with Applications, vol. 36, no. 1, pp. 702–715, 2009.
39. M. Ashrafzadeh, F. R. Mokhatab, and N. I. Mollaverdi, "Application of fuzzy TOPSIS method for selection of warehouse location: a case study," Interdisciplinary Journal of Contemporary Research Business, vol. 3, no. 9, pp. 655–671, 2012.
40. C.-C. Sun, "A performance evaluation model by integrating fuzzy AHP and fuzzy TOPSIS methods,"Expert Systems with Applications, vol. 37, no. 12, pp. 7745–7754, 2010.
41. C.-T. Chen, C.-T. Lin, and S.-F. Huang, "A fuzzy approach for supplier evaluation and selection in supply chain management," International Journal of Production Economics, vol. 102, no. 2, pp. 289–301, 2006.
42. M. Johanns and J. Craig, Pavement Maintenance Manual, NDOR, Grand Island, Neb, USA, 2002.

INDEX